Medical Microbiology

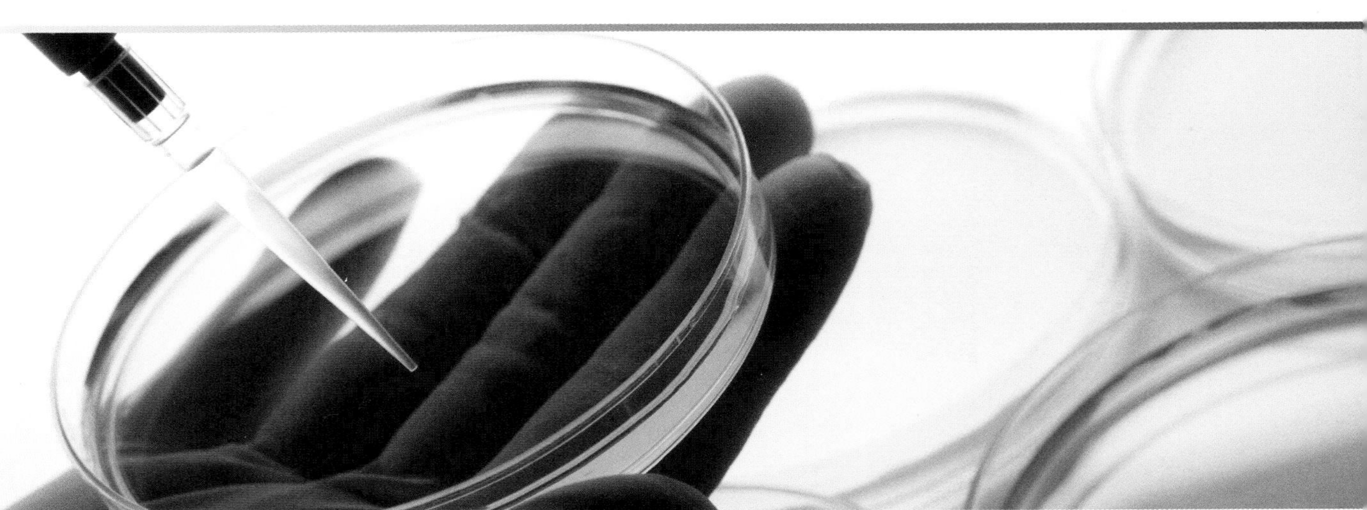

Edited by

Michael Ford

Microbiology Department, Freeman Hospital, Newcastle Upon Tyne

OXFORD
UNIVERSITY PRESS

Great Clarendon Street, Oxford OX2 6DP

Oxford University Press is a department of the University of Oxford.
It furthers the University's objective of excellence in research, scholarship,
and education by publishing worldwide in

Oxford New York

Auckland Cape Town Dar es Salaam Hong Kong Karachi
Kuala Lumpur Madrid Melbourne Mexico City Nairobi
New Delhi Shanghai Taipei Toronto

With offices in

Argentina Austria Brazil Chile Czech Republic France Greece
Guatemala Hungary Italy Japan Poland Portugal Singapore
South Korea Switzerland Thailand Turkey Ukraine Vietnam

Oxford is a registered trade mark of Oxford University Press
in the UK and in certain other countries

Published in the United States
by Oxford University Press Inc., New York

First edition published 2010

British Library Cataloguing in Publication Data
Data available

Library of Congress Cataloging in Publication Data
Data available

Typeset by MPS Limited, A Macmillan Company
Printed in Italy
on acid-free paper by
LEGO SpA – Lavis TN

ISBN 978-0-19-954963-4

3 5 7 9 10 8 6 4 2

Contents

An introduction to the Fundamentals of Biomedical Science series

Biomedical Scientists form the foundation of modern healthcare, from cancer screening to diagnosing HIV, from blood transfusion for surgery to food poisoning and infection control. Without Biomedical Scientists, the diagnosis of disease, the evaluation of the effectiveness of treatment, and research into the causes and cures of disease would not be possible.

However, the path to becoming a Biomedical Scientist is a challenging one: trainees must not only assimilate knowledge from a range of disciplines, but must understand—and demonstrate—how to apply this knowledge in a practical, hands-on environment.

The *Fundamentals of Biomedical Science* series is written to reflect the challenges of biomedical science education and training today. It blends essential basic science with insights into laboratory practice to show how an understanding of the biology of disease is coupled to the analytical approaches that lead to diagnosis.

The series provides coverage of the full range of disciplines to which a Biomedical Scientist may be exposed—from microbiology to cytopathology to transfusion science. Alongside volumes exploring specific biomedical themes and related laboratory diagnosis, an overarching Biomedical Science Practice volume provides a grounding in the general professional and experimental skills with which every Biomedical Scientist should be equipped.

Produced in collaboration with the Institute of Biomedical Science, the series

- understands the complex roles of Biomedical Scientists in the modern practice of medicine.
- understands the development needs of employers and the Profession.
- places the theoretical aspects of biomedical science in their practical context.

Learning from this series

The *Fundamentals of Biomedical Science* series draws on a range of learning features to help readers master both biomedical science theory, and biomedical science practice.

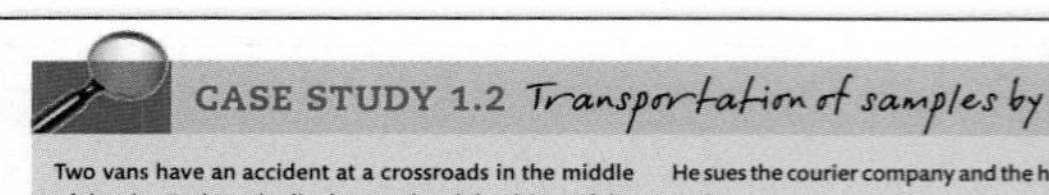
CASE STUDY 1.2 Transportation of samples by

Two vans have an accident at a crossroads in the middle of the city. Both are badly damaged and the driver of the specimen courier is unconscious. The emergency services arrive and find the contents of the van broken and all over the driver and the road. There is a lot of broken plastic/glass and lots of red and brown fluid everywhere.

The driver can't tell the ambulance staff what is all over

He sues the courier company and the h

each.

He dies of the infections 5 years late

precautions and labelling were not in p

If the samples had been in an appropr

been sealed and secured in the van,

Case Studies illustrate how the biomedical science theory and practice presented throughout the series relates to situations and experiences that are likely to be encountered routinely in the biomedical science laboratory.

Method boxes walk through the key protocols that the reader is likely to encounter in the laboratory on a regular basis.

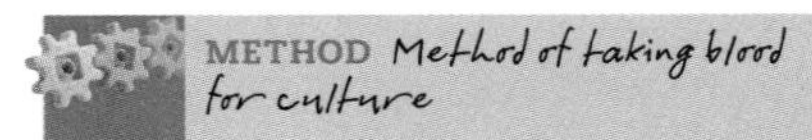

- Blood for culture may also be taken from central venous and midline catheters in some cases.
- The line connectors must be cleaned with chlorhexidine in 70% alcohol or aqueous povidone iodine before a syringe is used to withdraw blood.

Health & Safety boxes raise awareness of key health and safety issues related to topics featured in the series, with which the reader should be familiar.

HEALTH & SAFETY

Safety aspects of handling blood cultures

Blood cultures should be treated with care: breakages can occur and many systems use glass bottles which can cause cuts. Blood may harbour many dangerous pathogens including **HIV**, various types of hepatitis, and potentially haemorrhagic fever viruses, particularly in patients presenting with pyrexia of unknown origin (**PUO**). Any breakages should be handled with care using locally approved disposal and disinfectant methods, and the disinfectant used must have virucidal activity. Any blood stains on the outside of bottles should be wiped with an appropriate disinfectant, with the operator wearing suitable gloves.

Clinical correlations emphasize at a glance how the material sits in a clinical context.

CLINICAL CORRELATION

Because prompt administration of antibiotics can be life saving in cases of meningitis, patients may arrive at hospital having received antibiotics. In such cases organisms may be scanty or absent on Gram stain. The use of antigen detection kits in such situations can help identify the causative organism, as sufficient antigen will be present in the sample to give a positive reaction.

Further features are used to help consolidate and extend students' understanding of the subject

Key points reinforce the key concepts that the reader should master from having read the material presented, while **Summary** points act as an end-of-chapter checklists for readers to verify that they have remembered correctly the principal themes and ideas presented within each chapter.

techniques must be used to allow prompt and accurate identification and susceptibility testing of the infecting organism.

Key Point

It is essential the Gram stain result of a positive blood culture is telephoned as a matter of urgency to the clinician.

Because of the importance of obtaining rapid results with blood cultures we will see that this is one area of microbiology where automation is common place. In addition, con-

Key terms provide on-the-page explanations of terms with which the reader may not be familiar; in addition, each title in the series features a **glossary**, in which the key terms featured in that title are collated.

therapeutic index
Therapeutic index is the ratio between the toxic dose and the therapeutic dose of a drug, used as a measure of the relative safety of the drug for a particular treatment. A drug with little difference between toxic and therapeutic doses is referred to as having a narrow therapeutic index.

- To ensure that serum concentrations are high enough for clinical e
- To ensure that toxicity is minimized for those agents with a narro e.g. the aminoglycosides
- To monitor patient compliance, e.g. in the treatment of tuberculosi

Antibiotics that require monitoring primarily to minimize adverse e tobramycin, amikacin and vancomycin. In Tables 4.5 and 4.6 the norma is shown. For agents not commonly assayed, a service is generally pro reference laboratories.

Assay methods – microbiological assays

Self-check questions throughout each chapter provide the reader with a ready means of checking that they have understood the material they have just encountered; answers to self-check questions are presented at the end of each volume.

Disadvantages of HPLC:

- Costly
- Operator dependent
- New assay development is lengthy

SELF-CHECK 4.9

Which method of antibiotic assay would be the best for routine monitoring of antibiotics in patients' serum? Give reasons for your answer?

Cross-references help the reader to see biomedical science as a unified discipline, making connections between topics presented within each volume, and across all volumes in the series.

of the genital tract. Samples of urine taken directly from the bladder ood cultures may help confirm the diagnosis of renal candidiasis.

ve candidiasis

ntly taken to diagnose invasive candidiasis, although they may not

from other potentially infected body sites, such as the peritoneal sist with diagnosis. Isolation of *Candida* spp. from a normally sterile e candidiasis. However, isolation of *Candida* from sputum, urine or

Cross reference
See Chapter 6 for information on taking urine samples and interpretation of urine cultures.

Cross reference
Chapter 5 deals with blood culture samples.

Online learning materials

The *Fundamentals of Biomedical Science* series doesn't end with the printed books. Each title in the series is supported by an Online Resource Centre, which features additional materials for students, trainees, and lecturers.

www.oxfordtextbooks.co.uk/orc/fbs

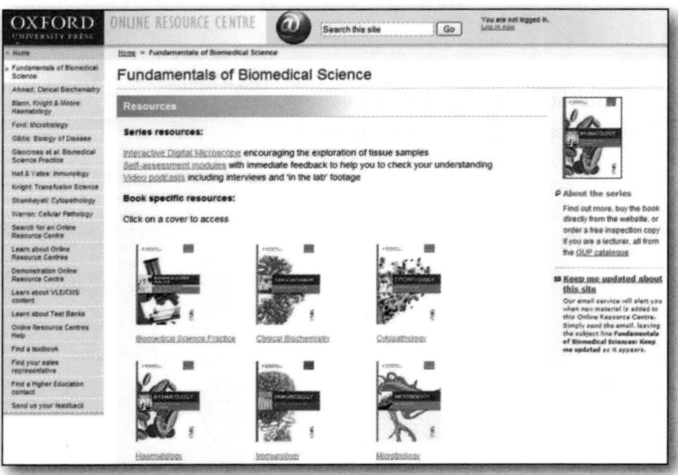

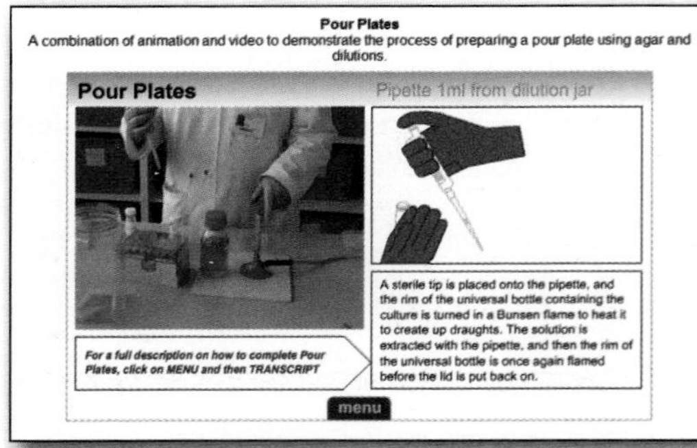

Guides to key experimental skills and methods

Multimedia walk-throughs of key experimental skills—including both animations and video—to help you master the essential skills that are the foundation of Biomedical Science practice.

Biomedical science in practice

Interviews with practising Biomedical Scientists working in a range of disciplines, to give you valuable insights into the reality of work in a Biomedical Science laboratory.

Jane Worthington, Specialist Biomedical Scientist in Microbiology at St Peters Hospital, Chertsey.

Digital Microscope

A library of microscopic images for you to investigate using this powerful online microscope, to help you gain a deeper appreciation of cell and tissue morphology.

The Digital Microscope is used under licence from the Open University.

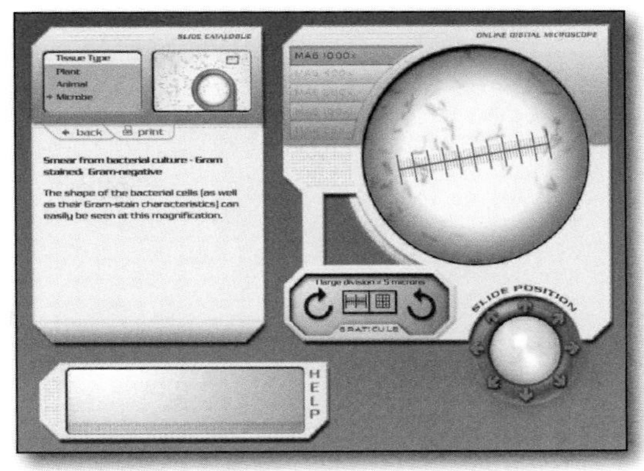

'Check your understanding' learning modules

A mix of interactive tasks and questions, which address a variety of topics explored throughout the series. Complete these modules to help you check that you have fully mastered all the key concepts and key ideas that are central to becoming a proficient Biomedical Scientist.

We extend our grateful thanks to colleagues in the School of Health Science at London Metropolitan University for their invaluable help in developing these online learning materials.

Lecturer support materials

The Online Resource Centre for each title in the series also features figures from the book in electronic format, for registered adopters to download for use in lecture presentations, and other educational resources.

To register as an adopter visit **www.oxfordtextbooks.co.uk/orc/ford** and follow the on-screen instructions.

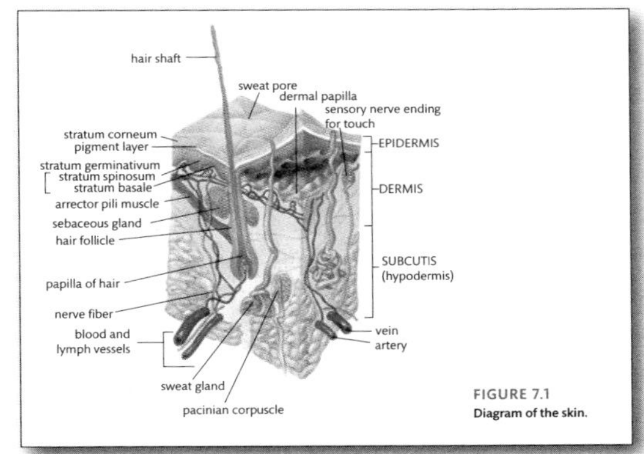

FIGURE 7.1
Diagram of the skin.

Any comments?

We welcome comments and feedback about any aspect of this series.
Just visit **www.oxfortextbooks.co.uk/orc/feedback/** and share your views.

Contributors

Jenny Andrews
Clinical Scientist, City Hospital NHS Trust, Birmingham

Steve Davis
Microbiology Department, Northern General Hospital, Sheffield Teaching Hospitals Foundation Trust

Dr Michael Ford
Microbiology Department, Freeman Hospital, Newcastle Upon Tyne

Dr Clive Graham
Consultant Microbiologist, Microbiology Department, West Cumberland Hospital, Whitehaven, Cumbria

Jayne Harwood
Chief Biomedical Scientist, Serology Department, Freeman Hospital, The Newcastle Upon Tyne Hospitals NHS Trust

Louise Hill-King
Team Leader, Microbiology Department, Frimley Park Hospital, Camberley

Dr Malcolm Holliday
Laboratory Manager, Microbiology Department, Freeman Hospital, The Newcastle Upon Tyne Hospitals NHS Foundation Trust

Dr Derek Law
Head of Microbiology, Microbiology Department, F2G Ltd, Eccles, Manchester

Sheila Morgan
Nurse Consultant, Microbiology Department, Freeman Hospital, The Newcastle Upon Tyne Hospitals NHS Foundation Trust

Kathy Nye
Consultant Microbiologist, Health Protection Agency, West Midlands Regional HPA, Birmingham Heartlands Hospital, Birmingham

Dr John Perry
Clinical Scientist, Microbiology Department, Freeman Hospital, The Newcastle Upon Tyne Hospitals NHS Foundation Trust

Lisa Tilley
Microbiology Department, Northern General Hospital, Sheffield Teaching Hospitals Foundation Trust

Mark Tovey
Microbiology Department, Northern General Hospital, Sheffield Teaching Hospitals Foundation Trust

Online materials developed by

Dr Sarah Atchia
Senior Lecturer Biomedical Sciences, Faculty of Life Sciences, London Metropolitan University

Dr Ian Hancock
Lecturer in Biomedical Sciences, Faculty of Human Sciences, London Metropolitan University

Sheelagh Heugh
Principal Lecturer Biomedical Sciences, Faculty of Human Sciences, London Metropolitan University

Dr Ken Hudson
Lecturer in Biomedical Sciences, Faculty of Human Sciences, London Metropolitan University

Dr Pamela McAthey
Academic Leader for Undergraduate Programmes, Faculty of Life Sciences, London Metropolitan University

European Committee on Antimicrobial
Susceptibility Testing

Abbreviations

4-MeU	4-methylumbelliferone
7-AMC	7-amido-4-methylcoumarin
AAFB	acid–alcohol fast bacilli
ABHG	alcohol-based hand gels
ABPA	allergic bronchopulmonary aspergillosis
ACDP	Advisory Committee on Dangerous Pathogens
ACV	acyclovir
ADH	arginine dihydrolyase
AFB	acid-fast bacilli
AIDS	acquired immune deficiency syndrome
API	analytical profile index
ASO	antistreptolysin O
ATCC	American Type Culture Collection
AUC	area under curve
AZT	azidothymidine
β-HS	β-haemolytic streptococci
BAL	bronchoalveolar lavage
Bcc	*Burkholderia cepacia* complex
BCG	Bacille Calmette Guerin
BCYE	buffered charcoal yeast extract
bDNA	branched DNA
BEAV	bile esculin azide vancomycin
BEB	bile esculin broth
BHI	brain heart infusion
BSAC	British Society for Antimicrobial Chemotherapy
BV	bacterial vaginosis
C390	9-chloro-9-(4-diethylaminophenyl)-10-phenylacridan
CAPD	continuous ambulatory peritoneal dialysis
CAT	chloramphenicol acetyltransferase
CDC	Centers for Disease Control
CF	cystic fibrosis
CFT	complement fixation test
CFTR	cystic fibrosis transmembrane conductance regulator
CFU	colony forming units
CLED	cysteine lactose electrolyte deficient
CLSI	Clinical Laboratory Standards Institute
CMV	cytomegalovirus
CNS	coagulase-negative *Staphylococcus*
COSHH	Control of Substances Hazardous to Health
CPE	cytopathic effect
CRP	C-reactive protein
CSF	cerebrospinal fluid
CSU	catheter stream urine
CT	computed tomography
CV	coefficient of variation
CVA	cerebrovascular accident
DAEC	diffusely adherent *E. coli*
DCA	deoxycholate citrate agar
DFA	direct fluorescent antibody
DGI	disseminated gonococcal infection
DNA	deoxyribonucleic acid
DNase	deoxyribonuclease
DTT	dithiothreitol
EAEC	enteroaggregative *E. coli*
EBV	Epstein–Barr virus
EDTA	ethylenediaminetetraacetic acid
EHO	Environmental Health Officer
EIA	enzyme immunoassay
EIEC	enteroinvasive *E. coli*
ELISA	enzyme-linked immunosorbent assay
EM	electron microscopy
EPEC	enteropathogenic *E. coli*
EPS	exopolysaccharide
ESBL	extended spectrum β-lactamase
ETEC	enterotoxigenic *E. coli*
E-Test	epsilometer test
EUCAST	European Committee on Antimicrobial Susceptibility Testing
FIC	fractional inhibitory concentration
FPIA	fluorescence polarization immunoassay
G6PD	glucose-6-phosphate dehydrogenase
GP	General Practitioner
GUM	genitourinary medicine
HA	haemagglutination
HACEK	group of CO_2 dependent organisms – *Haemophilus* sp, *Aggregatibacter actinomycetemcomitans* (previously named *Actinobacillus*), *Cardiobacterium hominis*, *Eikenella corrodens* and *Kingella* spp.

HAI	health-care-acquired infection
HAV	hepatitis A virus
HBV	hepatitis B virus
HCAI	health-care-acquired infection
HCV	hepatitis C virus
HEPA	high efficiency particulate absorption
Hib	*Haemophillus influenzae* type b
HIV	human immunodeficiency virus
HPLC	high-performance liquid chromatography
HPU	Health Protection Unit
HSV	herpes simplex virus
HVS	high vaginal swab
IA	invasive aspergillosis
IATA	International Air Transport Authority
ICU	Intensive Care Unit
IE	infective endocarditis
IFAT	indirect fluorescent antibody test
IgA	immunoglobulin A
IgG	immunoglobulin G
IgM	immunoglobulin M
ISTB	Iso-Sensitest broth
IUCD	intrauterine contraceptive device
KES	*Klebsiella–Enterobacter–Serratia*
KOH	potassium hydroxide
LAP	leucyl aminopeptidase
LCR	ligase chain reaction
LJ	Löwenstein–Jensen
LOS	lipooligopolysaccharide
LRTI	lower respiratory tract infection
MAC	*Mycobacterium avium* complex
MBC	minimum bactericidal concentration
MDR	multidrug resistant
MHB	Mueller–Hinton broth
MIC	minimum inhibitory concentration
MICE	minimum inhibitory concentration evaluation
MIRU-VNTR	mycobacterial interspersed repetitive units variable number tandem repeats
MRSA	methicillin-resistant *Staphylococcus aureus*
MSA	mannitol salt agar
MSU	midstream urine
MTB	*Mycobacterium tuberculosis*
NAAT	nucleic acid amplification test
NAC	*N*-acetyl-L-cysteine
NAD	nicotinamide adenine dinucleotide
NCTC	National Collection of Type Cultures
NEQAS	National External Quality Assurance Schemes
NHS	National Health Service
NNT	number needed to treat
NPA	nasopharyngeal aspirate
NSU	non-specific urethritis
O129	2,4-diamino-6,7-diisopropylpteridine phosphate
ONPG	*o*-nitrophenyl-β-D-galactopyranoside
PABA	para-aminobenzoic acid
PBP	penicillin-binding protein
PCP	*Pneumocystis jiroveci* pneumonia
PCR	polymerase chain reaction
PID	pelvic inflammatory disease
PMC	pseudomembranous colitis
***p*-NA**	para-nitroaniline
POC	products of conception
PPE	personal protective equipment
PUO	pyrexia of unknown origin
PVC	polyvinyl chloride
PVE	prosthetic valve endocarditis
PVL	Panton–Valentine leukocidin
RBC	red blood cell
RNA	ribonucleic acid
RPR	rapid plasma reagin
RSV	respiratory syncytial virus
SAH	subarachnoid haemorrhage
SCBU	Special Care Baby Unit
SDA	strand displacement amplification
SM-ID	*Salmonella* identification
SPA	suprapubic aspirate
SPS	sodium polyethanol sulphonate
STI	sexually transmitted infection
TB	tuberculosis
TCBS	thiosulphate citrate bile salt sucrose agar
TPHA	*Treponema pallidum* haemagglutination
TPPA	*Treponema pallidum* particle agglutination
TSS	toxic shock syndrome
TST	tuberculin skin test
TV	*Trichomonas vaginalis*
UBA	urine bacterial analysis
UCP	urine collection pad
UTI	urinary tract infection

UV	ultraviolet
VDRL	Venereal Disease Reference Laboratory
VHF	viral haemorrhagic fever
VP	Voges–Proskauer
VRE	vancomycin-resistant *Enterococcus*
VTEC	verocytotoxin-producing *E. coli*
VZV	varicella zoster virus
WBC	white blood cell
WD	washer disinfector
XLD	xylose lysine deoxycholate
ZN	Ziehl–Neelsen

1

Procedures for sample collection, transport and processing

Malcolm Holliday

Microbiology is a little like a computer - if you put the wrong information in, you may get a bad result out. This means that great care must be taken to make sure that the quality of samples for microbiological investigations are as good as possible. You could say that collecting the sample properly is the single most important stage of the whole process of microbiological examination. If the way the sample is collected isn't correct, contaminating organisms may be introduced which will confuse the result, and it is always possible that a poorly taken sample may not have actually collected anything from the site.

> **Key Point**
>
> **A poorly collected microbiology sample will produce a poor result no matter how good the laboratory is.**

This chapter will look at the important issues around sample collection, how the various sample types can be sent to the laboratory, the correct way of collecting them and transporting them to the laboratory, basic sample processing and some of the regulations and safety considerations concerned.

Learning objectives

After studying this chapter you should be able to:

- Discuss the principles of specimen collection for microbiological investigations
- Outline the processes necessary to ensure good quality samples are received in the laboratory
- Explain how contamination and improper transport and storage can affect the end result

- Describe the health and safety risks of collecting, transporting and processing microbiology samples
- List common types of sample and the correct collection and transport methods for each
- Discuss the principles behind the different types of transport media available
- Outline the legislation covering the collection and transport of microbiological samples
- Discuss basic microbiological sample processing methods
- Outline the methods of inoculation of culture media
- Describe the different organism hazard groups and containment levels used to process microbiological samples

1.1 The importance of sample collection

When a sample is taken for microbiology, it is a snapshot of what is happening at the sample site at the moment it is taken. What we really want to find out is what organisms were in the sample at that point.

However, because bacteria are living things, this begins to change right away. Some organisms start to die quickly, while others can multiply rapidly, depending on the conditions. This means that if the sample isn't collected properly and transported to the laboratory in the correct way and then processed appropriately, the results we get may bear little resemblance to the original organisms present at the site of infection. Incorrectly collected specimens can also be a common reason for samples being rejected by the laboratory as unsuitable for examination.

There are a number of principles that we can apply to microbiology specimen collection.

contaminant

A contaminant is a microorganism that wasn't present in the original sample, but somehow gets into it before or while it is cultured.

1. The sample must be collected from the correct infected site with a minimum of **contamination.**

   Example: Urine samples may be highly contaminated if they are not collected properly.

   Solution: Patients must be told exactly how to clean the surrounding area and collect the sample to minimize contamination.

2. The sample must be collected at the most appropriate time to recover the pathogens of interest.

   Example: In septicaemia, the maximum number of pathogens in the blood occurs when the patient's temperature rises significantly and blood cultures must be taken at this time. Taking blood cultures when the patient's temperature has settled to normal may not detect very low levels of pathogen.

   Solution: Clinical staff must be given clear instructions to take blood cultures during episodes of peaking temperature (pyrexia).

pus

Pus is a discharge made up of white blood cells and fluid which is present in infected wounds.

3. Enough sample must be collected to perform all the required tests.

   Example: A patient is found to have an infected abscess where **pus** is draining from a tonsil, and a swab of the discharge is sent to the laboratory. This may not allow all the different cultures required (including extended anaerobic and broth enrichment culture) to be done before all the material is wiped off the swab. A pus sample would have been better.

Solution: Clinical staff should be given clear instructions about the best types of sample to send. It's always better to send too much rather than too little.

4. The correct collection devices and specimen containers must be used for each type of sample.

   Example: A urine sample for culture is collected in an old medication bottle. Even if washed out, this may contain traces of the original medication, which may kill some pathogens and give a misleading culture result. Even if the medication had no antibacterial effect, it is unlikely that the container would be sterile and contaminants may be introduced to the sample.

   Solution: Provide suitable containers to all users, along with appropriate instructions, and reject any samples in incorrect containers.

5. Whenever possible, samples for microbial culture must be collected before antibiotics are given.

   Example: Throat swabs taken after antibiotics have been administered to the patient may not grow the infecting pathogen.

   Solution: Where possible, samples for microbiological culture should always be taken before giving antibiotics.

6. Specimen containers must be properly labelled and sealed before being sent to the laboratory and the details must match those on the request form.

   Example: A request for *N. gonorrhoeae* culture is sent from a Genitourinary Medicine Clinic with no label and no identification on the swab. The laboratory cannot be sure that this swab is from the patient whose details are on the separate request form. If the sample was not from that patient and *N. gonorrhoeae* was isolated, reporting it would cause the wrong patient unnecessary harm, inappropriate treatment and possible family problems. If this was later shown to be a mistake, the hospital could be sued.

   Solution: Provide clear instructions to all clinical staff regarding requirements for correct labelling of samples, and enforce a policy of rejecting and not culturing any samples that do not meet these requirements.

7. Specimens should be promptly transported to the laboratory.

   Example: A swab from an anaerobic abscess is taken at 08:00 am and is sent for culture. The sample is not collected from the ward until late afternoon and does not reach the laboratory until it is closed except for on-call samples. This swab will not be cultured until the next day, by which time many **anaerobes** could have died, and the pathogen may not be isolated.

   Solution: Ensure that porters, wards and clinical staff know that microbiology samples must be transported promptly to the laboratory. Contact the staff that do not comply and educate them appropriately.

Cross reference
Chapter 7, Section 7.7.

anaerobe
Anaerobes are organisms that only grow in the absence of oxygen.

The key to successful diagnosis depends on taking the appropriate sample correctly, at the proper time and transporting it to the laboratory without delay. This practice will help to ensure the best possible methods to ensure isolation of the target **pathogen(s)**.

pathogen
A pathogen is a microorganism that has the ability to grow in the human body and cause disease.

SELF-CHECK 1.1

What do you consider to be the main problems associated with sample collection and ensuring that the target organism is grown from the sample?

1.2 Overview of different sample types and specimen containers

There are lots of sample containers we can use to collect samples and transport them to the laboratory; however, we must be careful that we use the correct container for the job, or the results may be affected.

Sterile plastic or glass 'universal' containers

These are normally sterile bottles with a screw top which are used for urine, sputum, fluids, biopsies and tissues, catheter tips and almost any other small specimen.

Plastic containers tend to be used now, as glass containers can pose a safety hazard to staff if they break.

The 60-mL variety has a wider mouth and is sometimes preferred for sputum samples, as it's easier for patients to produce the sample directly into the container. Trying to get a sputum into a narrow-mouthed 30-mL container usually ends in a failure, with sample on the outside of the container as well as the inside.

Boric acid containers

These are sterile plastic universal containers with enough boric acid inside to give a concentration of 1.8% when 15 mL of urine is added and may be used when urine samples are likely to be delayed significantly before reaching the laboratory (for example overnight). The boric acid acts as a preservative and stops bacteria growing while the sample is transported. These containers are now rarely used due to improvements in sample transport.

Some bacteria may be killed if the volume of urine is insufficient.

Dipslides

These are sterile plastic universals with a large, flat plastic slide attached to the lid (Figure 1.1). This has bacterial culture media on each side. The slide is dipped in the urine after collection to inoculate it, the urine is then discarded, and the slide put back into the container and secured. It can then be sent to the laboratory where it can be incubated just like any other culture plate.

As no urine is received, no cell counts can be done.

Sterile plastic universals with spoons (faeces containers)

The spoon is attached to the container lid. It is used to collect a portion of faeces (stool) and transfer it to the container. It is not practical to attempt to fill a container directly while on the toilet, as the outside of the container is certain to become contaminated.

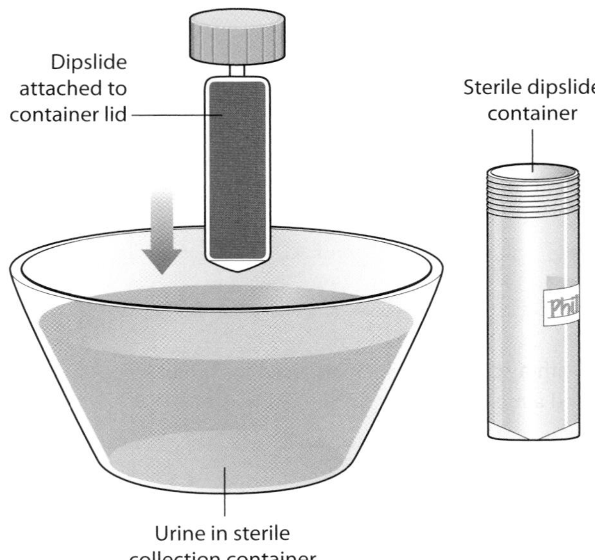

FIGURE 1.1
Dipslide urine sample.

Swabs

Swabs are available with different tips (cotton wool, fibre, serum coated), shafts (wood, metal, plastic) and transport media (with or without charcoal, Hanks, *Chlamydia*), and may be in tubes or be broken off into bijoux bottles (Figure 1.2).

- Serum coated swabs: these are dry swabs and are only used for hardy bacteria when there will be minimal delay in delivery to the laboratory (e.g. MRSA). Moistening the swab may increase the amount of material collected.
- Transport medium swabs: these are designed to maintain microorganisms alive while being transported to the laboratory and are preferable if there is any delay in transport.
- Fine-tipped swabs on flexible wire shanks: these are useful for sampling from difficult to reach areas where a normal swab is too large, or where the shaft has to bend slightly to allow the tip to reach the sample site (e.g. ear swabs or pernasal swabs).
- Plastic shaft swabs: these may be used where a wooden shaft would adversely affect the result of the investigation (e.g. *Chlamydia* investigation). Some plastic shaft swabs have a 'nick' in the shaft which conveniently snaps off, allowing the swab to fit into a bijou bottle.

Blood cultures

Cross reference
Chapter 5, Section 5.1.

Most blood culture bottles are designed for automated systems and consist of a bottle containing sufficient culture broth to remove the antibacterial effects of normal blood or antibiotics by dilution, or substances such as specialized resins, activated charcoal or sodium polyethanolsulfonate (SPS) which inactivate these substances.

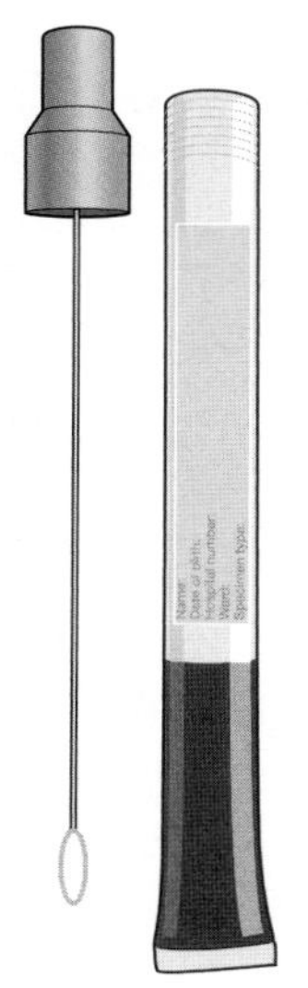

Pernasal swab with transport medium

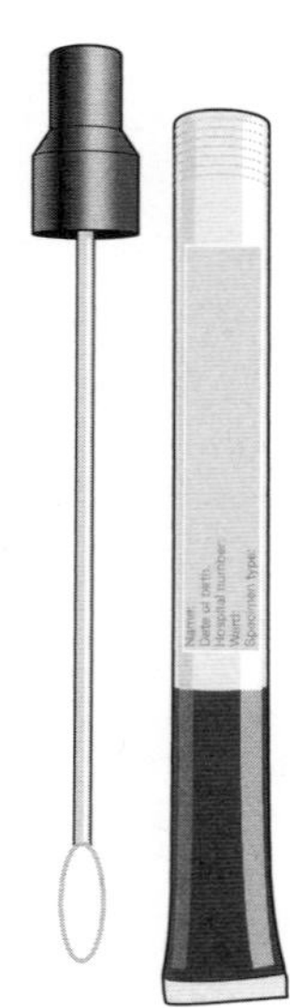

Plain swab with transport medium

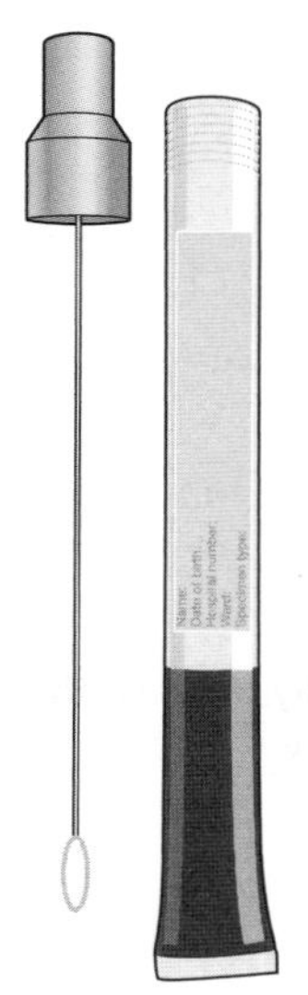

Ear swab with transport medium

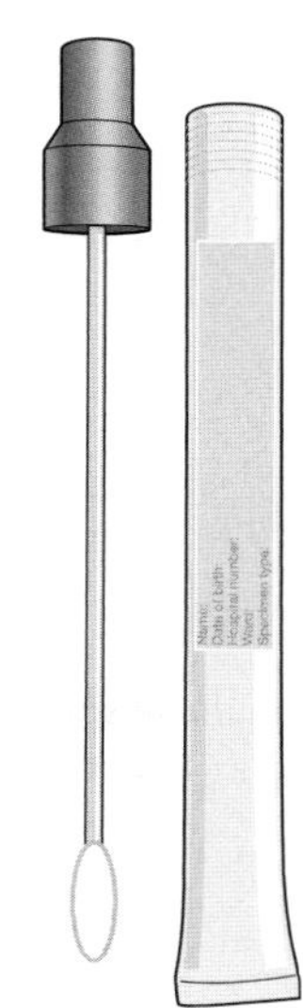

Plain swab with no transport medium

FIGURE 1.2
Swab types.

Systems where anticoagulated blood is lysed and centrifuged to concentrate any microorganisms before being cultured are also available.

SELF-CHECK 1.2

What are the main types of sample containers used for sample collection in microbiology?

1.3 Methods of sample collection

If microbiological specimens aren't collected properly, we can face all sorts of problems. These range from picking up contaminating organisms from around the site we want to sample, to missing the target altogether and sampling the wrong area. This is particularly relevant with children (who wriggle a lot and are a moving target, especially for swabbing etc.).

Whenever a sample is going to be taken, the procedure must be discussed with the patient beforehand. No one likes unpleasant surprises and having someone come up to you in hospital and start doing strange and often embarrassing things to you without any warning is not acceptable. Privacy must be ensured while the procedure is being carried out.

Samples for microbiological culture should be taken before antibiotic therapy is started if possible. This will avoid the results being distorted by the effects of the antibiotics.

All samples must be appropriately labelled before sending to the laboratory.

CASE STUDY 1.1 Put yourself in the patient's shoes

You have been sent to the hospital by your GP because you have some round, reddish areas on your back. You have taken your top off and been examined. The doctors have gone into a corner and muttered together, and now one has approached you with a scalpel and gone behind your back. You can feel something happening.

What would be your reaction here?

- **You have something dreadfully bad and the doctor is going to cut it out straight away.**
- **You have no idea what is happening, but the thought of someone you don't know behind you with a knife makes you very afraid.**
- **You imagine all sorts of things and the moment the scalpel touches you, you leap 2 feet in the air and scream.**
- **You trust the doctor and relax, safe in the knowledge you are in good hands.**

Hands must be washed thoroughly with bactericidal soap and water or alcohol hand gel and gloves must be worn when taking microbiological samples to avoid transferring organisms from one patient (or from your hands) to another patient. This is a key initiative to combat **health-care acquired infections** (HCAI).

health-care acquired infection
Health-care acquired infection (HCAI) occurs when pathogenic microorganisms are transferred from health-care workers, other patients or the hospital environment, to a previously uninfected patient, who then develops an infection.

Let's look at all the different sample types and how each should be collected to get the best quality sample.

Urine samples

Urine grows microorganisms quite well, therefore urine samples should be sent to the laboratory as quickly as possible to avoid overgrowth of contaminants. However cleanly a sample is taken, there is always the possibility that small numbers of contaminants may enter.

Mid-stream urine (MSU)

These are the commonest and easiest to collect. They are the most reliable non-invasive urine samples but may not be possible in some patients.

- Male: pull back the foreskin and clean the tip of the penis with soap and water or saline to remove contaminating organisms.
- Female: clean the outer part of the vagina and labia with soap and water or saline. Clean from front to back to avoid contamination from around the anus.

Ask the patient to urinate, directing the first and last part of the stream into the toilet, but collecting the middle part into a sterile container (bedpan, kidney bowl, bottle, etc.). This can then be transferred to a labelled, sterile laboratory specimen container which is then sent to the laboratory.

It is not a good idea to ask the patient to urinate directly into the laboratory container, as these are narrow, and there is a risk that the outside of the bottle may be contaminated with urine, which is unpleasant and hazardous for anyone then handling it.

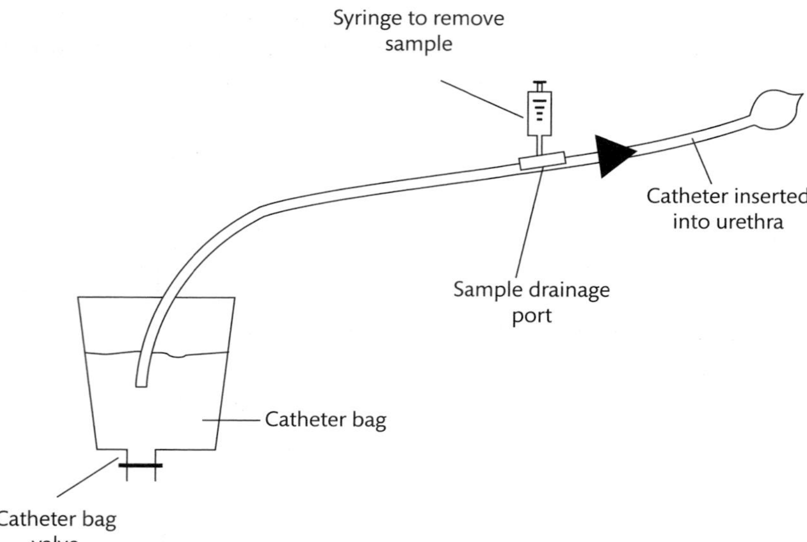

FIGURE 1.3
Catheter specimen of urine. (Courtesy of M. Ford.)

Catheter stream urine specimen (CSU)

When a patient has a catheter in place, it is not possible to collect an MSU sample. Catheter bags have a collection point provided (Figure 1.3). This should be cleaned using 70% alcohol and a sterile needle and syringe should be inserted into the collection point and sufficient urine withdrawn. This should be placed in a sterile sample container and be sent to the laboratory.

Where MSU or CSU samples can't be collected easily (e.g. from children and babies) the following techniques can be useful.

Bag urine

Choose the correct size of urine bag and remove the protective seal. Place the bag over the penis or vulva, working from back to front and sticking the bag to the skin. Watch the bag until urine is passed, then remove the bag, empty it into a sterile urine container and send to the laboratory.

The genital area should be washed after collection to prevent soreness.

Pad urine

The baby's groin should be cleaned with warm water and the urine collection pad (UCP) placed in the nappy, towards the front. The pad should be checked every 15 minutes to see if urine has been passed. Wear gloves and remove the wet pad. Put the tip of a sterile syringe into the pad and remove as much urine as possible. Place this in a sterile container and send to the laboratory.

Supra-pubic aspirate (SPA)

The abdominal skin over the bladder is cleaned thoroughly with alcohol-based disinfectant and allowed to dry, to ensure that possible contaminants are removed. A needle on a syringe is passed through the skin and directly into the bladder (which must be fairly full). A sample of the urine is withdrawn and placed into a sterile container and sent promptly to the laboratory. The needle is removed and safely disposed of.

Swab samples

Swabs are a convenient way to take samples for culture from superficial areas of the body. Samples may be taken from infected sites to look for the presence of pathogens, from any area to look for colonization with potential pathogens such as MRSA, or for specific organisms such as *Bordetella pertussis*, *Neisseria gonorrhoeae*, etc.

Wound swabs

Select the area of wound to be sampled and gently rotate the tip of the swab in the chosen spot. Collect as much material as possible, especially if there is pus present. Ideally, old pus should be removed to allow collection of the freshest sample possible. If not, contaminants may be present and this will interfere with isolation of the target pathogen. Using an **aseptic technique** place the swab back in the sample tube and send to the laboratory.

aseptic technique
Aseptic technique is the use of sterile equipment and procedures to avoid possible contamination.

Nasal swabs

If the swab is moistened beforehand it will help prevent discomfort to the patient. The patient should tilt their head backwards to make sample collection easier while the swab is placed into the nose, without touching the sides, and moved upwards into the tip of the nose. The swab should be gently rotated before withdrawing it and aseptically placed back in the sample tube.

Throat swabs

The patient should face the light. The patient's tongue should be depressed with a spatula to ensure that it does not rise up and contaminate the swab if the patient gags. Insert the swab into the mouth without touching the sides and quickly rub it gently over any inflamed area or any visible exudates at the back of the throat or the tonsil region before aseptically placing it back into the sample tube.

Eye swabs

The patient must be asked to try not to flinch when the sample is being taken. Moistening the swab with sterile water or saline may help. Ask the patient to look up, put on sterile disposable gloves and pull gently on the eyelid to expose the pink conjunctiva. Gently rub the swab across the lower eyelid from the inner corner outwards. Try to collect as much discharge as possible. Place the swab aseptically back in the sample tube.

Ear swabs

Use a fine-tipped swab with a flexible wire shaft. If any pus is present, sample this with the swab. Place the swab into the outer ear and gently rotate to collect any discharge. Do not push the swab too far into the ear to avoid damaging the eardrum. Place the swab aseptically back in the sample tube.

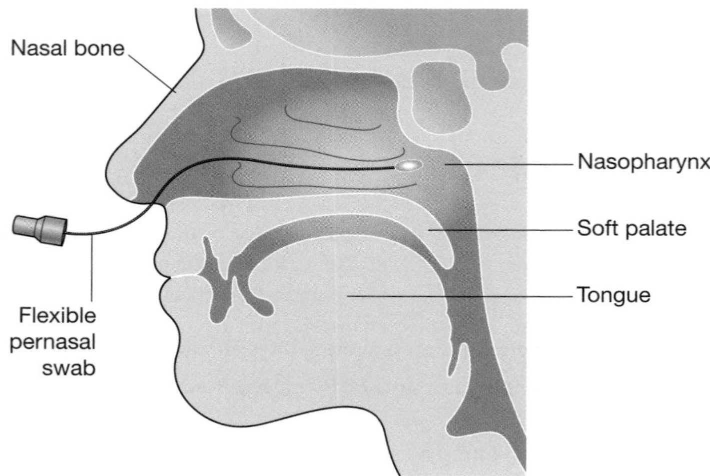

FIGURE 1.4
Method of sampling using a pernasal swab.

Deeper ear samples may be collected using a speculum; however, this should only be done by an experienced person to avoid causing damage.

The middle ear and inner ear cannot be sampled using a swab. Rarely, middle ear fluid can be aspirated if the eardrum is perforated, or by passing a fine needle through it.

Pernasal swabs (for Bordetella pertussis*)*

Use a fine-tipped swab with a flexible wire shaft and pass along the floor of the nasal cavity to the back wall of the nasal cavity (Figure 1.4). Rotate the swab gently, withdraw it carefully and place it aseptically back in the sample tube.

Blood samples

venepuncture
Taking blood via a needle placed into a vein.

Blood samples via a **venepuncture** should only be taken by suitably trained and qualified health-care workers (phlebotomists, nurses, doctors, etc.), as these are invasive samples and can pose significant risks to both the patient and the health-care worker if carried out incorrectly.

Blood cultures

It is particularly important not to introduce any contaminating bacteria into blood cultures, as these can cause misdiagnosis and unnecessary antibiotic therapy.

The tops of the blood culture bottles must be removed to expose the rubber septum through which blood is injected. The septum must be cleaned with alcohol and allowed to dry immediately before the sample is taken.

The patient's skin where the blood is to be taken must be cleaned thoroughly with alcohol-based disinfectant and allowed to dry to ensure that potential contaminants are removed.

METHOD Taking blood for culture

- **Blood for culture may also be taken from central venous and midline catheters in some cases.**
- **The line connectors must be cleaned with chlorhexidine in 70% alcohol or aqueous povidone iodine before a syringe is used to withdraw blood.**
- **The first 5 mL should be discarded.**
- **A sterile needle is attached to the syringe and the bottles inoculated as before.**
- **Separate syringes must be used for each culture bottle to prevent contamination.**
- **Forward blood cultures to the laboratory as soon as possible so the incubation process can begin.**

After withdrawing the blood, the syringe is removed and a new sterile needle is placed on the syringe. This removes any possibility of introducing skin contaminants on the outside of the needle into the bottle.

Insert the needle into the bottle through the septum and allow the appropriate amount of blood to enter. The volume of blood is critical, as overfilling the bottle can lead to false-positive results with some automated blood culture systems. Too little blood reduces the chance of growing a pathogen.

Syringe needles must be safely disposed of and must never be re-sheathed. Most blood culture systems require around 5 to10 mL of blood for adults, but often only 1 to 4 mL for neonates.

Many systems use two (or more) bottles, usually for aerobes and anaerobes, although systems for yeasts and mycobacteria are also available. Unless anaerobic infection is strongly suspected, the aerobic or general purpose bottles should be filled first. Special paediatric bottles are provided for neonates as the volume recovered is likely to be much lower than from an adult.

Blood culture bottles may have a vacuum inside, and if only a small blood sample is available, care must be taken to ensure that all the sample is not drawn into the first bottle, if more than one are required.

Inoculation of blood culture bottles should always be performed before putting blood into other containers for other laboratory tests, as these containers may not be sterile and accidental contamination may occur.

The bottles should be mixed gently by inversion and sent to the laboratory.

Blood samples for antibiotic levels

Because the result of antibiotic levels is required before the next dose can be calculated, these are often urgent.

METHOD Blood samples for antibiotic assay

- **Blood is collected by venepuncture and 10 mL placed in a clean blood tube. This may contain chemicals to aid blood clotting and serum separation, or they may not and allow blood clotting to occur naturally.**
- **The container is capped tightly and is sent to the laboratory as quickly as possible.**
- **Details of the drug, dosage and mode of administration must be supplied to the laboratory as well as whether the sample is a 'pre' or 'post' dose sample, to aid interpretation.**

Blood samples for bacterial and viral serology

These samples are taken in the same way as above. If only a small sample is available and many tests are required, it's important that the laboratory is informed which tests are the most important.

Blood samples for molecular investigations

Many of these require an **EDTA** blood sample; however, some investigations need clotted blood. Wards and clinical staff need to be aware of the specific requirements for the tests they are requesting.

EDTA
EDTA (ethylenediaminetetraacetic acid) is an anticoagulant to prevent blood samples from clotting.

Cerebrospinal fluid (CSF)

Cerebrospinal fluid is normally obtained by lumbar puncture, which is a specialized procedure performed by medical staff. The patient either lies or sits so that their back is arched and maximum widening of the spaces between the vertebrae is achieved.

cerebrospinal fluid
Cerebrospinal fluid is the fluid that fills the spinal column.

White and red blood cells in CSF can deteriorate or lyse rapidly, especially if refrigerated, so analysis is required as soon as possible and CSF should never be put in the refrigerator prior to microbiological examination.

METHOD CSF sample collection

- **The skin over the area to be punctured is cleaned with antiseptic and a spinal puncture needle is introduced between the second and third lumbar vertebrae and into the subarachnoid space (see Figure 8.2).**
- **The appropriate samples of CSF are obtained, after which the needle is withdrawn. Usually about 10 mL in total is required for microbiological and biochemical analysis. The usual volume is 2 mL per container.**
- **Samples for microbiology are placed into sterile universal bottles and transported to the laboratory promptly.**

Sterile body fluids (e.g. joint fluid, pleural fluid, etc.)

Skin over the site must be disinfected with chlorhexidine or alcohol. The sample must be obtained using a needle and syringe and placed directly into a sterile container, then transported to the laboratory.

Vesicle fluids for virology

The top of the vesicle or ulcer should be lifted and a swab from a viral transport kit should be used to swab the entire base of the vesicle firmly (without causing bleeding) to collect cellular material. Place the swab in viral transport medium and send promptly to the laboratory.

Catheter and line tips

The skin around the catheter must be cleaned with alcohol or chlorhexidine before the catheter is removed. Around 4 to 5 mm of the very tip is cut off with sterile scissors and placed into a sterile container, which is then sent to the laboratory.

Skin, hair and nails for dermatophyte fungi

These samples are sent for the isolation of dermatophyte fungi. For all samples, the site should be cleaned with 70% alcohol to reduce bacterial contamination, and especially if greasy ointments have been used. Samples are placed in folded black paper or universal containers for transportation to the laboratory.

Cross reference

Chapter 11, Section 11.4, Sample Collection.

Skin scrapings

Scrape outwards from the edge of the site with a blunt scalpel (or the back of the blade) or a glass slide and collect scales or flakes of skin into black paper. If there is no material being removed, a strip of Sellotape™ applied to the site and then placed on a slide can be useful. Slides must be transported inside a plastic slide holder.

Hair

Hairs should be plucked from the scalp with forceps so that the roots and follicles are attached. Ultraviolet light from a Wood's lamp may be used to detect infected hairs, which fluoresce.

Nail

Clippings should be taken from discoloured or brittle parts of the nail and cut back as far as possible from the edge. Scrapings should also be taken from beneath the nail.

SELF-CHECK 1.3

Why is it important for so many microbiological samples that the skin is cleaned first, before the sample is taken?

prosthesis

A prosthesis is a mechanical or electronic device placed in or on the patient's body to help it function. Examples include artificial knees, hips and pacemakers.

Tissue, prostheses, etc.

These are usually removed in theatre, and should be placed into a sterile container. A few drops of sterile saline will help stop tissues and biopsies drying out. Samples should be sent to the laboratory as quickly as possible.

Sputum

Sputum samples can easily be contaminated by passing through the mouth. Encourage the patient to cough deeply until a deep sputum specimen can be obtained (often best results are found first thing in the morning). This should be spat out into a clean (but not necessarily sterile) container such as a disposable kidney bowl.

The sample must be checked to ensure it is sputum, not saliva, and it can then be transferred to a sterile container and sent to the laboratory. Remember patients who are immunosuppressed may not produce a purulent sample.

Sputum can be induced by having the patient inhale 20 to 30 mL of 3 to 10% sterile saline using a nebulizer.

If sputum cannot be produced, a physiotherapist may be able to help the patient produce a satisfactory sample.

Nasopharyngeal aspirates (NPA) for virology

The patient's head should be reclined to a 70° angle. A container with a fine tube is attached to a suction pump and the tube is inserted gently into the nose. When the tube is at the very back of the nasal cavity the suction is switched on and the tube slowly withdrawn using a rotating movement. The tube should not remain in the nose more than 10 seconds. The sample should be transferred to a sterile container and transferred to the laboratory.

Sputum samples are unsuitable for viral investigations.

Faeces

The patient should pass faeces (stool) into a clean container such as a bedpan. The spoon on the faeces container lid is used to scoop up enough faeces to fill one-quarter to one-third of the container, and the lid (with spoon still attached) is closed securely. The sample is then sent to the laboratory. If parasites are requested it is important in some cases to transport to the laboratory as soon as possible.

If the sample is diarrhoea, using the spoon to fill the container is preferable to pouring from the bedpan, where there is a significant risk of contaminating the outside of the container.

Rectal swabs

When a faeces sample cannot be collected, a rectal swab may be used. This is obtained by passing a swab carefully through the anus into the rectum, rotating gently and withdrawing it. The swab is placed back in the sample tube and sent to the laboratory.

Key Point

Rectal swabs are not as good as faeces for microbiological examination.

Perianal swabs and Sellotape™ preparations

These are useful in diagnosing threadworm infections. A swab may be taken from the perianal (around the anus) region, or a piece of Sellotape™ may be stuck onto the skin around the anus and then removed and stuck onto a glass slide. These are done first thing in the morning and both techniques collect hookworm eggs laid by adult worms, which emerge during the night and lay eggs around the anus. Often in such cases, a diagnosis can be made clinically without the requirement to forward samples for analysis.

Vaginal swabs

To collect a high vaginal swab, a speculum is introduced into the vagina to separate the vaginal walls and a swab is inserted and rotated gently as high as possible in the vaginal vault. The outer areas of the vagina should be avoided. The swab should be placed into transport media and sent to the laboratory. If *Trichomonas vaginalis* is suspected, Diamond's or Feinberg's transport media can be used.

Cross reference
Chapter 7, Section 7.7, *T. vaginalis* infection.

Low vaginal swabs may also be sent to the laboratory. These are collected by inserting a swab into the vaginal entrance and rotating gently. This is not a suitable specimen for detecting vaginal pathogens as it will be heavily contaminated with organisms from the **perineum** area.

perineum
The perineum is the area between the anus and the vagina.

Cervical swabs

A speculum is introduced as above and a swab is gently inserted into the mouth of the cervix and rotated. If examination for *N. gonorrhoeae* is required, the swab should be placed into transport media containing activated charcoal to remove toxic substances and sent to the laboratory without delay. Cervical swabs are used for the diagnosis of *N. gonorrhoeae* and *Chlamydia*.

Penile (urethral) swabs

The prepuce (foreskin) is pulled back and a fine-tipped swab is placed 4 cm into the urethra and rotated gently. This is removed, placed in the sample tube and sent to the laboratory.

1.4 Methods of sample transport

In a hospital, most samples will be taken from wards and clinics to the laboratory by the portering services.

We need to realize that the porters will prioritize moving patients around, delivering meals and obtaining items such as oxygen cylinders before they transport laboratory samples, so don't be surprised if samples taken in the morning don't reach the laboratory until late in the day.

Many hospitals have air-tube systems, where samples are put into a plastic 'pod' which is placed into the sending station and is forced through a system of tubes to the chosen destination. This is often the fastest method of getting samples to the laboratory.

Often, however, wards may be discouraged from sending microbiology samples by this system because samples may break and there is an infection risk. This is illogical, especially for blood samples, when samples for haematology and biochemistry laboratories are sent routinely by the air tube system, but microbiology bloods from the same patients cannot be sent.

There may be fears that if a spillage occurs inside the system, a large quantity of infected specimen may be spread around the interior of the air tubes. These are very difficult to access, and even more difficult to disinfect.

Disinfection may be accomplished by closing the system for routine use and sending 'pods' with disinfectant-soaked cloth or sponge through the tubes to every location, sometimes repeatedly - a very long and complicated procedure.

For this reason, specimens known to be a high infection risk (hepatitis B, hepatitis C, HIV, *M. tuberculosis*, etc.) and anything likely to break easily should never be sent through an air-tube system.

Laboratory samples should always be transported in safe, leak-proof containers, so that if dropped, any breakages and spills will be contained.

Because the patient's personal and clinical details can be seen on the request form (and these may be extremely sensitive or confidential, such as HIV status), transport containers should be opaque to stop these being seen by other patients, staff, or members of the public. There are several types of carrier used, ranging from aluminium and plastic boxes to plastic bags.

Samples from GP surgeries are often collected by a regular courier service. Because these travel by road, they have to comply with the relevant road transport legislation. They must be transported in sealable, leak-proof boxes, which must carry the appropriate hazard labels, and should be securely fastened in the vehicle.

A risk assessment should have been carried out and all necessary measures put in place to minimize the dangers if there should be an accident. This includes driver training.

CASE STUDY 1.2 Transportation of samples by courier

Two vans have an accident at a crossroads in the middle of the city. Both are badly damaged and the driver of the specimen courier is unconscious. The emergency services arrive and find the contents of the van broken and all over the driver and the road. There is a lot of broken plastic/glass and lots of red and brown fluid everywhere.

The driver can't tell the ambulance staff what is all over him, there are no markings on the van or the contents to say what they are, and there is no contact number available. The driver needs mouth to mouth resuscitation. While doing this, the paramedic cuts himself on some of the glass. It later turns out that he has contracted HIV and hepatitis C from a broken high-risk sample.

He sues the courier company and the hospital for £5 million each.

He dies of the infections 5 years later. All because basic precautions and labelling were not in place.

If the samples had been in an appropriate box, which had been sealed and secured in the van, they wouldn't have scattered and broken. If the box had been correctly marked, the emergency services would have known what they were dealing with. If a laboratory contact number had been provided, advice could have been given regarding handling of the samples.

SELF-CHECK 1.4

What do you think is the best method of transporting microbiology samples and why?

1.5 Transport regulations

Microbiology samples and cultures are regarded as dangerous goods because they are infectious, and could contain a variety of **pathogens.** As such, they come under several different regulations including Control of Substances Hazardous to Health (COSHH), the Health and Safety at Work Act (1974), and, more specifically, the Carriage of Dangerous Goods and Use of Transportable Pressure Equipment Regulations (2004) and its amendment in 2005 which enacts the European Directive concerning the carriage of dangerous goods by road. International Air Transport Association (IATA) regulations cover dangerous goods transported by air.

pathogens
Pathogens are microorganisms that can cause disease in humans or animals.

Key Point

Infectious substances are those known or reasonably expected to contain pathogens.

Wastes from laboratories can also be infectious.

Infectious substances can be divided into two categories, A or B, based on how infectious they are likely to be.

If we are going to transport any infectious substance (samples or cultures) we have to decide which category it falls into and make sure that it is transported in a manner that complies with the regulations.

The substances that fall into Category A, which includes the higher risk infectious organisms, are shown in Table 1.1.

TABLE 1.1 **Category A samples.**

UN number and name	Microorganism
UN 2814 Infectious substances affecting humans	*Bacillus anthracis* (cultures only)
	Brucella abortus (cultures only)
	Brucella melitensis (cultures only)
	Brucella suis (cultures only)
	Burkholderia mallei – Pseudomonas mallei – Glanders (cultures only)
	Burkholderia pseudomallei – Pseudomonas pseudomallei (cultures only)
	Chlamydia psittaci – avian strains (cultures only)

(Continued)

TABLE 1.1 ***(Continued)***

UN number and name	Microorganism
	Clostridium botulinum (cultures only)
	Coccidioides immitis (cultures only)
	Coxiella burnetii (cultures only)
	Crimean–Congo haemorrhagic fever virus
	Dengue virus (cultures only)
	Eastern equine encephalitis virus (cultures only)
	Escherichia coli, verotoxigenic (cultures only)
	Ebola virus
	Flexal virus
	Francisella tularensis (cultures only)
	Guanarito virus
	Hantaan virus
	Hantaviruses causing hantavirus pulmonary syndrome
	Hendra virus
	Hepatitis B virus (cultures only)
	Herpes B (cultures only)
	Human immunodeficiency virus (cultures only)
	Highly pathogenic avian influenza virus (cultures only)
	Japanese encephalitis virus (cultures only)
	Junin virus
	Kyasanur Forest disease virus
	Lassa virus
	Machupo virus
	Marburg virus
	Monkeypox virus
	Mycobacterium tuberculosis (cultures only)
	Nipah virus
	Omsk haemorrhagic fever virus
	Poliovirus (cultures only)

(Continued)

TABLE 1.1 (*Continued*)

UN number and name	Microorganism
	Rabies virus
	Rickettsia prowazekii (cultures only)
	Rickettsia rickettsii (cultures only)
	Rift Valley fever virus
	Russian spring-summer encephalitis virus (cultures only)
	Sabia virus
	Shigella dysenteriae type 1 (cultures only)
	Tick-borne encephalitis (cultures only)
	Variola virus
	Venezuelan equine encephalitis virus
	West Nile virus (cultures only)
	Yellow fever virus (cultures only)
	Yersinia pestis (cultures only)
UN 2900 Infectious substances affecting animals only	African horse sickness virus
	African swine fever virus
	Avian paramyxovirus Type 1 - Newcastle disease virus
	Blue tongue virus
	Classical swine fever virus
	Foot and mouth disease virus
	Lumpy skin disease virus
	Mycoplasma mycoides - contagious bovine pleuropnemonia
	Peste des petits ruminants virus
	Rinderpest virus
	Sheep-pox virus
	Goatpox virus
	Swine vesicular disease virus
	Vesticular stomatitis virus

FIGURE 1.5
Example of a specimen transport label.

All samples not included in Category A come under Category B, and these must be packaged and transported according to Packaging Instruction 650 and labelled UN 3373.

This means that a leak-proof primary receptacle (normally the specimen container) must be placed inside an approved leak-proof secondary container containing absorbent material. This must be contained inside a rigid outer packaging, which is appropriately marked as in the example in Figure 1.5.

Category A samples must be packaged and transported under similar arrangements according to Packaging Instruction 620 and labelled UN 2814. Additional requirements include providing a leak-proof seal on the specimen container (e.g. heat seal, paraffin film), an itemized list of contents must be enclosed between the secondary and outer packaging and the words "*Suspected Category A infectious substance*" must be shown on this document.

1.6 Basic sample processing

The different parts of the body can have very different physical and chemical conditions. For example, the skin under your big toe is very different from the skin inside your ear, or the skin around your genitals. Conditions inside your lungs are different from the conditions in your bladder or intestines. This means that the bacteria that might live there as normal flora, and the ones that might enter as pathogens and cause disease, can be very different in each specific part of the body.

When we grow microorganisms in the laboratory, we usually try to put them onto culture media that provide similar conditions to the part of the body the sample came from originally. This means we try to make sure that the pH, nutrients such as sugars and proteins, gaseous conditions (aerobic, anaerobic, microaerophilic, increased CO_2, etc.) are the same as the body site we are interested in.

You can look at it as though we are trying to fool the pathogens into thinking they are still at the original body site, because if they were growing and multiplying happily there, then if we provide the same conditions in culture media, the pathogens should grow happily for us. This means that we have to provide many different culture media with different ingredients. There is no single 'best' culture medium that is suitable for all specimen types and sites.

Cross reference
Chapter 2

Specimens from sterile sites, such as cerebrospinal fluids (CSF), joint aspirates and blood cultures, can be cultured onto the most nutritious culture media available, because these samples have no resident normal flora that might overgrow any pathogens, and there may only be small numbers of the pathogens present.

In contrast, samples such as faeces, which come from areas with large numbers of normal flora, might require culture on media containing selective agents to help suppress some of the normal organisms and allow any pathogens to be detected.

Another way to achieve this is to use differential agents in the culture media, which will allow the pathogens to stand out from the normal flora, usually by being a specific, easily recognized colour.

In practice, many specimens will be cultured onto several different culture media (either culture plates or bottles of liquid media) to ensure that the widest possible range of pathogens can be grown.

Some samples, such as urine and CSF, will have microscopy performed on them before culture. This can help indicate whether the sample is infected, and may help in deciding whether antibiotic susceptibility tests should be put up directly.

Skin, hair and nail samples will have a portion of each type of sample placed into 30% potassium hydroxide (KOH) in a test tube and left for 10 to 120 minutes to digest the hard tissue.

A drop of this will then be placed on a slide, a cover slip is applied and the specimen is examined under the microscope to look for fungal cells.

Cross reference
Chapter 11, Section 11.4.

Microbiology specimens are usually processed in batches, keeping the different sample types together. All urines would be done together, in one section of the laboratory, while all blood cultures would be done in another section and all MRSA swabs in yet another. This allows maximum efficiency, as each section need stock only the culture media required for that particular type of specimen. As all the specimens will be processed using the same method, it is easier and more efficient for the person doing them.

Many microorganisms start to die as soon as they are taken away from the body, and it is necessary to process samples as soon as is reasonably possible, to get the best and most relevant result.

1.7 Inoculation of culture media

Because culture media will grow both contaminants and pathogens, it is important to use aseptic technique to avoid introducing contaminating organisms or allowing pathogens from the sample to come into contact with workers.

Culture media such as blood agar, which don't contain any inhibitors, should be inoculated first when possible. This prevents antibiotics and other selective substances killing microorganisms.

Sometimes, when only small numbers of pathogens may be present, a specific medium with inhibitors may be inoculated first (e.g. gonococcal culture plates).

If smears (wet preparations or stained slides) are needed, they should always be made after all culture media have been inoculated. This prevents contaminants that may be on the slides being transferred to the plates.

If smears must be made first (for example, if the result of the smear would guide the type of culture done) then a sterile slide must be used. These should be dipped in alcohol which is allowed to evaporate before use.

Liquid media (culture broths) should be inoculated after culture plates for swabs, but inoculated first with fluid samples.

Where possible, the following order should be followed:

- Media without inhibitors (e.g. blood agar, chocolate agar)
- Media with indicators (e.g. CLED agar, chromogenic agar)
- Media with inhibitors or selective agents (e.g. DCA agar, *B. cepacia* agar)
- Fluid media (e.g. nutrient broth, cooked meat broth)
- Smears for staining (e.g. Gram's stain, mycobacteria)

SELF-CHECK 1.5

Why is it best to inoculate non-selective media first?

Urine samples

Urine may be inoculated onto culture plates in several ways as follows:

- A sterile calibrated loop (1 μl, 2 μl or 10 μl) is dipped into the urine just below the surface and removed without tilting from vertical. This is then inoculated onto a culture plate.

A whole plate may be used, or two samples may be inoculated onto half a plate each. Four samples onto a quarter plate each has been used, but is not recommended.

- A sterile filter paper strip (available commercially, e.g. Bacteruritest) is dipped into the urine to the line on the strip. The strip is removed and any excess urine drained off onto the inside of the container. The strip is bent at the line and the inoculated area of the strip is then placed onto the culture plate, held for several seconds and removed. Ten to twelve urine samples may be inoculated onto a single culture plate using this method. More is not recommended.
- A number of urines (21, 36, 54 or 96) may be pipetted into sterile cups in a rack. A multipoint inoculator, which has an equivalent number of sterile pins, one of which goes into each cup, is used to pick up 1 μl of each urine. The multipoint inoculator pins are touched to the surface of an agar plate or agar in microtitre plates to inoculate spots with 1 μl each.

Some samples may also have antibiotic susceptibility tests performed directly from the sample, rather than from isolated colonies after culture. The process is similar for all samples. A sterile pipette, swab or loop is used to inoculate the sample on the centre of a plate, and the inoculum is spread all over the plate using a fresh swab, either from side to side by hand or in a circular pattern using a plate rotator, prior to placing antibiotic discs on the plate. Culture plates containing antibiotics can be inoculated directly with urine in the same way as above using a multipoint inoculator.

Swab samples

inoculum
The inoculum is the portion of the specimen that is put onto the culture plate or into the fluid medium.

Swabs are inoculated onto plates by rolling the tip over a quarter or a third of each plate to form a primary **inoculum.** Fluid media is inoculated by inserting the swab into the bottle and breaking (or cutting) it off near the tip.

Simply putting the swab in the fluid medium, agitating and removing the swab again is not sufficient as many organisms remain attached to the swab fibres and will be removed when the swab is, and so maximum inoculum will not be achieved.

Blood cultures

Cross reference
Chapter 5, Section 5.1.

Blood samples are inoculated into the blood culture bottles at the time the sample is taken. The bottles are incubated in the laboratory; often an automated system is used. When a bottle is suspected of being positive (i.e. having microorganisms present) it must be inoculated onto culture plates to grow the organism.

Most commercial systems supply a method for safely and aseptically removing a sample of the blood from inside the bottle. Often, this is a type of needle that can be pushed through the septum of the bottle and which allows a few drops of blood to escape when the bottle is inverted. These drops can be placed onto culture plates.

Fluids, pus and CSF

Fluid, pus and CSF samples may be centrifuged to concentrate small numbers of pathogens in the deposit, or may be inoculated uncentrifuged if the sample is thick or thought to contain lots of pathogens.

Either a sterile pipette or a sterile loop can be used to transfer around 10 μl of specimen to each culture plate or broth. If the sample is very thick, using a sterile swab to do this may be more convenient.

Tissue and biopsy samples

Tissue and biopsy samples are usually homogenized (ground up) in a little sterile saline using a sterile Griffiths tube, which is like a small enclosed pestle and mortar. This releases any organisms from inside the tissue sample.

A loopful or few drops from a sterile pipette are inoculated onto each plate or into fluid media.

Prostheses

These are usually made of metal or plastic and cannot be easily homogenized or sampled with a swab. Often the best way to culture these samples is to aseptically place them in a sterile container and add culture broth until the specimen is submerged. This can then be incubated.

Catheter tips

A catheter tip (or cannula) is rolled over the whole surface of the plate five times using a sterile swab, forceps or loop to move the tip, which is then removed. Any fluid or blood from the **lumen** of the catheter should be allowed to run onto the plate for culture.

lumen
The lumen is the inside of any hollow, tube-like structure.

Skin, hair and nails

A small flake of skin should be picked up with a sterile needle or forceps and pressed onto the surface of the culture plate or slope. This should be repeated for as many flakes as possible over the whole area of the plate.

Hair and nail samples should be picked up with forceps and pushed onto the surface of the culture plates in the same way.

Sputum

Sputum samples are normally quite sticky and mucoid, which makes picking a suitable portion of the sample to inoculate onto culture plates difficult. There are two common methods of inoculating sputum.

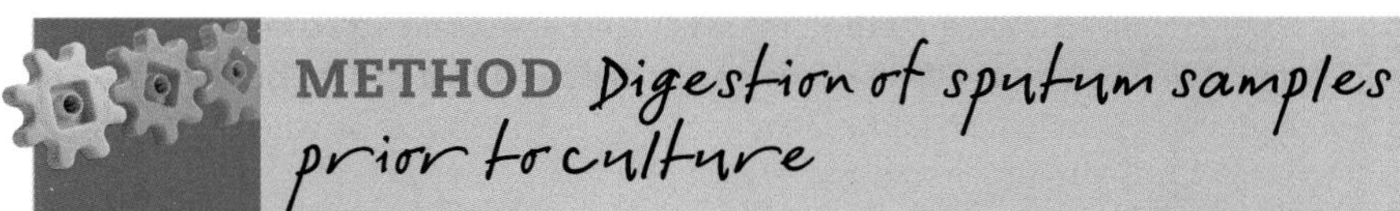

- An equal volume of a chemical that will break down the sticky mucus (e.g. Sputasol or *N*-acetyl-L-cysteine) is added to the sample.
- Mix well.
- The mixture is incubated at 37°C with periodic shaking until it is liquefied (around 10 min).
- Finally, the sample can then be inoculated onto culture plates with a sterile pipette or loop.

METHOD Decontamination of sputum samples

- An equal volume of 4% NaOH is added to the specimen and mixed well for 20 min. Continual mixing is best, but mixing several times may be used. TB is more resistant to the alkaline NaOH than contaminating bacteria.
- 0.2 mL of the digested mixture is inoculated onto culture media in slopes using a pipette. The culture slopes are at an acid pH, and adding the alkaline sample results in a neutral pH in the end.

OR

- An equal volume of 4% NaOH is added to the specimen and mixed well for 20 min. Continual mixing is best, but mixing several times may be used. TB is more resistant to the alkaline NaOH than contaminating bacteria.
- An equal volume of 2% HCl is added to the mixture to neutralize it
- 0.2 mL of the digested mixture is inoculated onto culture media in slopes using a pipette. The culture slopes are at a neutral pH, therefore the inoculum must be neutralized before being put on the slopes.

homogenized
Homogenized means liquefied and mixed, so that all parts of the whole sample are the same.

purulent
Purulent means containing a lot of pus.

Cross reference
Chapter 9, Section 9.2, Processing respiratory samples.

Some laboratories choose to dilute the **homogenized** sputum and inoculate 10 μl of the dilution onto culture plates. This helps to remove small numbers of contaminants picked up from the mouth and also to provide an estimate of the quantity of pathogens in the sample.

A **purulent** fleck of the sample is selected and is removed with a sterile swab or loop. This can be quite difficult to achieve due to the stickiness of sputum, and takes a little practice. The chosen fleck is inoculated onto culture plates using the loop or swab.

Sputum samples for tuberculosis (TB) culture need to be pre-treated with a digestion/decontamination stage to remove all non-TB microorganisms before culture. TB is slow-growing and any contaminants in the sample can hide the growth of TB. TB organisms are more resistant to chemicals than other bacteria. There are two methods of achieving this.

SELF-CHECK 1.6

Why are samples for TB treated with harsh chemicals before culture whereas other samples are not?

Faeces

Liquid faeces samples can be inoculated onto culture plates and broths using a loop, wooden stick or swab. These do not need to be sterile. Faeces is not sterile, and the pathogens that are looked for are not found as environmental contaminants, therefore it is unlikely that any contamination will cause a problem.

Solid faeces samples should be broken up using a loop, stick or swab and, if soft enough after this, it can be inoculated onto culture plates. If it is still solid, emulsifying it in a little saline or nutrient broth can be helpful.

Usually, several selective culture plates are inoculated for faecal pathogens.

Liquid media are also used for faeces culture, as there may only be very few pathogens present and enrichment may be necessary. These liquid media also contain selective agents to remove normal faecal organisms which may otherwise overgrow any pathogens present.

Wet preparations are made when looking for faecal parasites. These can be made in saline, or with a little fresh iodine added to help differentiation.

Smears and wet preparations for microscopy

Smears are made by rolling the swab across a circle marked with a diamond or a wax pencil on a slide. Smears of fluid or pus samples are made using a sterile pipette, loop or swab to transfer some of the sample to a marked circle on a slide.

If the sample is thick, a smear can be made from it directly; however, if it is very thin, it may need to be centrifuged first. The supernatant is then removed and the deposit resuspended in a small amount of liquid. This is used to make the smear.

Wet preparations may be useful for detecting parasites such as *Trichomonas* spp. or clue cells (indicating bacterial vaginosis) in genital samples, or faecal parasites from stools.

Wet preparations are made by dabbing a transport medium swab onto a slide and covering the removed material with a coverslip. The same effect can be achieved with a dry swab by placing a drop of saline on the slide and dabbing the swab into it.

A drop of a fluid sample is placed on the slide with a loop and covered with a coverslip. Solid faeces samples are emulsified in a drop of saline on the slide before the coverslip is added.

Spreading culture plates

In order to get the most out of bacterial cultures, the initial inoculum on a culture plate must be spread over the rest of the plate so that individual colonies grow. This allows mixed cultures to be easily detected and picked off for antibiotic susceptibilities or identification. This is known as 'spreading', 'plating out' or 'streak dilution'.

There are many different ways to achieve a good spread but the aim is to produce well isolated, single colonies after incubation. There are now instruments that can plate out swabs and achieve excellent results.

The initial inoculum should cover between a quarter and a third of the total area of the plate. If half plates are used, this should be reduced accordingly.

The initial inoculum is spread over the rest of the plate in several separate streaks using a sterile loop, as shown in Figure 1.6. A fresh loop may be used for each set of streaks if desired.

When quarter plates are used, it is normal for the initial inoculum to cover the entire segment of the plate.

1.8 Safety considerations

Why are samples sent to the microbiology laboratory in the first place? It is usually because the clinician thinks that the site the sample was taken from may be infected, but it is also a fact that there will always be patients or specimens that haven't been identified as infected,

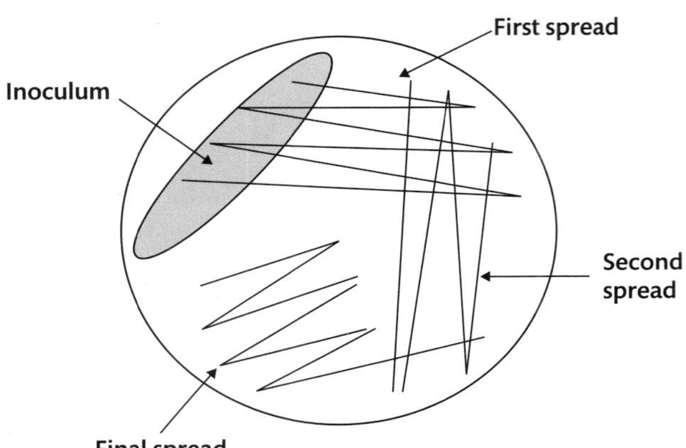

FIGURE 1.6
Method of spreading a culture plate.

but which turn out to be afterwards. All microbiology samples must therefore be regarded as potentially infected from the moment they are taken, and during transport, storage, sorting, processing and disposal. The correct specimen collection techniques must be observed, and samples must be collected into the correct laboratory approved container.

The container should not be overfilled. However generous the patient has been in supplying copious amounts of specimen, it doesn't all have to be sent to the laboratory. The rim and the outside of the container should not be contaminated when the sample is placed in it. If contamination occurs, the container should be discarded and a clean one used. The caps must be tightened correctly on all specimen containers to avoid leaking and spilling while in transit to the laboratory.

Anyone taking specimens is responsible for correctly completing all details on the sample container and request form, and must label any particularly high-risk samples (such as hepatitis or HIV-positive bloods) appropriately.

Samples should only be transported to the laboratory inside a sealed plastic bag. These may be attached to the request form, or may be separate, with a pocket to hold the request form. It is important to remember that porters and couriers who may transport specimens have little knowledge of microbiology, and may not understand the dangers of leaking or contaminated specimens. However, porters and couriers must be given appropriate training in handling specimens safely and how to deal with spillages.

All staff who handle specimens should be instructed to wear the appropriate, fastened protective clothing and to cover any cuts and grazes with waterproof dressings. Hands must be washed after each spell of duty and before breaks. Porters must wash their hands on entering and leaving each ward, and when leaving the laboratory.

In the laboratory, the following points should be considered at all times:

- Whenever specimens are handled, aerosol production should be minimized. Swabs should be removed from tubes gently and caps and lids of fluid samples should be opened slowly.
- Fluid samples may be under increased pressure, especially if microorganisms in the sample have been allowed to grow and produce gas during transport. If the top is removed too quickly, the contents may discharge explosively.
- Samples should not be mixed or shaken vigorously before opening.
- Disposable gloves must be worn when handling fluid samples and any high-risk specimens.

- If there is a danger of the specimen splashing during processing, eye protection (safety glasses, goggles or visors) must be worn.
- Leaking specimens should be assessed by a senior member of the technical staff to determine if they can be processed safely.
- Spills kits and disinfectants must be available, and staff must have appropriate training.
- All spills must be dealt with immediately and must be cleaned and disinfected using the most appropriate method.

All microorganisms in the laboratory are potentially infectious, but some pose a greater hazard than others. They are categorized into hazard groups as follows:

Hazard Group 1: A biological agent unlikely to cause disease.

Hazard Group 2: A biological agent that can cause human disease and may be a hazard to employees; it is unlikely to spread to the community and there is usually effective prophylaxis or effective treatment available.

Hazard Group 3: A biological agent that can cause severe human disease and presents a serious hazard to employees; it may present a risk of spreading to the community, but there is usually effective prophylaxis or treatment available.

Hazard Group 4: A biological agent that causes severe human disease and is a serious hazard to employees; it is likely to spread to the community and there is usually no effective prophylaxis or treatment available.

There are different containment levels that are appropriate for the different hazards of each group.

Microorganisms in Groups 1 and 2 are handled in Containment Level 1 or 2 laboratories. Level 2 is the normal standard for microbiology laboratories.

Microorganisms in Hazard Group 3 require Containment Level 3, which needs additional safety features such as a microbiological safety cabinet, **HEPA** filtration of exhaust air, a sealable laboratory for disinfection and access only to authorized people.

Hazard Group 4 microorganisms require Containment Level 4. This requires a separate laboratory with a fully sealed Class III safety cabinet, an airlock, an autoclave and extremely rigorous safe working practices over and above those for Level 3. There are very few of these laboratories in the UK.

SELF-CHECK 1.7

What are the highest and lowest containment levels? At which level are most routine microbiology samples processed?

CHAPTER SUMMARY

After reading this chapter you should know and understand:

- Sample quality is the single most important stage of microbiological examination.
- A poorly collected sample will produce a poor result.

- Some microorganisms start to die as soon as the sample is taken, while others can multiply rapidly.
- Insufficient or inappropriate samples may mean that the required investigations cannot be carried out.
- Wrongly collected, transported and labelled specimens should be rejected to ensure that incorrect results are not reported.
- Microbiology samples are potentially hazardous and must be handled and transported safely at all times.

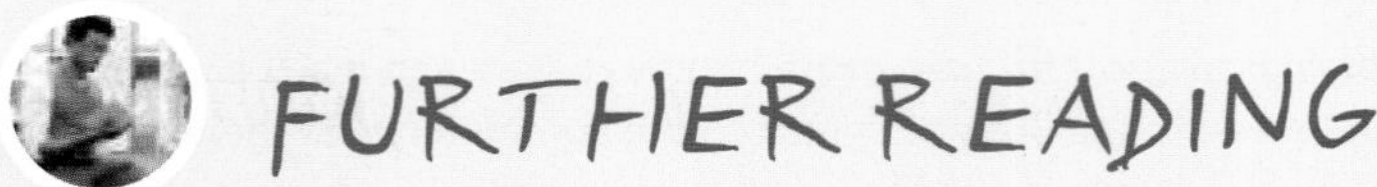

FURTHER READING

- **Health Services Advisory Committee. *Safe Working and the Prevention of Infection in Clinical Laboratories*. The Stationery Office, London, 1991.**

  Essential publication which covers the health and safety requirements for clinical laboratory work.

- **Hawkey P, Lewis D, eds. *Medical Bacteriology*, 2nd edition. Oxford University Press, 2004.**

  Excellent general text covering many aspects of sample processing and other areas of microbiology.

- **Mallett J, Bailey C, eds. *Manual of Clinical Nursing Procedures*, 6th edition. Blackwell Science, 2005.**

  Text which explains further the procedures for taking clinical samples.

- **Collins CH, Lyne PM, Grange JM, eds. *Collins and Lyne's Microbiological Methods*, 8th edition. Arnold, 2004.**

  This is a useful text which provides further detail on many aspects of this chapter. Essential laboratory handbook.

DISCUSSION QUESTIONS

1.1 Why do you think there are different types of swab available?

1.2 Why is it important to take blood cultures before blood for other laboratory tests and before antibiotic therapy is started?

1.3 Why is it important to transport microbiology samples correctly?

1.4 Why are microorganisms categorized into different groups and how does this affect laboratory handling?

2

Identification tests

John Perry

This chapter will concentrate principally on the biochemical reactions that allow for differentiation of clinically important bacteria. The application of other identification tests, including immunological tests, will be discussed.

Bacterial identification is typically based on a presumptive assignment to a genus or species based on morphological and cultural characteristics, followed by confirmation using biochemical tests. Bacteria are usually first isolated on agar-based culture media and this allows morphological and cultural features to be gathered first. These include: colony size, shape and colour; degree and type of haemolysis on blood-based media; size, shape and Gram reaction of cells on microscopy; atmospheric requirements for growth (e.g. failure to grow anaerobically). The assessment of these features will, in many cases, enable a presumptive assignment of the bacterial isolate to its family or genus.

For example, a Gram stain of colonies that only form under anaerobic conditions revealing spore-forming Gram-positive bacilli, would suggest that the isolate belongs to the genus *Clostridium*. Similarly, small α-haemolytic colonies confirmed as Gram-positive cocci and showing both aerobic and anaerobic growth would be presumptively regarded as *Streptococcus* sp. If there is any doubt about the likely family or genus to which an isolate belongs, two very simple biochemical tests for catalase and oxidase may be useful and results can be achieved within minutes.

Learning objectives

After studying this chapter you should be able to:

- Explain how bacterial hydrolases (e.g. glycosidases, peptidases, esterases) can be detected and give examples of specific hydrolases produced by pathogens
- Give examples of how carbohydrate metabolism by bacteria can be exploited in bacterial identification
- Give examples of how amino acid metabolism by bacteria can be exploited in bacterial identification
- Give examples of how susceptibility tests can assist microbial identification
- Explain the basis of simple immunological tests for the rapid identification of bacteria

2.1 Biochemical tests

Catalase test

Detection of catalase is a highly useful test in bacterial diagnostics and this enzyme is commonly found in a wide range of bacterial species. Catalase results in the decomposition of potentially toxic hydrogen peroxide into water and oxygen. Detection of catalase is most useful for discrimination between genera, for example, allowing differentiation between staphylococci (+) and streptococci (–) or listeria (+) and lactobacilli (–).

Oxidase test

The oxidase test is able to distinguish bacteria that possess cytochrome *c* as a respiratory enzyme (Figure 2.1). The test is simple and requires only a single reagent. Most commonly, the choice of reagent is *N*,*N*,*N*,*N*-tetramethyl-*p*-phenylenediamine dihydrochloride. This reagent is water soluble and colourless in solution. On exposure to oxidase-positive bacteria the reagent is rapidly oxidized to produce a purple–blue dye (Wurster's blue). As with catalase, detection of oxidase activity is useful for differentiation of bacterial families and genera. For example, all genera among the Enterobacteriaceae such as *Escherichia* or *Enterobacter* are oxidase negative (*Plesiomonas* is an exception) and this differentiates them from genera such as *Pseudomonas* and *Aeromonas* (oxidase positive). Among Gram-negative cocci, *Acinetobacter* (oxidase negative) may be differentiated from *Moraxella* and *Neisseria* (oxidase positive).

Although the oxidase test has its greatest use for differentiation of Gram-negative bacteria, a few Gram-positive species demonstrate oxidase activity and this can be useful, for example in the differentiation of *Staphylococcus sciuri* from most other species of staphylococci.

METHOD Catalase test

- **Using a plastic loop, sample a single colony and rub this inoculum onto a glass slide. The inoculum should be sufficiently large to be visible to the naked eye.**
- **Place a drop of 30% hydrogen peroxide onto a glass coverslip and place the coverslip onto the slide, exposing the bacterial inoculum to the hydrogen peroxide.**
- **Observe for immediate generation of oxygen bubbles (catalase positive).**
- **Discard the slide and coverslip into disinfectant.**
- **Precautions:**
  - **Avoid the use of metal loops or inoculating wires as these may cause false-positive results.**
  - **False-positive results are commonly obtained when colonies have been selected from blood agar as blood cells contain catalase.**
  - **Hydrogen peroxide is quite unstable and controls must be used to ensure optimal performance of the test.**

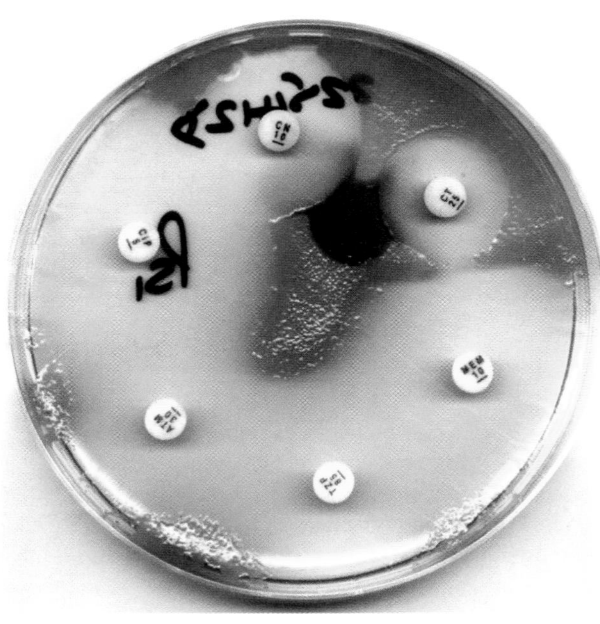

FIGURE 2.1
Oxidase reagent applied to a *P. aeruginosa* susceptibility test plate. The blue-violet coloration shows the organism is oxidase positive. (Courtesy of M. Ford.)

SELF-CHECK 2.1

What problems are associated with the catalase test and how can these be overcome?

Detection of bacterial hydrolases - glycosidases

Glycosides are sugar derivatives that occur naturally and are found commonly, for example, in plants. It has long been recognized that most clinically important bacteria produce glycosidase enzymes. These are enzymes that hydrolyse (or 'break down') glycosides to release the sugar part of the molecule, which is then used for generation of energy. Any non-sugar part of the molecule is referred to as an **'aglycone'** and detection of this free aglycone can be used to demonstrate the presence of a **glycosidase.** The nomenclature of the glycosidase enzyme is governed by the type of derivative it hydrolyses. So, for example, a galactosidase will hydrolyse derivatives of galactose (i.e. galactosides) and a glucosidase will hydrolyse glucosides, etc. It is important to remember that these enzymes do not hydrolyse the sugars themselves, but rather, sugar derivatives. So, for example, a glucosidase will not act on glucose and a glucuronidase will not act on glucuronic acid, etc.

The linkage between the sugar and the aglycone is also very important. This glycosidic linkage may have alpha or beta orientation and this will determine whether the enzyme is able to act. So, for example, a β-galactosidase will only hydrolyse substrates containing galactose that is linked to another molecule or group via a beta linkage.

In microbiology, one of the best known glycosides is esculin, a naturally occurring glucoside that can be extracted from the bark of the horse chestnut tree (*Aesculus hippocastanum*). Esculin is hydrolysed by a glycosidase known as β-glucosidase. Hydrolysis of esculin leads to the release of glucose and an aglycone known as esculetin. In the presence of iron salts (for example ferric ammonium citrate) esculetin forms a brown/black complex or 'chelate', which indicates that esculin has been hydrolysed and β-glucosidase is present. The structure of esculin and the complex formed after hydrolysis are shown in Figure 2.2a. Since 1926, the hydrolysis of esculin in the presence of bile has been recognized as a useful identification test to differentiate enterococci and group D streptococci from other streptococci (Figure 2.2b). The esculin hydrolysis test is also useful for differentiation of species within the Enterobacteriaceae.

aglycone
A molecule that is not a sugar but is linked to a natural sugar (e.g. galactose) via a glycosidic bond (example: nitrophenol in *ortho*-nitrophenyl galactopyranoside).

glycosidase
A glycosidase is an enzyme that hydrolyses compounds consisting of a sugar molecule linked to another molecule in a glycosidic linkage. The other molecule may or may not be another sugar. Natural substrates include esculin or disaccharides such as lactose.

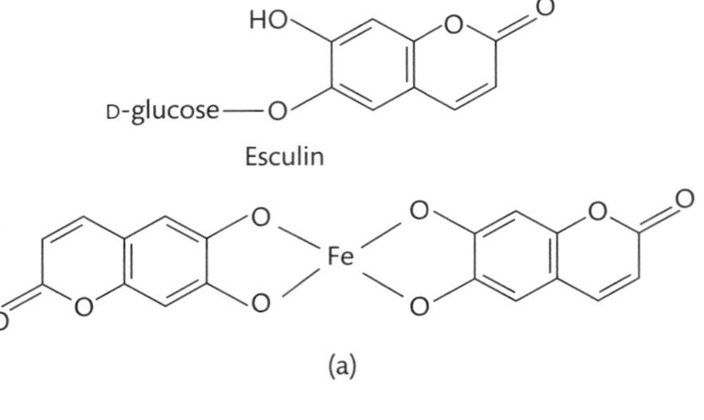

FIGURE 2.2
(*a*) The structure of esculin and the complex formed upon hydrolysis. (*b*) Colonies of *E. faecalis* showing hydrolysis of esculin on agar-based medium. (Courtesy of J. Perry.)

synthetic substrate
Synthetic enzyme substrates are not obtained from natural sources but are chemically synthesized in order to demonstrate enzyme activity. A synthetic substrate that releases a coloured dye when it is hydrolysed is referred to as a chromogenic substrate.

As a convenient assay for glycosidases, sugars may be linked to coloured dyes (via alpha or beta linkages) to form **synthetic enzyme substrates**. A key feature of such substrates is that when the dye is linked to the sugar, its colour is quenched. Hydrolysis of the substrate then leads to the release of the dye and the restoration of colour. This is, therefore, a very convenient and simple means of demonstrating bacterial enzyme activity. Such substrates that release coloured dyes when hydrolysed are known as 'chromogenic substrates' because a visible colour is produced and this can be interpreted by the biomedical scientist.

The best-known synthetic chromogenic substrate in clinical microbiology is *o*-nitrophenyl-β-D-galactopyranoside (ONPG), which comprises galactose in beta linkage with a yellow compound, *o*-nitrophenol. When in solution ONPG is virtually colourless, but the presence of bacteria that produce β-galactosidase causes hydrolysis, leading to the generation of a bright yellow colour. The ONPG test is a very useful test for differentiating *Salmonella* (–) from many other genera within the Enterobacteriaceae, for example *E. coli* (+) and *Citrobacter* spp (+). The reaction is illustrated in Figure 2.3.

fluorogenic substrate
A fluorogenic substrate may be natural or synthetic but is metabolized (usually by a hydrolase) to release a fluorescent compound.

As well as linking sugars to coloured dyes, they may also be linked to fluorescent molecules to form non-fluorescent compounds that are highly sensitive **'fluorogenic' enzyme substrates**. Hydrolysis leads to restoration of fluorescence, which may be measured by using a fluorimeter or simply observed under a suitable ultra violet light source. The most commonly used fluorescent molecule or 'fluorophore' is 4-methylumbelliferone (4-MeU), which emits a powerful sky-blue fluorescence under UV light. A commonly used **fluorogenic substrate** is 4-methylumbelliferyl-β-D-glucuronide, which is useful for the detection of β-glucuronidase produced by *E. coli*.

Other glycosidase enzymes useful in microbial identification include *N*-acetyl-β-D-glucosaminidase produced by *Candida albicans*, which distinguishes it from most other yeasts.

SELF-CHECK 2.2

What compounds are linked to carbohydrates and how are these used for bacterial identification?

Key Point

It is important to remember that the compounds released by bacterial enzyme action will readily diffuse through agar and it will be impossible to pick out the target organism in a mixture. Hence most of these tests are performed in tubes or other enclosed containers.

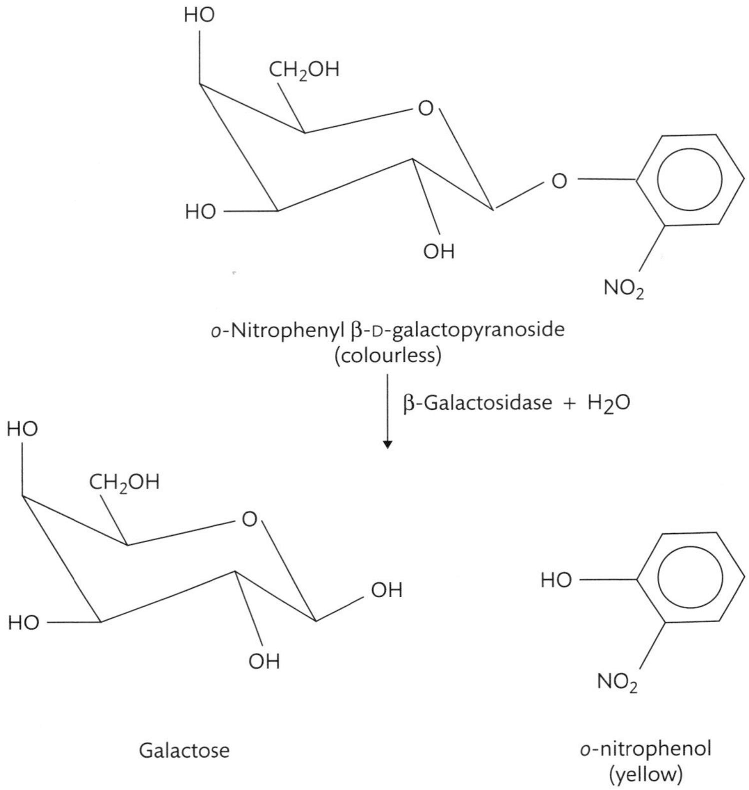

FIGURE 2.3
Hydrolysis of ONPG to release yellow *o*-nitrophenol. (Courtesy of J. Perry.)

Peptidases

Many bacteria produce a wealth of peptidases involved in the breakdown of the peptide linkages that exist between amino acids. Synthetic substrates have been devised that link an amino acid to a coloured dye to form a colourless enzyme substrate. Hydrolysis then releases the amino acid, which is used for metabolism, and the coloured dye. One such dye that is widely used is *p*-nitroaniline. This compound diffuses readily through agar and can also be quite toxic to bacterial cells.

The amino group of this molecule can be linked to the carboxyl group of various amino acids to generate aminopeptidase substrates. Whether or not the substrate is hydrolysed depends to a large extent on the nature of the amino acid. L-alanine-*p*-nitroanilide is a colourless substrate comprising L-alanine linked to *p*-nitroaniline. All Gram-negative bacteria demonstrate a strong L-alanyl aminopeptidase activity, resulting in hydrolysis of this substrate and release of a yellow dye. The enzyme is referred to as an 'aminopeptidase' because the amino group (NH_2) of the amino acid is free. An alternative name for the enzyme is L-alanyl aryl amidase.

L-Pyroglutamic acid may also be linked to *p*-nitroaniline to form a chromogenic substrate for detection of pyroglutamyl aminopeptidase (also known as pyrrolidonyl peptidase). Detection of this enzyme is highly useful for differentiation of enterococci (+) and group A streptococci (+) from other streptococci (–); and differentiation of *Salmonella* and *Shigella* (–) from many other genera of Enterobacteriaceae, for example *Citrobacter* (+) and *Hafnia* (+).

A number of other aminopeptidases are useful for biochemical identification including γ-glutamyl aminopeptidase that can discriminate *Neisseria meningitidis* (+) from other

Neisseria (–). Detection of prolyl aminopeptidase differentiates *Neisseria gonorrhoeae* (+) from other *Neisseria* (–) and is a useful identification marker for *Clostridium difficile*. β-alanyl aminopeptidase is a useful identification marker for *Pseudomonas aeruginosa*. Finally, leucyl aminopeptidase (LAP) is useful for the discrimination of *Leuconoctoc* spp. and *Aerococcus viridans* from the streptococci.

As well as chromogenic methods for the demonstration of aminopeptidase activity, fluorogenic substrates based on amino acids linked to 7-amino-4-methylcoumarin are commercially available and widely used. Like many other released compounds, the fluorogen will readily diffuse through agar and the whole plate will fluoresce and the target pathogen will be impossible to pick out.

We have shown that various aminopeptidase enzymes are useful for identification but other types of peptidase may also be valuable taxonomic markers. Hippuric acid (or in its salt form, sodium hippurate) comprises the amino acid glycine linked to benzoic acid in a peptide linkage. This substrate is found naturally as a product of excretion in the urine of horses and other herbivores. Many species of bacteria hydrolyse hippurate to form benzoic acid and glycine and the latter is usually detected by addition of ninhydrin reagent. This enzymatic activity is referred to as hippurate hydrolase (Figure 2.4) and is essentially a 'carboxypeptidase' activity as the carboxyl group of the glycine is free. Detection of this enzyme activity is useful for discrimination of group B streptococci (+) from most other streptococci (–); discrimination of *Campylobacter jejuni* (+) from *Campylobacter coli* (–) is also a useful marker for *Legionella* spp.

SELF-CHECK 2.3

Why do you think there are both chromogenic and fluorogenic substrates available?

Esterases

Coloured dyes such as *p*-nitrophenol and fluorophores such as 4-methylumbelliferone (see section on glycosidases above) may be derivatized with carboxylic acids ('fatty acids') to form esters such as acetate, butyrate, propionate, etc. Removal of the carboxylic acid by hydrolysis of the ester link is accomplished by bacterial esterases and results in generation of colour or fluorescence respectively. Short chain esters are easily hydrolysed and the vast majority of bacteria (if not all) hydrolyse acetate esters. Hydrolysis of butyrate esters is less common and is useful for discrimination of *Moraxella catarrhalis* from *Neisseria* spp. *Salmonella* spp. hydrolyse longer chain esters

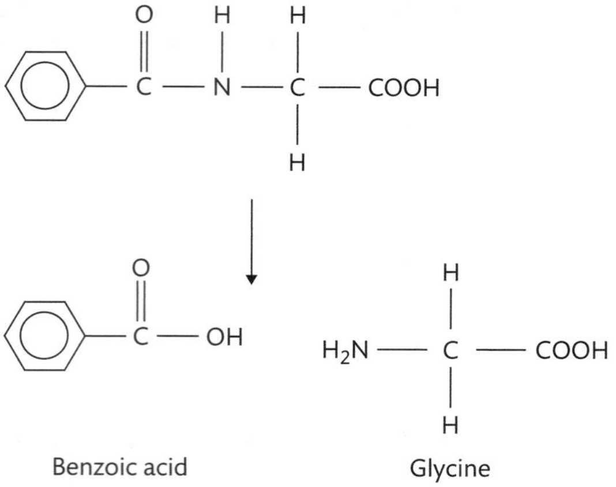

Benzoic acid

Glycine

FIGURE 2.4
Hydrolysis of hippuric acid to glycine and benzoic acid. (Courtesy of J. Perry.)

such as octanoate (C8) and nonanoate (C9) derivatives and this activity is an excellent diagnostic marker for the discrimination of *Salmonella* (+) from most other Enterobacteriaceae (–).

Phosphatase

Phosphatases are produced by a wide range of clinically important bacteria. They are most conveniently detected using phosphate esters of coloured or fluorescent dyes such as *p*-nitrophenol and 4-methylumbelliferone, respectively. Enzymatic hydrolysis liberates inorganic phosphate and restoration of colour or fluorescence. Phosphatase activity can be used as an identification marker for *S. aureus* but, depending on the choice of substrate and the length of incubation, the enzyme can be readily demonstrated in species of coagulase-negative staphylococci. Phosphatase is also a useful identification marker for *Clostridium perfringens* if an appropriate substrate with high specificity is used such as phenolphthalein diphosphate.

Deoxyribonuclease (DNase)

Deoxyribonuclease or DNase is an extracellular enzyme capable of hydrolysing DNA to release oligonucleotides. A heat resistant DNase or thermonuclease is produced by *S. aureus* and detection of this enzyme is a useful diagnostic marker as other staphylococci produce lower amounts of the enzyme or none at all. Traditionally, the enzyme has been detected by incorporating DNA into an agar-based medium and inoculating the test organism. After incubation, the medium is flooded with hydrochloric acid. The addition of acid results in precipitation of the DNA, which is visible in the agar; however, if the DNA is hydrolysed a clear zone will be apparent around bacterial colonies. Alternative assay formats have been developed to avoid the inconvenience of flooding agar plates with acid. One such adaptation of the method involves the inclusion of a metachromatic dye such as toluidine blue O. This dye complexes with intact DNA to produce a blue colour; however, when DNA is hydrolysed, this dye forms a pink colour with oligonucleotides. As well as being a useful identification marker for *S. aureus*, DNase is also produced by some streptococci (e.g. *S. pyogenes*), *Serratia marcescens* and *Corynebacterium diphtheriae*.

Hydrolysis of gelatine

Some bacteria demonstrate a proteolytic activity that results in digestion or liquefaction of gelatine, resulting in the generation of polypeptides and/or amino acids. There are various assay formats involving both solid and liquid media. In one of the most effective assays, finely powdered charcoal particles are bound to a small amount of solid gelatine. Digestion of the gelatine by bacterial enzymes results in the release of the charcoal particles into suspension, this forming a clearly visible black precipitate. Hydrolysis of gelatine is a useful diagnostic marker for several species of Enterobacteriaceae including *Proteus* spp. and *Serratia* spp. Gelatinase is also produced by members of the Vibrionaceae, for example *Vibrio cholerae* and *Aeromonas hydrophila*, as well as being useful for the differentiation of non-fermentative Gram-negative rods and for differentiation between *Clostridia* spp.

Amino acid metabolism

Amino acids may be metabolized by bacteria in a variety of ways depending on their ability to produce certain enzymes. Lysine and ornithine are atypical amino acids in that they each contain two amino groups and a single carboxyl group. Some Gram-negative bacteria decarboxylate either lysine or ornithine, or both, to generate diamines (cadaverine and putrescine respectively), which are very alkaline. This activity can be conveniently detected using an appropriate pH indicator such as bromocresol purple or phenol red. Pyridoxyl phosphate may be included as a coenzyme to enhance activity. Detection of lysine or ornithine decarboxylase activity provides excellent discrimination between species of Enterobacteriaceae (see Table 2.1).

Some bacteria are able to split the amino acid arginine via a complex dihydrolase reaction, also generating ammonia as an end-product, which may be detected using a pH indicator. The demonstration of arginine dihydrolase (ADH) is a useful marker for some species of Enterobacteriaceae including *Enterobacter cloacae* and *Salmonella* spp. and also related bacteria such as *Aeromonas hydrophila*. It is also useful for differentiation within the staphylococci, enterococci and streptococci.

When testing for either decarboxylase or dihydrolase activity it is preferable to perform the tests in tubes in which the test medium is overlaid with mineral oil to create anaerobic conditions for growth. In the presence of air, bacteria readily oxidize peptone to form alkaline end-products which may lead to false positive results.

TABLE 2.1 A simple biochemical test scheme for identification of Enterobacteriaceae from urine samples.

Biochemical profile						
Lysine	**Ornithine**	**Glucose**	**Indole**	**Cellobiose**	**Urea**	**Identification**
–	+	+	+	+	–	*Citrobacter koseri*
–	–	+	–	–	–	*Citrobacter freundii*
+	+	+	–	+	–	*Enterobacter aerogenes*
–	+	+	–	+	–	*Enterobacter cloacae*
+	+ or –	+	+	–	–	*Escherichia coli*
+ or –	+	+	+	–	–	*Escherichia coli*
+	–	+	+	+	+ or –	*Klebsiella oxytoca*
+	–	+	–	+	+ or –	*Klebsiella pneumoniae*
–	+	+	+	–	+	*Morganella morganii*
–	+	+	–	–	+	*Proteus mirabilis*
–	–	+	–	–	+	*Proteus penneri*
–	–	+	+	–	+	*Proteus vulgaris*
–	–	+	+	–	–	*Providencia* species
+	+	+	–	–	–	*Serratia* species

As well as the removal of the carboxyl group of amino acids, an amino group may be removed by deaminase enzymes. Deamination of aromatic amino acids such as tryptophan and phenylalanine leads to the release of ammonia and the generation of organic acids (indole pyruvic acid and phenylpyruvic acid respectively) which may be detected by their formation of coloured complexes in the presence of ferric ions. Deaminase activity is largely restricted to the three related genera *Proteus*, *Providencia* and *Morganella*, and it is therefore an excellent diagnostic marker for this group.

In addition to being a target for deaminases, certain species are able to split tryptophan to form pyruvate, ammonia and indole. The latter is detected by specific reagents such as Kovac's or James' reagent to generate a red coloration. Generation of indole is a very useful identification test for differentiation of Enterobacteriaceae and as a marker for many other species of Gram-negative bacteria such as *Aeromonas hydrophila*, *Pasteurella multocida* and *Fusobacterium necrophorum*. The ability to generate indole from tryptophan is rare amongst Gram-positive bacteria but it is useful for identification of certain species, for example *Clostridium tetani*.

Carbohydrate oxidation/fermentation

Most species of clinical importance utilize sugars via either oxidation or fermentation to generate organic acids. This activity is generally detected by use of a suitable pH indicator such as phenol red or bromothymol blue. The ability to acidify glucose under anaerobic conditions (fermentation) is a fundamental test that can differentiate major groups of bacteria, for example distinction of all members of the family Enterobacteriaceae (+) from non-fermentative genera such as *Pseudomonas* (–), *Acinetobacter* (–) and *Stenotrophomonas* (–), or the differentiation of staphylococci (+) from micrococci (–).

Not only is the ability to ferment glucose of fundamental importance to bacterial identification, so is the way in which it is fermented. Fermentation of glucose may lead to the formation of different end-products and this varies between species. For example, when *Klebsiella* spp. and *Enterobacter* spp. ferment glucose, 2,3 butanediol is a major end product of metabolism whereas the closely related *E. coli* does not generate this compound. These differences in metabolism are exploited by the Voges–Proskauer (VP) test. In the VP test bacteria are allowed to ferment glucose and, after incubation, 40% potassium hydroxide and 5% α-naphthol (dissolved in ethanol) are added. These reagents react with acetoin (the precursor of 2,3 butanediol) to form a pink coloration indicating a positive VP reaction. The VP test is highly useful for differentiation of species within the Enterobacteriaceae and also within staphylococci and streptococci.

For differentiation of very closely related species within a genus (e.g. *Staphylococcus* or *Enterococcus*) it is often necessary to test a range of appropriate sugars to obtain a biochemical profile. A wide range of compounds are acidified by bacteria and these include pentoses (e.g. L-arabinose, D-xylose), hexoses (e.g. D-glucose, L-sorbose), deoxysugars (e.g. L-rhamnose), disaccharides (e.g. sucrose, lactose), trisaccharides (e.g. melezitose, raffinose), polysaccharides (e.g. inulin, starch), glycosides (e.g. esculin, salicin), and sugar alcohols (e.g. adonitol, D-mannitol).

For most bacteria, such tests are generally straightforward and consist of peptone, 1% sugar and a suitable pH indicator, for example phenol red. When testing fastidious bacteria, such as *Neisseria* spp., it may be necessary to include 2% serum to allow growth. Serum should be heat-treated to ensure that potentially interfering enzymes (e.g. maltase) are inactivated.

SELF-CHECK 2.4

Having read the above, what tests would you use to differentiate between the following?

(a) *Pseudomonas* and Enterobacteriaceae

(b) *Proteus mirabilis* and *E. coli*

(c) *Salmonella* spp. and *Shigella* spp.

(d) *Staphylococcus* spp. and *Streptococcus* spp.

2.2 Miscellaneous biochemical tests

Urease

This enzyme is found in a range of diverse species and causes the hydrolysis of urea into carbon dioxide and ammonia. Ammonia is a strong base and consequently there is a notable rise in pH caused by hydrolysis of urea. This is usually detected with a suitable indicator such as phenol red (Figure 2.5). Detection of urease is useful for the differentiation of genera within the Enterobacteriaceae. A strong urease reaction is generated by *Proteus* spp. and weaker reactions are common in *Klebsiella* spp. and occasionally in *Serratia* spp. On media for the isolation of enteric pathogens, *Proteus* often resembles *Salmonella* and *Shigella* and a rapid test for urease can be useful to identify *Proteus* and thus exclude these two pathogens.

Helicobacter pylori produces a urease with exceptional activity that may be detected in minutes and detection of this enzyme is highly useful in the diagnosis of gastritis.

Hydrogen sulphide production

Some bacteria are able to generate hydrogen sulphide from compounds containing sulphur, including sodium thiosulphate and sulphur-containing amino acids such as cysteine or methionine. The release of hydrogen sulphide is detected using a suitable metal ion that forms an insoluble black precipitate with hydrogen sulphide. Commonly, this involves incorporation of an iron salt into the growth medium (e.g. ferric ammonium citrate) or the use of filter paper soaked with lead acetate to detect hydrogen sulphide liberated into the atmosphere above the growth medium. The test is mainly applied to the differentiation of Enterobacteriaceae and few species are positive including *Salmonella* spp., *Proteus* spp. and *Citrobacter freundii*.

Reduction of nitrate

Nitrate may be reduced by a wide range of bacteria to produce either nitrite or nitrogen gas. To test for nitrate reductase, bacteria are incubated in a nutrient broth supplemented with 0.1% potassium nitrate. After incubation, two reagents are added, *N*,*N*-dimethyl-α-naphthylamine and sulphanilic acid, and these will cause the development of a red diazonium compound if nitrite is present. A red coloration therefore indicates a positive result, that is nitrate has been reduced. Lack of colour upon addition of reagents does not necessarily indicate a negative result as denitrification may have led to decomposition of nitrite and generation of nitrogen gas. This can be resolved by addition of powdered zinc to the test medium, which will cause reduction of intact nitrate to nitrite, resulting in a red coloration in the presence of the

FIGURE 2.5
Urease test showing positive (*P. mirabilis*: red) and negative (*S. enteritidis*: colourless) results. (Courtesy of Paul Kibby: Oxoid.)

reagents. If the test medium remains colourless after addition of zinc, it can therefore be concluded that nitrate has been completely reduced resulting in formation of nitrogen gas, that is the test is positive.

Most bacteria encountered in the clinical laboratory reduce nitrate and the test is therefore a useful marker for species that do not. These include a number of Gram-negative non-fermenters such as *Acinetobacter* spp., *Alcaligenes faecalis* and *Sphingobacterium* spp. The nitrate reduction test is also highly useful for division of the mycobacteria where only a minority of species, including *Mycobacterium tuberculosis*, are able to reduce nitrate.

Coagulase

Outside of the urinary tract, *Staphylococcus aureus* is the most predominant pathogen encountered in clinical microbiology. Detection of staphylocoagulase or 'free' coagulase using

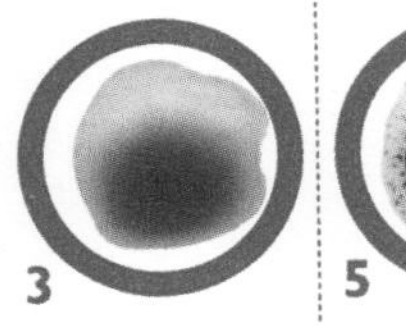
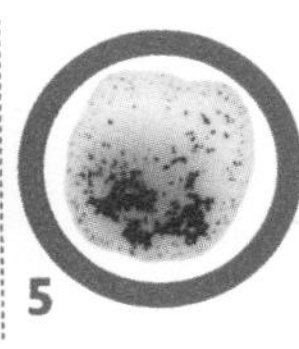

FIGURE 2.6
Slide coagulase test showing negative control (*S. epidermidis*: left) and positive result (*S. aureus*: right). (Courtesy of M. Ford.)

the tube coagulase test is a highly reliable means of confirming the identification of *S. aureus*. Among other staphylococci, coagulase is only produced by rarely encountered species such as *S. schleiferi* and *S. intermedius*. *S. aureus* strains lacking staphylocoagulase are uncommon.

The biochemical mechanism of the coagulase enzyme is complex and the subject of much debate. The enzyme reacts with a thermostable thrombin-like substance present in plasma, resulting in the formation of a fibrin clot. Human or rabbit plasma are recommended for the test and suitable preparations (containing anticoagulant, e.g. citrate) are available from a number of commercial sources. To perform the test, the plasma is mixed with an equal volume of an overnight broth culture of the test strain in a test tube; alternatively a loopful of growth from solid media may be directly suspended in the plasma to produce an equivalent turbidity. The tube is incubated at 37°C and examined after 1, 4 and 18 h incubation. If staphylocoagulase is present, a visible clot is normally apparent within 4 h of incubation.

As well as producing free staphylocoagulase most strains of *S. aureus* possess a bound coagulase or 'clumping factor' that may be detected in slide tests (the tube method detects both types). A loopful of growth from solid medium is mixed in physiological saline to produce a heavy suspension. The suspension must be homogeneous and show no evidence of autoagglutination after thorough mixing. Autoagglutination is common if strains are selected from media containing a hypertonic salt concentration, for example mannitol salt agar, and this will invalidate the test. If the saline suspension is suitable, a loopful of plasma may then be added and mixed into the suspension. The presence of *S. aureus* will result in immediate, coarse clumping of the plasma. Any delayed reactions (i.e. > 20 seconds) should be confirmed by the tube coagulase test (see above). For both tests, positive and negative controls are essential and must give expected reactions before interpretation is made.

Chromogenic and fluorogenic substrates have been developed for detection of staphylocoagulase. Examples include those based on the chemical formula *N*-*t*-boc-Val-Pro-Arg-X where X is a fluorescent or coloured label such as 7-amino-4-methylcoumarin (7-AMC) or *p*-nitroaniline (*p*-NA), respectively. Prothrombin also needs to be incorporated in the assay. The presence of staphylocoagulase results in release of the label, which may be detected visually as colour or fluorescence or by a suitable optical device. A number of commercially available assays are available that employ chromogenic or fluorogenic methods. An example is the Staphychrom II, which is able to detect staphylocoagulase with high specificity in a 2-h test.

SELF-CHECK 2.5

Why are two types of coagulase test available to identify *S. aureus*? What are the differences and when would you use them?

Lecithinase (phospholipase C)

A major component of lecithin is phosphatidylcholine, which may be hydrolysed by bacteria to generate phosphorylcholine and insoluble digylceride. Egg yolk is a rich source of lecithin and this may be incorporated into agar media in order to demonstrate lecithinase activity. Colonies that produce lecithinase produce a halo of opalescence due to the precipitation of diglyceride which is released from lecithin by enzyme activity.

Demonstration of lecithinase is characteristic of a number of species of *Clostridium* including *C. perfringens*, *C. sordellii*, *C. novyi* and *C. bifermentans*. It is also produced by a number of *Bacillus* spp. including the pathogenic species *B. cereus* and *B. anthracis*. Most commonly, the lecithinase reaction is used for the identification of *C. perfringens*. The α toxin of *C. perfringens* is responsible

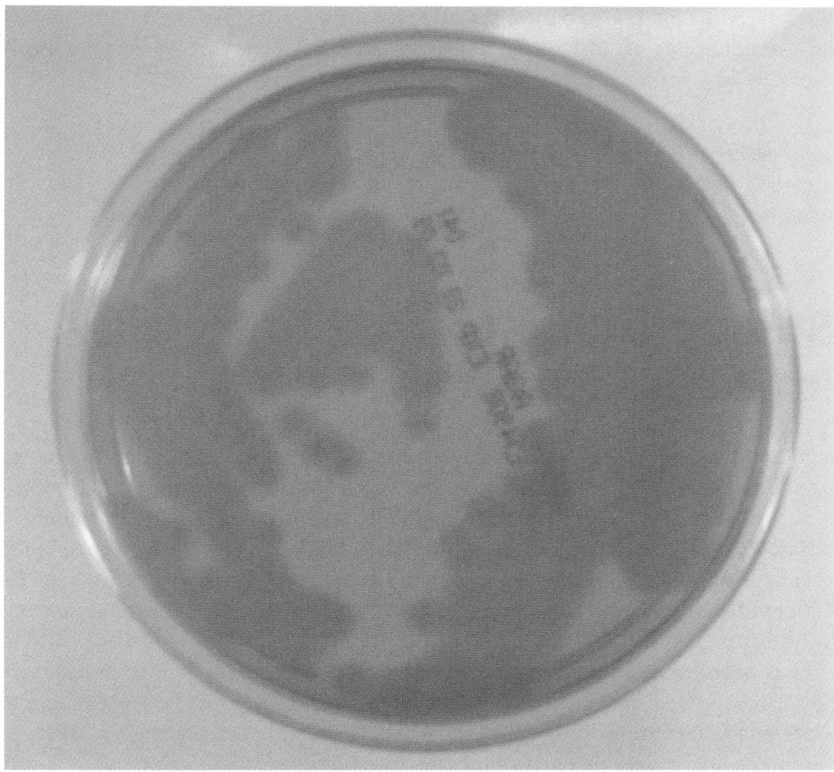

FIGURE 2.7
Clostridium perfringens growing on Nagler egg yolk medium. Note the diffuse opacity due to production of lecithinase C. This opalescence is neutralized with specific antitoxin. (Courtesy of M. Ford.)

for lecithinase activity and its reaction on egg yolk media can be neutralized using specific antitoxin. This is known as the Nagler reaction (Figure 2.7). To test for the Nagler reaction, an agar plate medium containing egg yolk is prepared and specific antitoxin is spread on the surface of half of the plate. A suspect strain of *C. perfringens* is then streaked across the medium so that both halves of the plate are inoculated. After incubation *C. perfringens* shows a clearly visible opalescence, which is inhibited on one side of the plate due to neutralization by specific antitoxin.

2.3 Carbon source utilization tests

These are used widely for bacterial identification and are particularly useful for relatively 'unreactive' bacteria, for example those that do not acidify sugars. The best known carbon source utilization test is the citrate test. Bacteria are incubated with various salts including ammonium phosphate (as a sole nitrogen source) and citrate (as a sole carbon source). Bacteria that are able to utilize citrate are able to grow and visible growth is used as an endpoint to indicate a positive test. Alternatively, bromothymol blue may be included and an alkaline (blue) reaction indicates growth due to the release of ammonia from ammonium phosphate. Classically, the citrate test is used in the differentiation of the Enterobacteriaceae and can help differentiate *E. coli* and *Shigella* spp., which in contrast to most other genera are unable to utilize citrate as a sole carbon source.

A range of other organic acids may also be used in carbon source utilization tests to help differentiate species. These include sodium salts of acetate, benzoate, lactate, oxalate, propionate, pyruvate, succinate, tartrate, malate, gluconate and malonate. When performing such tests care must be taken to ensure that nutrient media (containing additional sources of carbon) are not accidently included with the bacterial inoculum and the inoculum itself should generally be low.

SELF-CHECK 2.6

Why are carbon source utilization tests used and not carbohydrate fermentation tests for some organisms?

2.4 Susceptibility to antibiotics and chemical inhibitors

The susceptibility of bacteria to antimicrobial agents can provide useful information to assist the identification process. *Staphylococcus saprophyticus* is a urinary tract pathogen that possesses an intrinsic resistance to novobiocin, thus differentiating it from most other commonly encountered species of staphylococci, although some other species (e.g. *S. xylosus*) are also resistant. Among Gram-negative bacteria, resistance to colistin can be a useful identification marker, for example for distinction of *Burkholderia cepacia* complex (always resistant) from *Pseudomonas* species (usually susceptible). Susceptibility to bacitracin has been used for many years for the differentiation of *Streptococcus pyogenes* from other β-haemolytic streptococci. Finally, susceptibility to metronidazole can help to differentiate anaerobes from non-anaerobes. The use of antibiotic resistance markers as tools in identification should be used with care as acquired resistance can emerge during antimicrobial chemotherapy. Nalidixic acid was widely recommended as an identification marker in the 1980s to differentiate *Campylobacter laridis* (resistant) from other thermophilic campylobacters. However, widespread use of quinolones in subsequent years led to a large proportion of *Campylobacter jejuni* expressing resistance to nalidixic acid. Despite these limitations, an experienced biomedical scientist will always review the biochemical identification of a pathogen in the light of all other available information, including its antibiogram.

Susceptibility to non-antibiotic substances may also be of diagnostic use. Optochin (ethyl hydrocupreine hydrochloride) is a highly useful test to differentiate *Streptococcus pneumoniae* from other α-haemolytic streptococci. 2,4-diamino-6,7-isopropylpteridine phosphate (O/129) is an agent that inhibits the growth of *Vibrio* spp. providing differentiation from closely related genera (e.g. *Aeromonas* spp.). Resistance to 9-chloro-9-(4-diethylaminophenyl)-10-phenylacridan (C390) is a highly useful marker for *Pseudomonas aeruginosa* among non-fermentative Gram-negative bacteria, although around 4% of strains are susceptible. Tolerance to various salts has also been used as an identification marker. For example, the ability to grow in the presence of 6.5% sodium chloride is a characteristic of enterococci but not streptococci and growth in the presence of potassium cyanide (0.0075%) has been shown to be useful for differentiation of species of Enterobacteriaceae. Finally, the ability to grow in the presence of various dyes (e.g. thionin, basic fuchsin and safranin O) has been used to distinguish between the three main pathogenic species of *Brucella*.

SELF-CHECK 2.7

Which do you consider to be the better tests to differentiate between organisms—biochemical or substance susceptibility testing? What are your reasons?

2.5 Tolerance to environmental conditions

Clinically important bacteria grow at pH 7 at 37°C, although that may not be their optimum growth temperature or pH. The ability to withstand deviations in temperature and pH varies between different groups of bacteria and tests for tolerance to such deviations can be useful for identification. For example, it has been recognized for many years that enterococci are particularly robust and will typically grow at temperatures between 10 and 45°C and at a pH as high as 9.6. They are also able to maintain viability after being heated at 60°C for 30 min and these features may contribute to their differentiation from closely related genera. *Pseudomonas aeruginosa* is able to grow at 42°C unlike the closely related *Pseudomonas fluorescens* and *Pseudomonas putida*. *Listeria* spp. are able to grow at 4°C thus distinguishing them from *Erysipelothrix* and *Lactobacillus* spp.

As well as the ability to grow or survive at extremes of temperature, the behaviour of bacteria may vary according to the temperature of incubation. This is not surprising, as 37°C is higher than the optimal temperature for a number of pathogenic bacteria. For example, *Listeria* spp. show characteristic tumbling motility at 22°C but are sluggishly motile or non-motile at 37°C. *Yersinia pseudotuberculosis* is also motile at 22°C but not at 37°C. Biochemical characteristics of bacteria may also vary and this is classically seen with *Yersinia* spp. For example, *Yersinia enterocolitica* is much more likely to demonstrate motility, ornithine decarboxylase, β-galactosidase and a positive VP reaction at 22°C rather than 37°C. The ability to grow in different atmospheric conditions may also be highly useful in bacterial identification.

2.6 Requirement for growth factors

The growth requirements of bacteria can be exploited for the identification of species as best illustrated by the requirements of *Haemophilus*. On primary isolation, all species of *Haemophilus* require either X factor (haemin or haematin) or V factor (nicotinamide adenine dinucleotide), or both, for growth. These requirements can be demonstrated by inoculating a suspect strain of *Haemophilus* onto a basic medium (without blood) and placing paper discs impregnated with X, V and X + V onto the surface of the medium. Nutrient agar has traditionally been used for X and V tests but brain heart infusion agar, trypticase soy agar and Mueller Hinton agar have also been recommended. Following incubation, growth should only be visible around the discs that provide the factors necessary for growth. For example, *H. influenzae* is the most frequent *Haemophilus* spp. isolated in clinical laboratories and requires both X and V factors to grow. Growth will therefore only occur around the disc containing both growth factors. Identification of the requirement for X and V assists with discrimination between species. The requirement for growth factors (particularly the requirement for X) can be lost during serial subculture and this has been well documented with certain species, for example *H. aphrophilus*. As with any test for growth factor dependence, great care should be taken to use a modest inoculum and to avoid the transfer of nutrients that may contaminate the bacterial inoculum.

2.7 Identification schemes

We have discussed the principles of a range of biochemical tests. To assign a bacterial strain to a particular species, it is often necessary to perform a battery of tests to obtain a 'biochemical profile'. The choice of which biochemical tests to perform will be based on the results of preliminary observations such as colonial appearance, interpretation of Gram stain, etc. Table 2.1 shows a simple test scheme for identification of a suspect strain of Enterobacteriaceae from a urine sample using six biochemical tests for fermentation of glucose and cellobiose, hydrolysis of urea, production of indole and decarboxylation of lysine and ornithine. For most commonly encountered species, a reliable identification is provided. Problems arise when uncommon urinary pathogens such as *Salmonella* spp. or *Kluyvera* spp. are encountered and these are likely to be misidentified.

To ensure accurate identification of a wider range of species, including those that are rarely encountered, a wider range of biochemical tests must be employed and, in general, the more tests that are used, the more accurate the identification. This is facilitated by the availability of commercial identification kits and automated identification systems.

2.8 Commercial kits

A wide range of commercial kits are available from a number of manufacturers for the identification of bacteria. Typically, these consist of a plastic strip with compartments or wells containing dehydrated growth media and metabolic substrates. Manual inoculation of a bacterial suspension of known density is then required to re-hydrate the contents of the wells. Following incubation, the wells of the strip are observed for growth and/or colour changes. Manual addition of reagents is often required for some of the tests to be interpreted. In 1971, the first evaluation of the API 20 E was reported. This strip comprised a gallery of 20 biochemical tests including nine distinct sugar fermentation tests. Following addition of reagents to three of the tests, and interpretation of colour changes in all 20 wells, a seven-digit profile is generated. This profile number is then computed using dedicated software to give the most likely identification of the strain. These strips allow for the discrimination and identification of over 100 species and are widely used today.

The last 30 years have seen the development of API galleries designed for the identification of most other clinically important species. Some groups of bacteria do not readily acidify sugars and the most useful way to identify these is with the use of 'assimilation tests'. In such tests, a minimal growth medium in which the bacteria do not grow is supplemented with a single substrate, which offers the bacteria its only significant source of carbon. The ability of the bacteria to metabolize this substrate will result in growth and therefore a positive test. Carbon source assimilation tests are a major feature of the API 20 NE (Figure 2.8) (for non-enteric Gram-negative bacteria, e.g. *Pseudomonas* spp.) and the API Campy strip (for *Campylobacter* spp.).

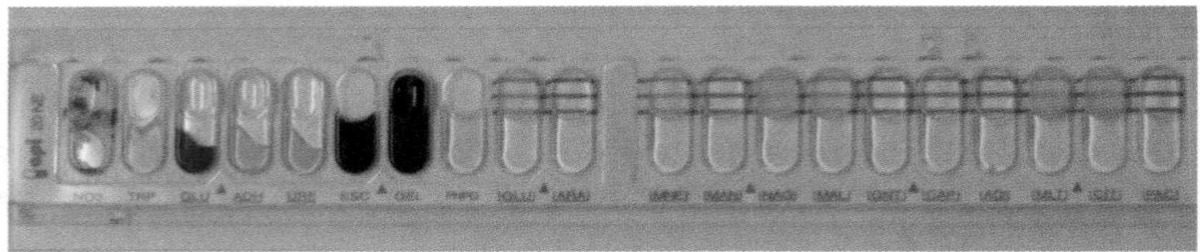

FIGURE 2.8

Identification of *Stenotrophomonas maltophilia* using the API 20 NE. The strain shows reduction of nitrate, hydrolysis of esculin, liquefaction of gelatin and production of β-galactosidase. In addition, the strain is able to utilize any of several substrates as a sole source of carbon, including glucose, mannose, *N*-acetyl glucosamine, maltose, malic acid and citrate. (Courtesy of J. Perry.)

In addition to the API strip, a wide range of identification systems based on similar principles are available from a number of manufacturers. Examples of kits for the identification of Enterobacteriaceae include the GN2 MicroPlate (Biolog Inc.), the Crystal E/NF (Becton Dickinson), the Microbact (Oxoid) and the Micro-ID (Remel). There are ample comparisons of the performance of these kits published in microbiology journals to allow potential users to assess their relative attributes and limitations.

SELF-CHECK 2.8

What are the advantages of using commercial kit systems such as API for the identification of microorganisms?

2.9 Automated systems

Biochemical tests that do not require addition of reagents after incubation lend themselves to automation. Cards containing panels of miniaturized biochemical tests (Figure 2.9) may be inoculated with a standardized bacterial suspension and then monitored during incubation by sophisticated optical readers for changes in turbidity or development of colour or fluorescence due to hydrolysis of enzyme substrates. Since the early 1980s, automated instruments have been available for both bacterial identification and antimicrobial susceptibility testing. A number of systems are available for automated identification including the Microscan WalkAway (Siemens), Vitek 2 (bioMérieux), BD Phoenix (Becton Dickenson), Trek Sensititre (Sensititre) and the GN2 MicroPlate (Biolog). For some rapidly growing bacteria (e.g. strains of Enterobacteriaceae) some automated systems may generate an identification result in as little as 2–3 h.

A somewhat different approach is employed in the GN2 MicroPlate system (Biolog). The system utilizes a 96-well microtitre plate and each test well contains a different carbon source. The test strain is inoculated into all wells, which also contain a tetrazolium dye. If the strain is able to utilize the carbon source, microbial growth causes reduction of the dye and generation of a purple colour. This results in the generation of a carbon source metabolic 'fingerprint'. The trays can be used manually with visual recording of the results or as part of a fully automated system. Identification of at least 526 species or taxa is possible including Enterobacteriaceae, non-fermenters and fastidious organisms. For some species, identification can be achieved in 4 h.

Scientific evaluations of commercially available products may show a wide range of accuracy, most often due to differences in study design, and it is prudent to review a number

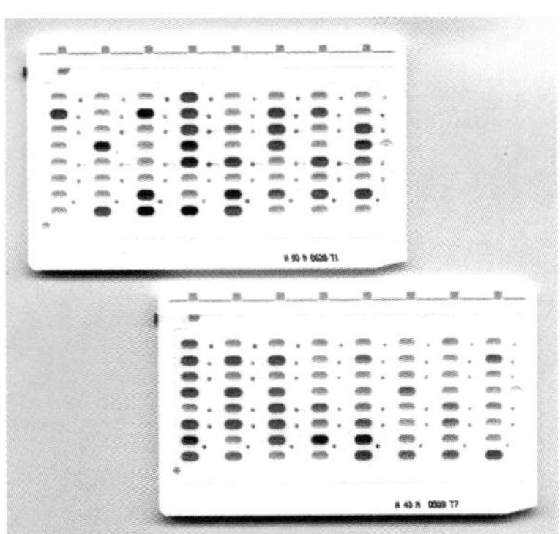

FIGURE 2.9
Identification cards used in the Vitek 2 automated system for the identification of bacteria. The 64-well card includes 47 biochemical reactions which are all colorimetric and are monitored by the optical readers of the instrument. Chromogenic substrates for detection of hydrolases include those based on *p*-nitrophenol and indoxyl (for detection of glycosidases and esterases) and *p*-nitroaniline and dichloroaminopenol (for detection of peptidases). The identification profile that is generated for Gram-negative organisms allows differentiation of 55 genera and 130 species of Enterobacteriaceae and non-fermentative Gram-negative bacilli. (Courtesy of M. Ford.)

of published evaluations. There are a number of systems with essentially equivalent performances and other factors such as cost, patient population, throughput and workflow play an important part in selection. A detailed description of some of these systems has recently been reviewed (see Further Reading).

2.10 Immunological tests

The surface of a bacterial cell contains a variety of antigens that may be targeted by antibodies so that a signal is generated when a specific antigen–antibody reaction occurs. In the simplest immunological tests, a bacterial suspension is mixed on a glass slide with a suspension of unlabelled antibody and the mixture is gently agitated and observed for visual agglutination. This results from an antibody, which has two identical variable regions that can interact with antigen, interacting with more than one bacterial cell and causing clumps of organisms bound to antibody. Such agglutination tests are widely used for the detection of somatic polysaccharide ('O' antigens) present in the Gram-negative outer membrane. Agglutination tests are widely used for differentiation of species of *Shigella*, for typing within the genus *Salmonella* and for differentiation of *E. coli* that cause diarrhoea (e.g. *E. coli* O157). Typing of *Salmonella* flagella (H) is also routinely sought using agglutination tests. Cross reactions between *Salmonella* antibodies and other species of Enterobacteriaceae are common and it is necessary to confirm the identity of *Salmonella* and *Shigella* spp. using biochemical tests as well as agglutination tests.

The interaction of antibody and antigen resulting in agglutination can be magnified by using a latex agglutination test. In such a test, latex particles of approximately 1 μm in diameter are coated with antibody and then mixed with a bacterial suspension. If specific antigens are present on the cell surface agglutination is readily visible as the latex particles leave the milky suspension and form clumps due to the antigen–antibody reaction. Commercially available latex reagents have become widely used for the detection of Gram-positive pathogens such as *S. aureus* and *S. pneumoniae*. Latex particles for detection of *S. aureus* may be coated with a number of distinct monoclonal antibodies targeting cell surface antigens and may also be coated with fibrinogen to detect 'clumping factor'. Some other species of staphylococci can generate specific agglutination reactions such as *S. intermedius* and *S. lugdunensis* although these are relatively uncommon.

The latex agglutination test is also widely used for identification of haemolytic streptococci of groups A, B, C, F and G. The letters designate carbohydrate antigens present in the cell wall and these may be readily extracted using simple enzymatic methods or chemicals such as nitrous acid. Some bacteria are capable of causing apparent clumping of latex particles that is unrelated to specific antigen–antibody reactions. For this reason it is essential to perform a negative control, usually consisting of latex reagent without specific antibody, to exclude non-specific latex agglutination.

2.11 Pitfalls and quality issues with biochemical identification tests

The most fundamental point for any identification test or kit is the necessity for pure cultures. Ideally, the inoculum for identification tests should originate from a single well-isolated colony but this is not possible for some commercial kits as they require a high inoculum. Some selective culture media will inhibit bacterial growth but do not kill bacteria and it is possible that viable bacteria that are not visible may be inadvertently included in the inoculum for identification tests.

When bacteria are inoculated into identification media and incubated overnight, a suitable non-selective culture plate should always be inoculated and incubated concomitantly. Before the results of any biochemical tests are interpreted, the purity of the inoculum must be confirmed by examination of the subculture (often termed a 'purity plate'). This safeguard is compromised when rapid identification tests are performed and results are reported in 4–6 h, before purity can be verified. Some automated systems may be able to detect the likelihood of a mixed inoculum but this cannot be guaranteed and if there is any question about the purity of the inoculum, this should be resolved before any conclusions are drawn.

Biochemical identification is also compromised by bacteria that behave atypically. For example, auxotrophic mutants may arise that are dependent on growth factors, for example thymidine-dependent *S. aureus* or *E. coli*, and such strains may not produce expected reactions in identification tests. With highly atypical or rarely encountered strains the identification process can seem more like an art form than a scientific process and the microbiologist must assimilate all of the available information about the strain and use all of his or her experience. In addition to the results of biochemical tests, all of the following may prove useful to point the microbiologist in the right direction: colony morphology, Gram stain, atmospheric growth requirements (e.g. requirement for carbon dioxide), antimicrobial susceptibility pattern, ability to grow on a range of media (e.g. requirement for blood) and seemingly extraneous information such as the site of isolation (blood, urine, etc.), clinical details of the patient, antibiotic therapy, etc.

Reagents for performing identification tests such as catalase and oxidase reagents or latex agglutination reagents should be tested with control organisms that give positive and negative results to ensure adequate performance, and results should be documented. This should be performed either daily or whenever a test is carried out. Identification media should also be inoculated with controls. Strains obtained from recognized culture collections with well-defined characteristics should be used as control organisms.

After reading this chapter you should know and understand:

- The importance of biochemical tests in the identification process
- The most important biochemical tests used in microbiology and how these tests work and are performed
- How susceptibility patterns can aid in the identification of specific organisms or groups of organisms
- How tests are performed routinely by use of commercial kit systems, e.g. API kits
- The role and methodology of immunological tests for bacterial identification
- The advantages and disadvantages associated with automated identification of microorganisms
- Problems associated with identification of bacteria

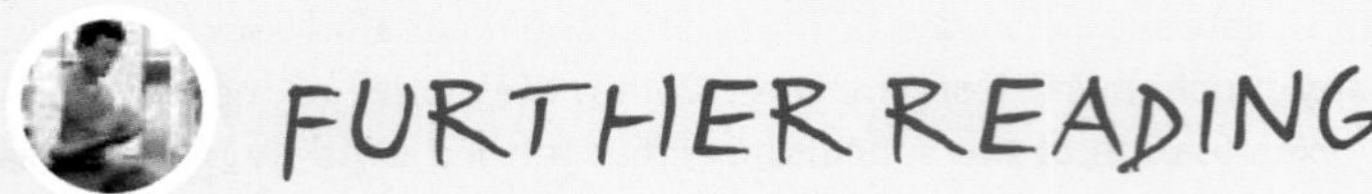

FURTHER READING

- Barrow GI, Feltham RKA, eds. ***Cowan and Steel's Manual for the Identification of Medical Bacteria***, **3rd edition. Cambridge University Press, 2008.**

  A recommended, comprehensive overview of biochemical tests for identification of medically important bacteria.

- MacFaddin J. ***Biochemical Tests for Identification of Medical Bacteria*****. Williams & Wilkins, 1999.**

  Another recommended, comprehensive overview of biochemical tests for identification of medically important bacteria.

- O'Hara CM. ***Manual and automated instrumentation for identification of Enterobacteriaceae and other aerobic Gram-negative bacilli.*** *Clinical Microbiology Reviews* **2005**: **18**: **147–162**.

  A useful review for further information on automated systems for microbial identification and susceptibility testing.

- Orenga S, James AL, Manafi M, Perry JD, Pincus DH. ***Enzymatic substrates in microbiology (review).*** *Journal of Microbiological Methods* **2009**: **79**: **139–155.**

  A detailed overview of enzyme substrates used in microbiology including automated instruments.

DISCUSSION QUESTIONS

2.1 Do you believe it is necessary to identify all organisms isolated from culture? If not, which organisms do you need to identify? Provide reasons for your answer.

2.2 Do you believe automated identification systems will replace biochemical and immunological identification tests? If not, why not?

3

Culture media

John Perry

In 1890, Robert Koch published four criteria designed to establish a causal relationship between a microbe and a disease. One of these stipulated that the microorganisms must be isolated from the diseased individual and grown in pure culture. Despite over 100 years of advances in diagnostic microbiology, the use of culture media remains central to the diagnosis of most bacterial and fungal infections in any clinical microbiology laboratory.

Pathological specimens are referred to the laboratory for diagnosis and are typically inoculated onto one or more culture media. Following incubation, microbial colonies are identified to detect pathogenic bacteria thus assisting the diagnosis of infection. A wide range of media are available for assisting the isolation and identification of numerous pathogenic species and this chapter will review the types of media available, how they work and how they are prepared.

Learning objectives

After studying this chapter you should be able to:

- Identify the basic growth requirements of bacteria including some fastidious bacteria
- Identify the basic ingredients of culture media and the contribution of specific supplements
- Give examples of selective, differential and enrichment media and what they are used for
- Explain the principles of chromogenic media and provide examples of their use
- Understand the principles and pitfalls of media preparation
- Explain how media are assessed for quality and performance

3.1 Microbial requirements for growth

A bacterial culture medium must provide all of the nutritional requirements for the bacterial species it is expected to recover. All bacteria require a range of elements in order to grow; these include carbon, oxygen, hydrogen and nitrogen which are constituents of amino acids; sulphur for the synthesis of amino acids such as cysteine and methionine; phosphorus for nucleic acids. Potassium, magnesium and calcium are major cellular cations and iron is an important component of cytochromes. Trace elements such as manganese, colbalt and zinc are also important as co-factors for enzymes.

Bacteria encountered in the clinical laboratory are heterotrophic and therefore need to be supplied with organic compounds in order to grow. Their requirements may vary widely, for example some species such as *Pseudomonas aeruginosa* are able to grow on a culture medium containing only a single source of carbon, such as acetate. This is remarkable when one considers the vast diversity of carbon-containing compounds that *Pseudomonas* spp. will need to synthesize in order to grow. Other bacteria may be more demanding and require other **growth factors** such as vitamins, purines and pyrimidines, or specific amino acids. Bacteria that require a range of growth factors are termed **fastidious**. Occasionally, strains of bacteria lose their innate ability to synthesize a compound essential for growth. Such strains are termed auxotrophic and therefore rely upon the addition of this compound as a growth factor.

growth factor

A growth factor is a specific compound that is required for growth of a particular organism. As it cannot be synthesized it must be added to culture media in order to allow growth.

fastidious

Fastidious organisms require a number of growth factors.

SELF-CHECK 3.1

Why do microorganisms require so many different types of nutrients for growth?

3.2 Ingredients of culture media

Media may be 'chemically defined', which means that the composition has been formulated using known amounts of pure chemical compounds. Such media are invaluable for studying the growth requirements of bacteria and for studies of bacterial metabolism. In practice, culture media used by diagnostic laboratories are not defined as they contain extracts of complex biological materials of which the exact composition is unknown. The reasons for this are convenience and the fact that bacteria generally grow better on complex undefined media. Some common constituents of complex culture media are listed below.

Water

A culture medium typically comprises >90% water. Water makes up 80–90% of the bacterial cell and is the solvent in which all metabolic processes take place. Deionized water is usually recommended for media preparation as tap water may have a variable mineral content and some constituents can be harmful to microorganisms.

Protein derivatives

Digests or extracts of protein are major constituents of most culture media. 'Peptones' are very commonly used and these consist of the water soluble products generated when a suitable protein source is digested with proteolytic enzymes, for example trypsin. Suitable protein sources include meat, soya flour and casein. The products of digestion include a variety of compounds to support bacterial growth including peptides, amino acids, phosphate, various metal ions in trace amounts and potential growth factors such as nicotinic acid and riboflavin. A good peptone can be dissolved in water, and without other additives can provide a good medium for growth of many commonly encountered bacteria, for example *E. coli* and *S. aureus*.

Another common source of nutrients is provided by extracting the water soluble components of protein. For example, using methods originally devised by Liebig, lean finely divided beef may be infused with boiling water to extract its water soluble components. Following removal of fats, these components may be concentrated to produce a powdered 'meat extract' or 'meat infusion'. This is a rich source of nutrients and includes nitrogenous compounds such as amino acids, purines and glutathione as well as salt and other minerals and a range of potential growth factors. The combination of peptone and meat extracts is frequently used in the formulation of commercially produced culture media.

Key Point

Media used in clinical microbiology are not strictly defined in terms of their chemical composition and are mostly based on extracts or digests of animal or plant tissues that are chemically complex.

Yeast extract

Cells of brewer's or baker's yeast may be lysed by gentle heat or chemical methods and the insoluble cellular material (e.g. cell walls) removed by filtration. The resulting product is a rich mix of a wide range of amino acids, carbohydrate (e.g. glycogen), minerals and growth factors (especially of the vitamin B group).

Gelling agents

In order to separate different bacterial species from a polymicrobial inoculum it is necessary to obtain isolated colonies on a solid medium. Agar is by far the most widely used and convenient gelling agent for solidifying culture media. Agar is a product of various seaweeds and is extracted as an aqueous solution at temperatures above 100°C. Its attributes include the fact that it is relatively inert, it has convenient solidifying and melting temperatures (38°C and 84°C respectively) and has high gel strength at a relatively low concentration (1%). Agar may contain unwanted, or even toxic, substances including inorganic salts and long chain fatty acids. These are normally removed during processing to produce a final product with low toxicity and high clarity. Typically, microbial culture media are poured into sterile Petri dishes, normally 90 mm in diameter. Media can also be poured as slopes so organism can be stored for long periods or onto plastic carriers which are used as dipslides.

Other solid media that do not contain agar have been used for the isolation of pathogenic bacteria. Gelatine was used as a solidifying agent to create some of the earliest solid culture media and serum or egg can be coagulated by inspissation to create solid media. Löwenstein–Jensen medium is the most widely used egg-based medium and is an effective and economical option for the culture of mycobacteria, including *Mycobacterium tuberculosis*. Whole egg is aseptically removed from its shell and added to a sterile solution of minerals and malachite green. This suspension is then dispensed into sterile bottles and heated at 80–85°C for 45 min using an inspissator. This procedure is used to solidify the medium rather than sterilize it.

Buffers

All bacteria have an optimal pH for growth and a pH range within which they are able to grow but this can vary widely for different bacteria. During growth bacteria may generate acid from carbohydrate or alkali from oxidation of peptone resulting in deviation from optimal pH. For some species this may adversely affect growth or even cause loss of viability.

Phosphate buffers are most commonly used but a wide range is available and the choice of buffer is determined chiefly by the starting pH of the medium. Typically, general-purpose agars are formulated to give a pH of between 7 and 7.5, but use of a different pH may be advantageous. For example, *Vibrio cholerae* is susceptible to acid but grows readily under alkaline conditions. Hence, media for *V. cholerae* are often formulated to give a pH of around 8.5 and this limits the growth of competitors. In contrast, media for the isolation of pathogenic yeasts, such as Sabouraud agar, have a pH of around 5.6.

SELF-CHECK 3.2

Why do you think most culture media have a pH of 7 to 7.5 and why is it important to maintain this pH during bacterial growth?

Blood and serum

Blood from various mammals (including man) may be collected aseptically and used to supplement or enrich sterilized media to improve the growth of fastidious bacteria. Horse blood (mainly) or sheep blood are most commonly used. Blood is normally defibrinated to prevent it from clotting and this is preferable to the addition of anticoagulants that may interfere with bacterial growth. It is also important, as with any culture ingredient derived from biological material, to ensure that the blood is free from infective agents and antibiotics. Addition of blood or serum improves the growth of a range of fastidious bacteria and allows the demonstration of haemolysis, which may be useful for identification.

SELF-CHECK 3.3

What are the advantages of using horse blood as the main blood supplement in culture media? Why is human blood not routinely used?

Charcoal

Activated charcoal is porous and has a very large surface area. It has a tendency to adsorb a range of molecules to its surface by both physical and chemical means and it is used in culture media to bind compounds that may be toxic to growth. Some bacteria have a tendency to generate toxic products such as peroxide free radicals that can inhibit growth. Activated charcoal has been shown to be beneficial when used in culture media for pathogenic Gram-negative rods such as *Bordetella pertussis*, *Campylobacter* spp., *Haemophilus ducreyi* and *Helicobacter pylori* and it is indispensable for growth of *Legionella* spp. Because of the non-toxic nature of pure activated charcoal, it can be used in general-purpose broth media for the culture of blood and can assist in the adsorption of antimicrobial compounds. Other substances included in media, such as starch and blood, may also reduce the effect of potentially toxic products of metabolism.

Other additives for fastidious bacteria and anaerobes

Glucose may be added to culture media as a ready source of energy and promotes the growth of a number of species. For example, it is included in agar media for *Brucella* sp. and in brain heart infusion broth, which is used for the cultivation of fastidious organisms. For fastidious bacteria, growth factors may need to be added. If the exact growth requirements are known, specific compounds can be added, for example nicotinic acid is added to culture media for the isolation of *Bordetella pertussis*. Alternatively, commercial companies supply nutritional supplements that consist of a concentrate of vitamins and other potential growth factors. Examples include Vitox and IsoVitaleX and these are particularly useful for inclusion in media for the isolation of *Neisseria gonorrhoeae*.

Cultivation of anaerobic bacteria can be particularly challenging as some Gram-negative species are fastidious in their growth requirements and oxygen may be toxic to some anaerobes, although tolerance varies between species. To address these needs specialized media have been developed that are supplemented with growth factors and reducing agents to maintain a low redox potential in the culture medium. Growth factors include haemin, menadione and succinate for fastidious Gram-negative rods such as *Porphyromonas* spp., *Prevotella* spp. and *Bacteroides* spp. Reducing agents may include L-cysteine, dithiothreitol and sodium thioglycolate.

Isolation of spore-bearing anaerobes may also be difficult, but for different reasons. *Clostridium difficile*, as implied by its name, is a case in point. Spores may take some 'persuasion' to germinate on synthetic media and compounds that initiate spore germination can substantially increase the yield of *C. difficile* from stool samples. Examples of such compounds include lysozyme and certain bile salts such as sodium taurocholate.

3.3 Selective media

Antibiotics or other chemical inhibitors of growth can be added to general-purpose media to inhibit the growth of commensal bacteria and thus increase the likelihood of isolating target pathogens. *Staphylococcus aureus* is the most commonly encountered pathogen associated with infections of skin and wounds. Some wounds, particularly from the lower body, may be dirty and contain faecal flora including strains of Enterobacteriaceae such as *Escherichia coli* and *Proteus* spp. The large colonies formed by Enterobacteriaceae and the swarming growth of *Proteus* sp. may mask the presence of *S. aureus* colonies on general-purpose culture media such as blood agar. It can therefore by advantageous to include a **selective agent** to inhibit the growth of Gram-negative bacteria and increase the likelihood of recovering *S. aureus*. A useful selective agent for this purpose is aztreonam, which inhibits Gram-negative bacteria including most strains of Enterobacteriaceae and *Pseudomonas aeruginosa* (Figure 3.1). Studies have shown that the use of blood agar containing up to 32 mg/L of aztreonam can increase the isolation rate of *S. aureus* from wound swabs compared with blood agar alone.

selective agent
A selective agent is an antimicrobial substance that is incorporated into culture media to inhibit the growth of unwanted commensal bacteria thus assisting the isolation of target pathogens.

The growth of Enterobacteriaceae may also mask the presence of anaerobic bacteria such as *Bacteroides fragilis* and anaerobic streptococci. The inclusion of 30 mg/L nalidixic acid inhibits

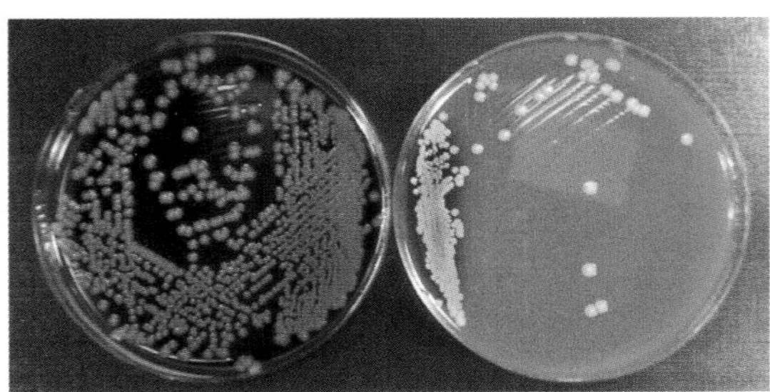

FIGURE 3.1
Culture of a leg ulcer swab on blood agar with an apparent pure growth of coliform (left). Culture of the same swab on blood agar plus aztreonam (right) reveals the presence of *Staphylococcus aureus*. (Courtesy of M. Ford/J. Perry.)

the growth of most Enterobacteriaceae and allows the growth of these anaerobes. Nalidixic acid is generally inhibitory to strains of clostridia, and if *Clostridium* infection is suspected on clinical grounds, a more appropriate selective agar would be blood agar supplemented with neomycin. The commonest pathogen from upper respiratory tract infection is *Haemophilus influenzae* and this is typically diagnosed by culture of sputum. If *H. influenzae* is present, it is typically there in large numbers and isolation on chocolate agar is normally straightforward. In some cases, sputa may be heavily contaminated by oral streptococci and these may complicate the isolation of small numbers of *H. influenzae*. Many laboratories therefore employ chocolate agar supplemented with an anti-Gram positive antimicrobial, bacitracin, to increase the isolation rate of *H. influenzae*.

In some cases, a cocktail of antibiotics may be employed to drastically reduce the amount of normal flora and therefore optimize the chances of recovering target pathogens. Stool samples typically contain a wide variety of Gram-negative and Gram-positive flora and consequently the isolation of *Campylobacter* spp. would be almost impossible without the use of selective agents. Traditionally, a cocktail of agents has been used, for example, Skirrow's formulation includes vancomycin, polymyxin and trimethoprim for inhibition of commensal faecal flora. A similar cocktail of agents is applied in the isolation of *Neisseria gonorrhoeae* from urethral and endocervical swabs. For example, Modified New York City medium contains vancomycin, colistin, trimethoprim and amphotericin B. Vancomycin may be inhibitory to a small proportion of *Neisseria gonorrhoeae* strains and the inclusion of lincomycin, in place of vancomycin, is preferred by some laboratories. The activity of trimethoprim is reduced or abolished in the presence of thymidine, which may be present in small amounts in nutrient media. It is therefore customary to include either lysed or laked blood in selective media that contain trimethoprim. Erythrocytes from horses contain an enzyme, thymidine phosphorylase, which can inactivate thymidine thus allowing trimethoprim to exert its activity.

SELF-CHECK 3.4

What would be useful antimicrobial agents for the selective isolation of (a) Gram-negative bacteria and (b) Gram-positive bacteria?

3.4 Differential media

Ideally, for dedicated media that target a specific pathogen, the growth of all other bacteria and fungi would be suppressed using selective agents. In such a scenario only growth of the pathogen would occur and biochemical identification would be unnecessary. In practice, it is extremely difficult to inhibit all unwanted bacteria using selective agents and particularly those that are closely related to the target pathogen. In such cases it may be useful to incorporate a 'detection system' into the culture medium that will highlight a particular metabolic activity thus providing differentiation of the target pathogen from other species that are able to grow on the medium. Such media are described as **differential media**.

By far the commonest means of highlighting a metabolic feature is the use of carbohydrates, which may be used by bacteria to generate acid and this can be highlighted using a suitable pH indicator. A suitable example is cystine lactose electrolyte deficient (CLED) medium, which is a non-selective medium used principally for the cultivation of urinary tract pathogens. The inclusion of lactose, and bromothymol blue as a pH indicator, allows the differentiation of species and provides clues to their likely identity. So for example, *Proteus* spp. produce medium-sized, blue colonies due to an inability to acidify lactose whereas enterococci form much smaller yellow colonies due to acidification of lactose. The swarming of *Proteus* spp. is inhibited by the absence of electrolytes (e.g. sodium chloride) in the medium. Some brands of CLED medium include Andrade's indicator in addition to bromothymol blue to provide a greater range of colours produced by lactose fermenters (red–yellow) and thus provide enhanced differentiation of colony types.

Key Point

Differential media allow the distinction of different species based on their biochemical reactions. This results in different species forming different coloured colonies, usually as a result of their ability to produce acid from a sugar in the presence of a pH indicator.

In practice, most differential media incorporate selective agents and may therefore be described as selective and differential. A suitable example is mannitol salt agar, which highlights *S. aureus* as large yellow colonies due to mannitol fermentation in the presence of phenol red indicator. This provides differentiation from some other species of staphylococci and most notably the commonest commensal of the skin, *Staphylococcus epidermidis*, which fails to ferment mannitol (Figure 3.2). However, mannitol salt agar is also selective due to the inclusion of a high concentration of sodium chloride (up to 7.5%) resulting in the inhibition of most Gram-negative bacteria and some Gram-positive species.

The use of media that are both selective and differential is employed for the detection of enteric pathogens such as *Salmonella* and *Shigella*. Traditional media for isolation of *Salmonella* have relied on the fact that *Salmonella* (a) does not usually ferment lactose and (b) typically generates hydrogen sulphide from sulphur-containing compounds such as thiosulphate. One of several media that exploits these attributes of *Salmonella* is desoxycholate citrate agar (DCA). This medium employs sodium desoxycholate and sodium citrate as selective agents and also contains a pH indicator (neutral red) plus lactose as well as chemicals for the demonstration of hydrogen sulphide production (sodium thiosulphate/ferric citrate). Most species of Enterobacteriaceae ferment lactose and therefore grow as pink/red colonies; *Shigella* are unreactive on the medium and are therefore colourless or very pale pink due to late fermentation of lactose. Salmonella are colourless due to the absence of lactose fermentation but colonies usually show a black centre due to generation of hydrogen sulphide, which subsequently reacts with iron to form the black precipitate (Figure 3.3). This classical methodology has been widely used for detection of *Salmonella* and *Shigella* but is relatively non-specific. For example, *Proteus* spp. are common in stool samples, are always non-fermenters of lactose and frequently generate hydrogen sulphide. They are therefore frequently confused with

FIGURE 3.2

Differentiation of *Staphylococcus aureus* (left) from *Staphylococcus epidermidis* (right) using mannitol salt agar. (Courtesy of J. Perry.)

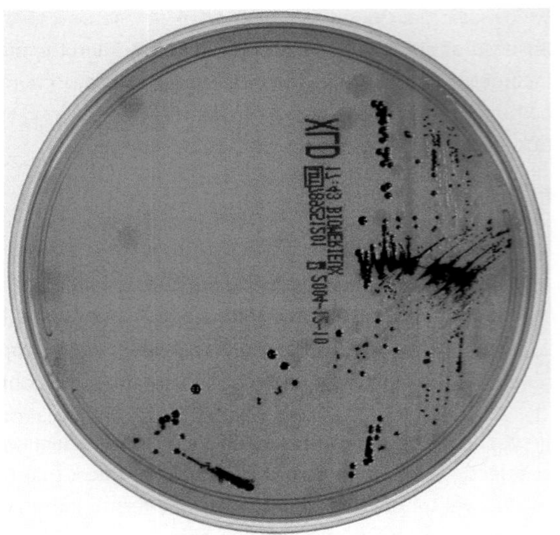

FIGURE 3.3
Growth of *Salmonella enteritidis* showing typical black colonies on XLD. *Shigella sonnei* (pale colonies) is also shown for comparison. (Courtesy of bioMérieux.)

Salmonella or *Shigella* and require other tests such as biochemical screening to differentiate them from *Salmonella* in particular.

A much more ingenious approach is utilized in xylose lysine deoxycholate medium (XLD), which can detect up to five distinct metabolic activities and can therefore detect *Salmonella* and *Shigella* with higher specificity. This medium contains three carbohydrates, xylose, lactose and sucrose, any of which may be fermented to generate acid and thus cause bacterial colonies to appear yellow in the presence of phenol red. The vast majority of Enterobacteriaceae, except *Shigella* spp., ferment at least one or more of these three sugars and therefore appear as yellow colonies. *Salmonella* spp. typically ferment the small amount of xylose present to generate acid but then proceed to decarboxylate lysine to cause a rise in pH to at least 7.0. At this pH, the generation of hydrogen sulphide is possible and *Salmonella* therefore generate black colonies. Other common producers of hydrogen sulphide (e.g. *Proteus* spp. and *Citrobacter freundii*) do not possess the enzyme to decarboxylate lysine and are not able to demonstrate hydrogen sulphide production at low pH. Some species of Enterobacteriaceae (e.g. *Providencia* spp.) may not undergo any of these reactions and consequently may resemble *Shigella* spp.

> Key Point
>
> **In practice, most differential media contain selective agents and may also be classed as selective media.**

3.5 Enrichment media

Enrichment media are broth media that are useful for recovery of small numbers of target pathogens that may be present in clinical samples alongside larger numbers of non-target commensal bacteria. Some broth media are completely non-selective and their function is simply to amplify the numbers of any bacteria that are able to grow. Brain heart infusion broth (BHI) is one such medium and is useful for supporting the growth of fastidious bacteria. Robertson's cooked meat broth is another example and is helpful for the isolation of anaerobes. Non-selective broth media that can support fastidious bacteria (e.g. BHI and trypticase soy broth) are used for the culture of blood to detect bacteraemia and these are discussed in more detail in Chapter 5.

Cross reference
Chapter 5, Section 5.1.

Enrichment media are broth media that are designed to favour the growth of target pathogens and discourage the growth of commensals. By achieving this, small numbers of pathogenic bacteria are able to multiply more rapidly than competitors and may then be readily isolated following subculture of the broth onto solid media. A good example is alkaline peptone water, which is used for the isolation of *V. cholerae* from stool samples. Alkaline peptone water has a high pH (8.4) and includes sodium chloride (0.5–2%) to encourage rapid growth of *Vibrio* spp. and suppress the growth of other flora (e.g. most Enterobacteriaceae). The alkaline peptone water can then be subcultured after 6–8 h of incubation onto agar media such as thiosulphate-citrate-bile-sucrose (TCBS) to isolate colonies and confirm identification. Selenite F broth is an enrichment medium that is very useful for the isolation of *Salmonella* spp. from stool samples and various modifications are available to optimize the growth of *Salmonella* spp.

3.6 Chromogenic media

In Chapter 2, the principles of chromogenic enzyme substrates were described. Chromogenic substrates may be incorporated into agar-based culture media to form **chromogenic media**. Such media often contain selective agents and are therefore both selective and differential. A limited range of chromogenic substrates may be used in an agar based medium because it is critical that the coloured dye released by hydrolysis does not diffuse in agar. By ensuring that the colour generated remains highly localized on bacterial colonies, this allows differentiation of bacterial colonies producing a particular enzyme from those that do not. For agar media the most widely used substrates are those based on indoxyl (or its halogenated derivatives). Indoxyl may be derivatized to form a phosphate, a glycoside (sugar-derivative) or an ester with a fatty acid (e.g. indoxyl octanoate). Removal of these groups by hydrolysis by a specific enzyme yields free indoxyl, which, in the presence of oxygen, forms an insoluble blue-grey dye. Compounds closely related to indoxyl produce different coloured dyes, for example, 6-chloro-indoxyl forms a pink coloured dye, 5-bromo-6-chloro-indoxyl forms a magenta dye and 5-bromo-4-chloro-indoxyl forms a green dye. Figure 3.4 shows the structure of indoxyl and some halogenated derivatives and all of these may be derivatized to form substrates that are suitable for use in agar media.

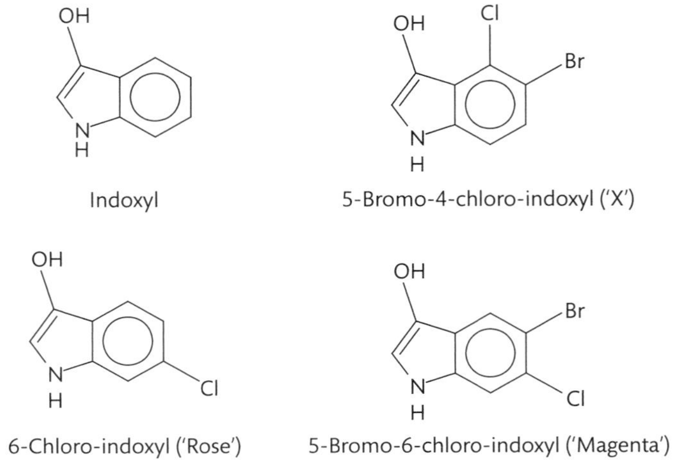

Indoxyl

5-Bromo-4-chloro-indoxyl ('X')

6-Chloro-indoxyl ('Rose')

5-Bromo-6-chloro-indoxyl ('Magenta')

FIGURE 3.4
Structure of commonly used indoxylic dyes. (Courtesy of J. Perry.)

Detection of *Salmonella* and *E. coli* O157 using chromogenic media

One of the first chromogenic media applicable to clinical microbiology was Rambach agar and its performance was first reported in 1990. Rambach agar relies on the fact that the vast majority of *Salmonella* strains encountered in clinical specimens do not produce β-galactosidase. The medium therefore employed a chromogenic substrate for β-galactosidase to enable most common strains of Enterobacteriaceae (e.g. *E. coli*, *Klebsiella* spp., *Enterobacter* spp., *Citrobacter* spp.) to form blue colonies. In order to differentiate *Salmonella* from the remaining flora (e.g. *Proteus* spp., *Pseudomonas* spp.), the medium also contained neutral red indicator and propylene glycol. The latter was fermented by *Salmonella* to generate pink–red colonies. SM-ID agar was designed at around the same time and worked on identical principles except that glucuronic acid was used in place of propylene glycol.

Over the last 18 years, chromogenic media for detection of *Salmonella* have evolved and a variety are now available from different manufacturers. Most of these media rely on the fact that *Salmonella* can hydrolyse long chain esters, for example octanoate or nonanoate derivatives of chromogenic dyes. Only a few other species of Enterobacteriaceae produce such esterases and these can be made to produce a different coloration by incorporation of a second chromogenic substrate (e.g. for β-glucosidase). Esterases are also produced by non-fermentative Gram-negative rods and yeasts but their growth can be limited by incorporation of selective agents, for example cefsulodin and amphotericin, respectively.

The main advantage of chromogenic media for isolation of *Salmonella* is their high specificity. Coloured colonies are usually confirmed as *Salmonella* and, when compared with the use of conventional (i.e. non-chromogenic) media, there are far fewer species that resemble *Salmonella*. A disadvantage of **chromogenic media** for *Salmonella* is that, unlike deoxycholate agar and XLD medium, they do not accommodate the isolation of *Shigella* spp. and therefore two different media need to be used if both pathogens are sought.

chromogenic media
Chromogenic media are sophisticated differential media. Target pathogens are distinguished on the basis of colour, and the colour originates from hydrolysis of a chromogenic enzyme substrate.

E. coli O157 is a cause of haemorrhagic colitis and its presence is sought in diarrhoeal stool samples. Traditional methods rely on the fact that, unlike most *E. coli* strains, this serotype fails to ferment sorbitol. *E. coli* O157 also fails to produce β-glucuronidase and a chromogenic substrate for this enzyme may be used for differentiation. Several media have been developed and marketed using this approach.

SELF-CHECK 3.5

What are the advantages of using chromogenic media for the isolation of *Salmonella* spp. from stool samples?

Detection of *Staphylococcus aureus* (including MRSA) with chromogenic media

The presence of *S. aureus* is routinely sought in swabs from skin and soft tissue infection and there is widespread demand in many hospitals for screening patients and staff for colonization with methicillin-resistant strains (MRSA). Such demand has led to the development of chromogenic media both for the specific detection of *S. aureus* and the specific detection of MRSA. Such media exploit selective agents for the inhibition of Gram-negative bacteria and enterococci. If only methicillin resistant strains are sought, the medium is further supplemented with a β-lactam antibiotic (e.g. oxacillin or cefoxitin) to inhibit methicillin-sensitive strains of staphylococci. There is now good evidence to show that cefoxitin is the preferred of these two options. A chromogenic substrate is then used for the differentiation of *S. aureus* from other staphylococci. Substrates for phosphatase activity or α-glucosidase activity have been shown to be suitable for this purpose.

Cross reference
Chapter 12, Section 12.2.

Chromogenic media for urinary tract pathogens

E. coli is the predominant species associated with urinary tract infections and other species of Enterobacteriaceae and enterococci are also commonly isolated. Non-selective chromogenic media have been designed to isolate and differentiate all common pathogens associated with urinary tract infection. A number of media are available and most rely on the same principles. They include chromogenic substrates for the detection of two enzymes, β-galactosidase and β-glucosidase, as well as tryptophan with iron salts in order to detect deaminase activity in the *Proteus–Providencia–Morganella* (PPM) group of species. Production of β-galactosidase generates colonies with a red–pink coloration due to hydrolysis of the chromogenic substrate, whereas production of β-glucosidase generates blue or green colonies. Deamination of tryptophan in the presence of iron results in a diffuse brown coloration.

Approximately 99% of *E. coli* strains produce β-galactosidase, but not β-glucosidase, and they therefore grow as pink or red colonies. The *Klebsiella–Enterobacter–Serratia* (KES) group of species typically produce both of these enzymes and colonies appear predominantly blue. Enterococci also have a strong β-glucosidase activity and also appear blue but are smaller in size. Colonies of the PPM group produce a brown coloration. The reactions shown by representatives of these species are shown in Figure 3.5. Several brands of chromogenic media are available that target the same enzyme activities including UriSelect 4, Chromogenic UTI medium and CHROMagar Orientation. Additional biochemical tests may be used to confirm the identification of colonies on these chromogenic media including a spot indole test to confirm the identity of *E. coli*. *Citrobacter freundii* is infrequently found in urine samples but, when present, it often grows as pink colonies due to its strong β-galactosidase activity and weak or absent β-glucosidase activity. A rapid indole test can be used on pink colonies to exclude *C. freundii* and thereby provide reliable confirmation of *E. coli* (indole positive).

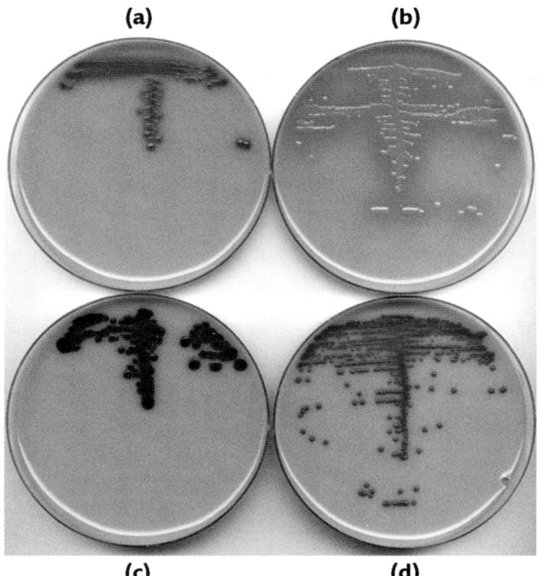

FIGURE 3.5

Differentiation of common urinary tract pathogens using a chromogenic medium. (a) *E. coli* (producing β-galactosidase), (b) *Proteus mirabilis* (producing tryptophan deaminase), (c) *Klebsiella pneumoniae* (producing β-galactosidase and β-glucosidase) and (d) *Enterococcus faecalis* (producing β-glucosidase). (Courtesy of J. Perry.)

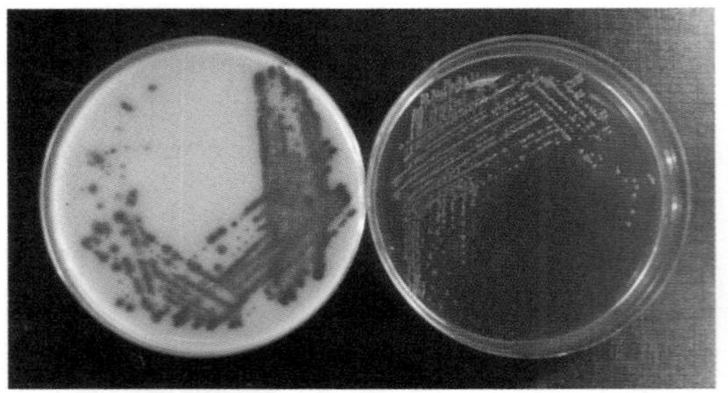

FIGURE 3.6
A contaminated urine specimen cultured on UriSelect 4. On CLED medium (right), a mixture of large and small lactose-fermenting colonies are visible. On the chromogenic medium (left) differentiation of *E. coli* (pink), *Klebsiella pneumoniae* (large blue) and *Enterococcus faecalis* (small turquoise) is significantly easier. (Courtesy of J. Perry.)

The CPS ID brand of media (CPS ID2, CPS ID3) operate on slightly different principles in that a chromogenic substrate for β-glucuronidase is used in place of a substrate for β-galactosidase. This strategy has advantages and disadvantages. Firstly, β-glucuronidase is largely restricted to *E. coli* among urinary tract pathogens - the enzyme may be found in *Salmonella*, *Shigella* and *Yersinia* but these are only rarely encountered in urine. This means that only *E. coli* produce pink colonies and there is arguably no requirement for further confirmation of identity. The disadvantage is that only 94% of *E. coli* strains from urine produce β-glucuronidase and appear as pink colonies and the remaining strains form colourless colonies requiring identification. Members of the KES and PPM groups, as well as enterococci, are distinguished as on the media described above.

The most useful aspect of all of these media is their ability to differentiate mixed cultures, which are commonly encountered in urine samples and are often an indication of contamination rather than true polymicrobial infection. Figure 3.6 shows a urine sample that has been cultured simultaneously on cystine–lactose–electrolyte-deficient (CLED) medium and on a chromogenic medium, UriSelect 4. The sample contains *E. coli*, *K. pneumoniae* and *Enterococcus faecalis*. All of these grow as lactose-fermenting (pink) colonies on CLED medium and their respective numbers are difficult to estimate due to their similar appearance. On UriSelect 4 medium, colony differentiation is much easier due to the different colours generated. Several large studies comparing chromogenic media with conventional media (e.g. CLED) have demonstrated that mixed cultures may be detected with significantly higher frequency when using chromogenic media, thus allowing a more accurate diagnosis to be performed.

Key Point

Chromogenic media for urines allow for the specific identification of *E. coli* and the differentiation of other species, thus assisting the recognition of mixed cultures.

Chromogenic media for detection and differentiation of yeasts

A range of chromogenic media is available for detection of yeasts. Such media enable microbiologists to differentiate between different species unlike conventional media such as Sabouraud agar. There are sound clinical reasons for this, as some *Candida* spp. are intrinsically resistant to some antifungal agents. On conventional media it is virtually impossible to determine if more than one species of *Candida* is present.

A principal feature of these media is the inclusion of a chromogenic substrate for *N*-acetyl-β-D-glucosaminidase, thus allowing the differentiation and identification of the

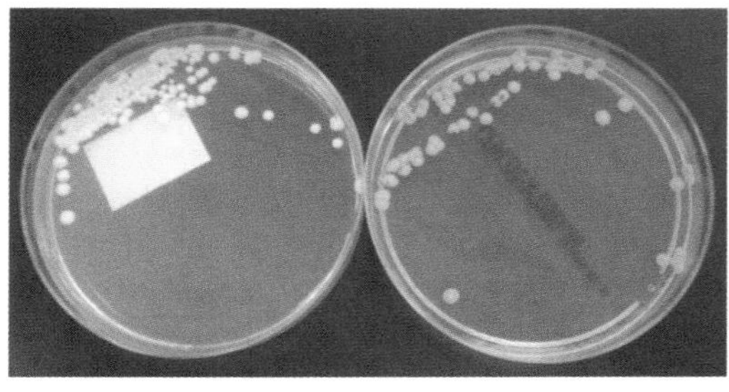

FIGURE 3.7
The use of chromogenic *Candida* agar to detect mixed yeast cultures. The plate on the left is Sabouraud's dextrose agar and on the right CAN2 agar (bioMérieux) which shows pink colonies of *C. tropicalis* and blue colonies of *C. albicans*. This mixture is very difficult to detect on non-chromogenic media. (Courtesy of M. Ford.)

most frequent and clinically important species, *Candida albicans*. Use of such a substrate (e.g. 5-bromo-4-chloro-3-indolyl-β-D-*N*-acetyl-glucosaminide) results in *C. albicans* forming characteristic green-blue colonies, although *Candida dubliniensis* cannot be differentiated.

Some chromogenic media for yeasts include a second chromogenic substrate (e.g. for phosphatase activity) and enable the recognition of species other than *Candida albicans*. CAN2 agar (Figure 3.7) employs two substrates and enables the differentiation of *C. albicans* as green colonies and also the identification of *Candida tropicalis* (blue colonies). Other species form either pink or white colonies. *Candida krusei* may be recognized by the formation of characteristic pink, flat colonies.

SELF-CHECK 3.6

Why would it be clinically important to be able to determine that a culture contains more than one species of *Candida*?

3.7 Media for antimicrobial susceptibility testing

As discussed earlier in the chapter, media used in clinical microbiology are not strictly defined in terms of their chemical composition and are mostly based on extracts or digests of animal or plant tissues that are chemically complex. In spite of the great deal of knowledge that has been accumulated by microbiologists regarding the growth requirements of bacteria, it has not proved possible to produce chemically defined media that match the performance of complex undefined media. This is not a problem media provided that the performance of different batches of media is consistent. However, complex media are problematic for antimicrobial susceptibility testing. The action of various antibiotics may be adversely affected by media components. For example, thymidine is antagonistic to the action of trimethoprim and the various cations may impact upon the activity of aminoglycosides such as gentamicin.

One solution has been to design culture media that are 'semi-defined' so that complex components are kept to the minimum required for good growth, the presence of antagonists is minimized and batches of agar may be produced with reproducible performance. Isosensitest agar (or broth), which is widely used in UK laboratories, is specifically designed for susceptibility testing and comprises only around 1.4% (w/v) of undefined content. Over 25 other chemically defined ingredients are included in the medium including glucose, buffer, minerals and vitamins to ensure satisfactory growth of most commonly encountered bacteria. Fastidious organisms such as streptococci may require the addition of 5% defibrinated horse blood in order to grow well and *Haemophilus* spp. require the further addition of 20 mg/L NAD.

Mueller-Hinton agar is less defined than Isosensitest agar and contains some 30% (w/v) solids derived from infusion of beef. Historically, the medium has shown problems with batch-to-batch variation in its content of divalent cations and thymidine leading to problems with its consistency of performance. However, Mueller-Hinton agar remains the standard medium for antimicrobial susceptibility testing recommended by the Clinical Laboratory Standards Institute (CLSI) and is used widely in many countries including the United States. Problems have been overcome by ensuring that batches conform with strict performance criteria published by the CLSI, before distribution from manufacturer to laboratory. Wilkins-Chalgren agar has been recommended for the susceptibility testing of anaerobic bacteria. It contains a range of growth factors to support the growth of anaerobes (see Section 3.2) without requiring the addition of blood.

Cross reference
Chapter 4, Section 4.3.

3.8 Preparation and sterilization of culture media

Although an increasing number of diagnostic laboratories rely on the purchase of ready-made culture plates from commercial suppliers, many laboratories continue to prepare their own culture media. Invariably, culture media are prepared from fully formulated, dehydrated powders supplied by commercial manufacturers. Therefore, for simple media, laboratory personnel simply have to add water to the powder, sterilize and dispense the medium into Petri dishes. More complex media may require a number of additives. Some general principles of media preparation are outlined below.

Commercially supplied, dehydrated media should be stored and prepared in strict accordance with manufacturers' instructions. An expiry date will be supplied for each batch of media and this should be strictly adhered to. Dehydrated media are adversely affected by heat and should be stored away from sunlight and other sources of heat such as autoclaves and hot air ovens. They are hygroscopic and therefore highly susceptible to moisture. They must therefore be kept in tightly closed containers and not exposed to the atmosphere unnecessarily. If the medium is no longer 'free-flowing' (i.e. shows evidence of clumping due to absorption of moisture) or becomes discoloured, it should be discarded.

Dehydrated media commonly contain hazardous materials, often incorporated as selective agents, and examples include bile salts, lithium chloride and various dyes. Any hazardous ingredients will be detailed in the Material Safety Data Sheet accompanying the product. Weighing out fine powders is hazardous and requires the use of a weighing station, in which the powder is removed from the air by filters. Personal protective equipment such as a mask and gloves should also be used and laboratory personnel should be familiar with relevant COSHH assessments. To reduce the potential problem of inhalation of toxic substances, some media suppliers offer granulated media containing a lower proportion of fine powder.

Media should be rehydrated with either distilled or deionized water in glassware that has been meticulously cleaned. Thorough mixing is essential to ensure that none of the powder remains adherent to the glassware or floating on the surface of the water. Bottles of media should be clearly labelled using heat sensitive tape that will change colour if the medium has passed through the autoclave, although this is not an indication of sterilization.

Sterilization

This is normally achieved by autoclaving using a standard sterilization cycle, for example 116°C for 20 min or 121°C for 15 min. Bottles of media should be autoclaved with the lids loosened and personal protective equipment should be employed when sterilized media are removed. The autoclave should be dedicated to media preparation and should not be used, for example, for sterilization of bacterial cultures or clinical samples. The temperature of the autoclave should be monitored during the sterilization cycle and records of performance should be kept. After removal from the autoclave, agar media should be cooled to 50°C in a water-bath and dispensed as soon as possible. Prolonged exposure to heat is very damaging to culture media resulting in autooxidation and the formation of superoxides.

The use of semiautomated media preparators is advisable for sterilization of large batches of agar-based media (>1 litre). Such instruments perform continuous stirring of the media and therefore allow excellent heat dispersal and minimal overheating. Care must be taken, however, to ensure meticulous rinsing of the drum which holds the medium and the tubing used for dispensing, particularly if different types of media are being prepared using the same instrument. Cross contamination of antimicrobials between batches occurs very readily and is a common source of error. It is therefore advisable to have dedicated instruments (or at least dedicated tubing) for selective media.

Although all culture media will suffer from prolonged heating, some ingredients are more sensitive than others and sugars are particularly susceptible to breakdown. When preparing some selective agars, such as XLD and TCBS, it is sufficient to boil the medium briefly to dissolve all of the constituents and then dispense the medium into Petri dishes. This does not constitute sterilization but the exposure to 100°C combined with the selectivity of the medium obviate any problems of contamination.

For non-selective broth media that do not withstand sterilization via autoclaving, a process known as tyndallization has been traditionally used. This process involves heating the medium at 100°C for 30 min and then incubating overnight at 37°C. This process is then repeated twice. The heating process kills all vegetative bacteria but may not kill all spores. The incubation period is designed to allow the vegetation of any spores, which will subsequently be killed by a further round of heating, and so on. This process is not ideal as it allows for the germination of spores and subsequent bacterial growth that may alter the constituents of the medium. For this reason, and for its general lack of convenience, the method is now infrequently used.

SELF-CHECK 3.7

What types of media supplements are heat sensitive and how is this problem overcome?

Dispensing media

Once the culture medium has been sterilized by an appropriate method, heat sensitive additives such as blood, serum, egg yolk or antimicrobials may be added as sterile supplements to cold broth media or molten agar at 45–50°C. When adding to agar, supplements should be warmed up to at least room temperature. If supplements are added, these may alter the pH of the medium and checking and/or adjustment of the pH may be necessary. After addition of heat-sensitive supplements, media should be dispensed without delay.

Media should be dispensed into sterile containers (e.g. Petri dishes) in a dedicated, thoroughly clean preparation area using all aseptic precautions. Individual culture plates must be labelled or coded with a unique batch number so that a user can easily tell the exact type of media and its expiry date. Generally, culture plates may be stored for a week at 4°C without loss of performance, or longer if they are sealed to avoid loss of moisture. Some ingredients, such as antibiotics, are relatively labile and it is important for laboratories to establish the 'shelf life' of any culture media. For every batch of media, a record should be available to access the lot numbers of its ingredients as well as details of the sterilization cycle and, if applicable, the automated preparator used for dispensing the media (Figure 3.8).

FIGURE 3.8
An automated media preparator for production of culture plates. The autoclave (left) sterilizes a batch of agar, which is then automatically dispensed at 42°C into sterile Petri dishes. (Courtesy of J. Perry.)

3.9 Quality control

Laboratories have an obligation to assess the performance of culture media and they should not be used for diagnostic procedures until performance criteria have been met. The physical attributes of the media may be assessed by simple visual examination and measurement of pH. If these are acceptable, sterility testing and assessment of microbiological performance are then required.

Sterility testing

A random sample of each batch of culture plates and broth media should be incubated at 30°C for at least 24 h and closely examined for microbial colonies. This does not guarantee sterility as some microbes may take days to appear, but most contamination events will be detected by these measures. Any indication of contamination on the surface of the medium or within the medium should prompt disposal of the batch.

Choice of bacterial strains

Microbiological assessment involves the inoculation of culture media with microbial strains of known characteristics to examine their growth and colony appearance. Strains should be derived from a recognized culture collection that is accessible to others such as the National Collection of Type Cultures (NCTC) or the American Type Culture Collection (ATCC). If problems are encountered, testing can then be reproduced by other laboratories or commercial suppliers using the same strains. Such strains are often referred to as 'control strains' and they are generally susceptible to antimicrobial agents. When strains are first obtained from a culture collection, they should be stored to provide a 'seed lot' that can be accessed for subculture at least once a month. The use of control strains that have undergone serial subculture for many weeks should be avoided as strains may lose their characteristics or become contaminated.

Microbiological assessment should be designed to test the major characteristics of the medium and to ensure that media are 'fit for purpose'. So, for example, if a medium is differential, microbiology tests should be performed to demonstrate its ability to differentiate between relevant species and if a medium is selective then the ability of the medium to inhibit relevant species should be demonstrated and so on. So, for example, a general purpose non-selective medium such as blood agar could be inoculated with control strains of *Staphylococcus aureus*, *Escherichia coli* and *Streptococcus pyogenes*. This will demonstrate the growth of commonly encountered pathogens (both Gram-positive and Gram-negative) as well as the ability of *S. pyogenes* to demonstrate typical β-haemolysis. CLED medium could be inoculated with

E. coli, *Proteus mirabilis* and *Enterococcus faecalis* thus assessing the growth of three commonly encountered urinary tract pathogens (both Gram-positive and Gram-negative) and showing the ability of the medium to differentiate between *E. coli* and *P. mirabilis* (based on lactose fermentation) and prevent the swarming growth of *Proteus*. As well as inoculation of test media, it is important to inoculate a medium of proven performance (i.e. from an in-use batch) in order to demonstrate the viability and inoculum level of the test bacteria.

Inoculum

The most common pitfall associated with the assessment of culture media is the use of high inocula. When attempting the isolation of a bacterial pathogen from a pathological sample (e.g. a cerebrospinal fluid) success may depend upon the isolation of only a handful of bacterial colonies. It is therefore of little relevance to assess the performance of a culture medium by inoculating a loopful of growth containing $>10^8$ colony forming units (CFU); however, such practice is commonplace. An ideal inoculum for control strains should deliver no more than 100 CFU onto the culture medium. This will allow a colony count to be performed and this count can be monitored to allow batch to batch comparison of media in order to detect deterioration of performance. Various methods may be used to achieve such inocula and most commonly this involves the preparation of a standardized suspension by measurement of turbidity followed by serial dilution. This is cumbersome and prone to errors and some products are now commercially available to deliver a controlled inoculum onto culture media. Examples include lenticule discs and vitroids, which consist of a small inoculum of microorganisms preserved in a water soluble solid matrix.

There are occasions when it may be appropriate to use a higher inoculum (e.g. at least 1000 CFU) of a control strain. For example, this may be useful for demonstrating the ability of a selective agent to inhibit growth of unwanted bacteria. Similarly, an enrichment broth is most usefully assessed using a low inoculum of target pathogen (100 CFU) along with a higher concentration (1000 CFU) of an unwanted commensal. For example, selenite broth could be inoculated with 100 CFU of *Salmonella enteritidis* and 1000 CFU of *E. coli*. After incubation, subculture onto a medium that demonstrates lactose fermentation (e.g. CLED or McConkey agar) should demonstrate a predominance of *Salmonella* colonies (non-lactose fermenting). This illustrates that the selenite broth has performed effectively as an enrichment medium for *Salmonella*.

Batch failure

There are a number of potential sources of error that can lead to unacceptable quality or performance of culture media. Assuming that the sterilization cycle has been monitored effectively, most contamination events are due to the inadvertent addition of non-sterile additives. For example, batches of sterile blood may become contaminated during storage and this is usually detected after overnight incubation of culture media. The inclusion of an excessive amount of a selective agent, due to a dilution error, is also a potential source of error. In the event of a failure, all media from the batch should be discarded and an audit conducted to examine all of the steps and ingredients involved in production of the batch.

Once a batch has passed quality assessment it can be released for use in the laboratory. Batch rotation is essential to ensure that media are used in chronological order.

SELF-CHECK 3.8

Why is quality control of media so important?

CHAPTER SUMMARY

After reading this chapter you should know and understand:

- Why culture media are essential in the diagnosis of infection
- How culture media are produced and the ingredients essential for optimal bacterial growth
- How the various types of media such as selective media are used in routine microbiology and the types of selective agents used
- How and why enrichment media are used in the routine laboratory
- How and why susceptibility test media are used and why they are so different to other culture media
- The importance of chromogenic media over other more conventional media, and also the advantages and disadvantages of using such media
- How media are sterilized and the processes involved
- The importance of quality control of media and how batches of media can fail quality control and what action should be taken

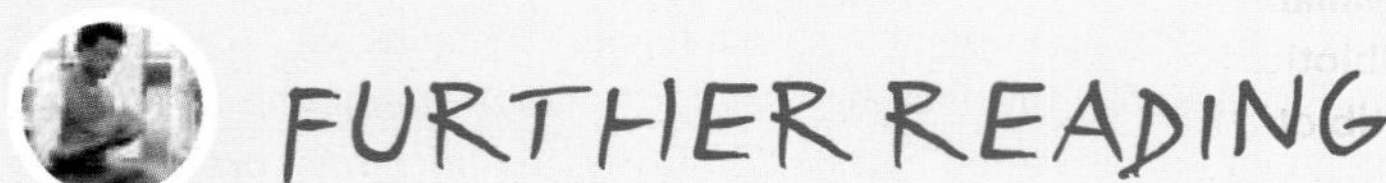

FURTHER READING

- ***The Oxoid Manual*. Available from Oxoid Ltd, Basingstoke, UK.**

  A comprehensive source of references for conventional culture media and an excellent resource in itself. Contains detailed descriptions of a wide range of media for clinical and food microbiology.

- **Perry JD, Freydiere AM. *The application of chromogenic media in clinical microbiology.*** *Journal of Applied Microbiology* **2007; 103: 2046–2055.**

  Provides a review of the chromogenic media available for detection of clinical pathogens and how they work.

DISCUSSION QUESTIONS

3.1 Chromogenic media are gradually replacing more conventional media for the isolation of specific pathogens from clinical samples. Why is this change happening and what are the advantages of this change?

3.2 Do you think the future of microbiology will be to move away from traditional methods of culture and towards more rapid detection methods such as PCR? If not, why not?

3.3 Many laboratories now purchase all media used in their laboratory – do you think this is a good or bad practice? Explain your reasoning.

4

Susceptibility testing and antibiotic assay

Jenny Andrews

Antibiotics have played a major role in the treatment of infectious disease since the discovery of penicillin in the 1920s. Since then hundreds of antibiotics have been made or discovered with a wide range of different properties and activities. Although some antibiotics have been rendered useless by the development of bacterial resistance, there are still a large number available for clinicians to use for the treatment of infection. However, whilst numerous antibiotics are available, laboratory support is required to determine if this is the correct antibiotic for the treatment of a particular infection. Throughout this chapter you will be shown the importance of the laboratory in determining the susceptibility of an organism to an antibiotic, and in the monitoring of the patient's serum for the correct concentration of antibiotic.

antibiotic
An antibiotic is classed as a chemical compound that has the ability to kill or inhibit the growth of microorganisms. They are sufficiently non-toxic to be administered to the host.

The primary purpose of an antimicrobial susceptibility test performed in the laboratory (*in vitro*) is to assist clinicians in the choice of appropriate agents they can use for therapy. As such, it is an essential component of the management of patients that antimicrobial susceptibility is performed and reported in the most timely manner. Antibiotics are one of the most widely prescribed drugs in the world and almost everyone will undergo at least one course of antibiotics in their lifetime. The importance of antimicrobial therapy is enormous, from both a clinical and laboratory perspective.

Learning objectives

After studying this chapter you should confidently be able to:

- Understand why antimicrobial susceptibility tests are performed
- Discuss current national/international guidelines and methods available
- Describe what is meant by intrinsic resistance and in which pathogens it occurs
- Have a thorough knowledge of commonly used susceptibility test media
- Assess the significance of MIC and MBC
- Understand the difference between bacteristatic and bactericidal agents
- Understand why combination therapy is used and common synergistic combinations
- Give clinical reasons for combination therapy and examples

- Understand procedures for testing for antibiotic synergy between two different agents
- Understand how resistance markers may prove useful in detecting more widespread resistance to that class of antibiotic (e.g. ESBLs)
- Know the types of antibiotics that require patient's serum level monitoring
- Understand procedures for monitoring serum antibiotic levels and the principles of the methodology
- Understand acceptable ranges of peak and trough serum levels for the commonly monitored antibiotics
- Understand the role of reference laboratories

4.1 Rationale for antimicrobial susceptibility testing

It is essential that for the management of patients with an infection the organism isolated is tested for susceptibility against a relevant set of antimicrobial agents. Without susceptibility test results, the clinician would be uncertain as to whether an infection was being treated with the correct therapy. However, it is not quite as clear cut as this because if a patient is admitted to hospital then the clinician has no choice but to prescribe the most appropriate antimicrobial in the hope that the patient will respond (**empiric antibiotic therapy**). The results of susceptibility tests will follow and hopefully confirm that the correct therapy was administered.

empiric antibiotic therapy
Empiric antibiotic therapy is antibiotic therapy commenced before the identification of the causative microorganism is available. Typically, full identification and susceptibility testing of bacteria from clinical specimens is not available for 48–72 h after collection of the specimen from the patient.

Routine laboratory susceptibility testing, in addition to providing a result to the clinician, also provides essential current and relevant accumulated data that are used in the formulation of hospital prescribing policies, the selection of the most suitable agents for empirical use, surveillance of resistance studies and the future development of new agents.

Key Point

A prescribing policy is where clinicians are guided in the use of antimicrobials rather than prescribing any drug.

In a clinical setting, agents are commonly used empirically and the laboratory test is used to confirm susceptibility, explain treatment failures and to provide suitable alternative agents. Just consider GP surgeries for a moment; a patient showing signs of infection would have an antimicrobial agent prescribed. The patient wouldn't be happy if they had to return to the practice several days later for treatment, while their symptoms continued, waiting for the laboratory to isolate the organism and test it against the antibiotic prescribed.

4.2 Commonly used antibiotics

A wide variety of antibiotics are used for patient therapy and a number of different classes of antibiotics exist. Some antibiotics are active only against Gram-positive organisms and some only against Gram-negative strains or anaerobes. Whilst it would be impossible to list them all, the commonest antibiotics used are as follows:

- β-lactams - active against bacterial cell walls and include penicillins and cephalosporins.
- Macrolides - often used as alternatives to β-lactam agents if the patient is allergic, particularly for respiratory and soft tissue infections. Examples include erythromycin and clindamycin.
- Glycopeptides - active virtually exclusively against Gram-positive organisms and resistance is uncommon. Examples are vancomycin and teicoplanin.
- Aminoglycosides - these are **broad-spectrum antibiotics** which are normally administered IV. Common examples include gentamicin and tobramycin.
- Quinolones/fluoroquinolones - these are broad-spectrum antibiotics with Gram-negative organisms showing good susceptibility. A common example is ciprofloxacin.

broad-spectrum antibiotic
Broad-spectrum antibiotics are those that show activity against a wide range of organisms, typically both Gram-negative and Gram-positive species.

4.3 Antimicrobial susceptibility testing

It is essential that tests are performed in such a way that accurate and reproducible results are obtained because of the clinical implications to individual patients. The need for standardization of *in vitro* testing was recognized by the World Health Organization (WHO) in the 1950s when a first expert committee was convened in Geneva. Since that time, several national committees have developed standardized methods, taking into consideration all the components of the test. This includes media constituents, concentration of antimicrobials used in discs, rate of organism growth, supplements required, atmospheric conditions and length of incubation.

Cross reference
Chapter 3, Section 3.7.

It is important that laboratories are aware that standardized methods of susceptibility testing have to be followed precisely and that the procedure cannot be modified. Alterations in the procedure may result in inaccurate interpretation of susceptibility.

Disc susceptibility testing

The most widely used method of determining antimicrobial susceptibility is by the use of disc diffusion tests. The method involves inoculation of an organism onto a specific medium and application of paper discs containing antimicrobials of known strength. All of the above must be standardized. The discs used for susceptibility testing may contain either the antibiotic that is to be used for therapy or **indicator antibiotics**.

indicator antibiotic
Indicator antibiotics are agents that are generally the least microbiologically active members of a family of antibiotics and are therefore more likely to detect low-level resistance. Examples of agents that have been used for this purpose include oxacillin for the detection of penicillin resistance in *Streptococcus pneumoniae* and methicillin resistant *S. aureus*.

Disc susceptibility testing - inoculum required

The method involves the application of paper discs containing different antibiotics to the surface of an agar plate containing a known inoculum of the organism to be tested. Several methods are available for this process including those recommended by the Clinical Laboratory Standards Institute (CLSI) or the British Society for Antimicrobial Chemotherapy (BSAC), both of which require standardization of inoculum, media, disc content and incubation time, the results of which determine if the organism is termed sensitive, intermediate or resistant to the antibiotic(s) tested. It is appropriate to test more than one antibiotic - routinely up to six antibiotics are tested against any organism. This saves the need for further testing if the organism is resistant to one particular antibiotic and results can be reported after overnight incubation.

FIGURE 4.1
Examples of a range of inoculum densities for disc susceptibility testing with the ideal being semi-confluent growth (middle plate: BSAC methodology). (Courtesy of M. Ford.)

Essentially, the target pathogen is inoculated into an appropriate medium (e.g. Iso-Sensitest broth) with a standardized inoculum, made by reference to McFarland standards. Suspensions are accurately prepared using a spectrophotometer, or more commonly a bench-top nephlometer, to ensure the inoculum is correct. According to the methodology used, a dilution of the suspension may be required before the plate is inoculated. The plate is then inoculated and spread to achieve the required 'lawn'; this can be confluent or semiconfluent. For the disc diffusion test, the inoculum density has been shown to be one of the most important factors in ensuring the correct result is obtained. Studies have shown that intra- and inter-laboratory reproducibility is affected most by an inoculum which was considered 'too heavy', whereas a lighter inoculum gave less variation. Figure 4.1 shows the types of lawn, from too light to confluent.

Depth of agar

In the case of disc susceptibility testing, the depth of agar affects performance; very thin medium results in pronounced increase in zone sizes and *vice versa* with very thick media. The medium should have a depth of 4 mm (± 0.5 mm). It is therefore essential that proper quality control is carried out on all media, especially susceptibility test media.

SELF-CHECK 4.1

What would happen to the zone sizes if the plate was (a) too thin and (b) too thick? What consequences could this have for the patient?

Application and storage of discs

Discs are applied using sterile forceps to the plate or more conveniently by use of disc dispensers. The discs generally hold six different antibiotics and are 'stamped' onto the agar plate by the operator. The plates are then incubated under the required conditions and examined for the zones around the various antibiotic discs. The following points are important:

- Discs must be firmly applied to the surface of the agar.
- Discs should be stored correctly as potency of the antibiotic is lost following exposure to heat, light and moisture.
- Discs must be stored in sealed containers with a desiccant and protected from light - this is important for light-susceptible agents such as metronidazole, chloramphenicol and the quinolones.

- Stock discs should be stored at -20°C unless otherwise advised by the manufacturer.
- To prevent condensation, allow discs to warm to room temperature before use.
- Discs must be discarded on, or ideally before, the expiry date shown on the side of the container.

Incubation and atmospheric conditions

Generally, incubation of disc susceptibility tests is in air at 35-37°C for 18-20 h. However, there are exceptions to these recommendations. There is much debate about the effect of atmospheric conditions on the *in vitro* activity of antibiotics. In the case of the respiratory pathogens *Haemophilus influenzae* and *Streptococcus pneumoniae* there are two diametrically opposed opinions, one stating that for optimum growth of newly isolated clinical strains, an atmosphere enriched with 4-6% CO_2 is needed. Other opinions advise against the addition of CO_2 because of its effect on the macrolide antibiotics such as erythromycin - agents that are used to treat respiratory infections. The argument for incubating in 4-6% CO_2 is based on observations that 20% of *H. influenzae* and 40% of *S. pneumoniae* isolates from clinical samples required an atmosphere enriched with CO_2 to give semiconfluent growth. Also, it has been suggested that *in vivo* the acid within the respiratory tree may reduce the efficacy of the macrolide/azalide antibiotics.

The detection of glycopeptide resistance in enterococci has proved unreliable by disc testing methods. However, detection is improved if the plates are incubated for at least 24 h before examination, which allows time for the resistant microcolonies to be visibly detectable.

When using cefoxitin as an indicator antibiotic for the detection of methicillin resistance in *S. aureus* the temperature should not exceed 36°C because this will result in the unacceptable merging of the susceptible and resistant populations. When methicillin or oxacillin is used to detect resistance, plates must be incubated at 30°C to ensure the detection of the minority resistant population. Incubation conditions for all disc susceptibility testing are available on the BSAC web site (http://www.bsac.org.uk) or in CLSI publications.

Key Point

It is essential that all parameters are standardized in order that the zone size result can be interpreted correctly.

SELF-CHECK 4.2

Why are temperature and atmospheric conditions so important in disc susceptibility testing?

What do zones of inhibition mean - what do they relate to?

Just because an organism shows a zone around an antibiotic disc doesn't mean the organism is susceptible to it. Is a zone of 15 mm sensitive, or is a zone of 30 mm? If so, does this mean you can reliably treat the patient? Because of this, it is important that you understand how zone sizes relate to the concentration of antibiotic that is required in the body to inhibit growth

minimum inhibitory concentration (MIC)
Minimum inhibitory concentration is the lowest concentration of an antibiotic needed to inhibit bacterial growth.

breakpoint
The breakpoint refers to when only certain antibiotic concentrations are tested, normally two. These concentrations would be appropriate for differentiating between susceptible, intermediate and resistant microorganisms.

or kill the infecting organism. There have been several approaches to obtain a relationship between **minimum inhibitory concentration (MIC)** and zone of inhibition, so that critical zone diameters can be obtained that separate the susceptible, intermediate and resistant populations. One approach uses regression analysis where zone diameters (arithmetic scale plotted on the *x* axis) and corresponding MICs (logarithm to base 2 plotted on the *y* axis) for 100 to 150 representative clinical isolates with an MIC evenly distributed over the range studied are tested simultaneously by MIC and disc diffusion (Figure 4.2).

By applying the MIC **breakpoint** the critical zone diameters can be calculated, keeping the number of major discrepancies (organism resistant by MIC, yet susceptible by disc testing) as low as possible. A disadvantage of this method is that it relies on an even distribution of MIC over the range tested. This is not always possible, particularly when new agents are introduced.

In view of the inconsistencies of regression analysis, Metzler and DeHaan devised a bivariate 'error minimization' or 'error rate bounded' classification scheme for relating MIC values to zone diameters which is not concerned with regression analysis. Figure 4.3 illustrates a scattergram where the zone diameter breakpoint is calculated by taking into consideration the MIC

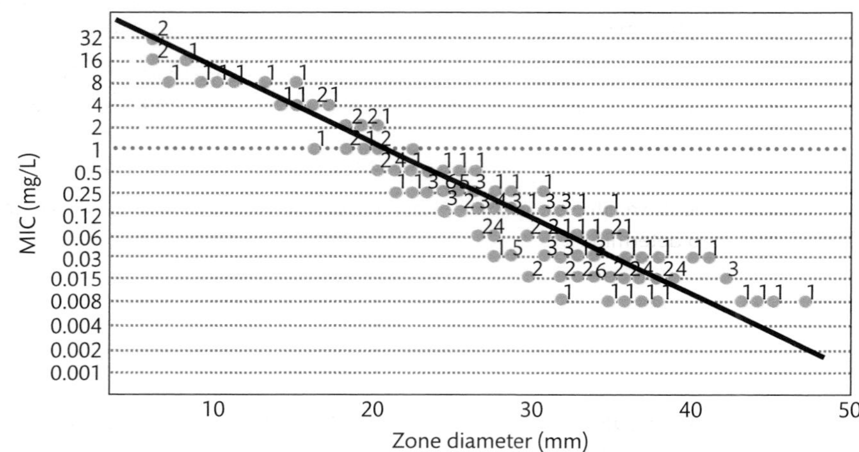

FIGURE 4.2
Example of a regression line correlating the zone diameters to agar dilution MICs. (Courtesy of J. Andrews.)

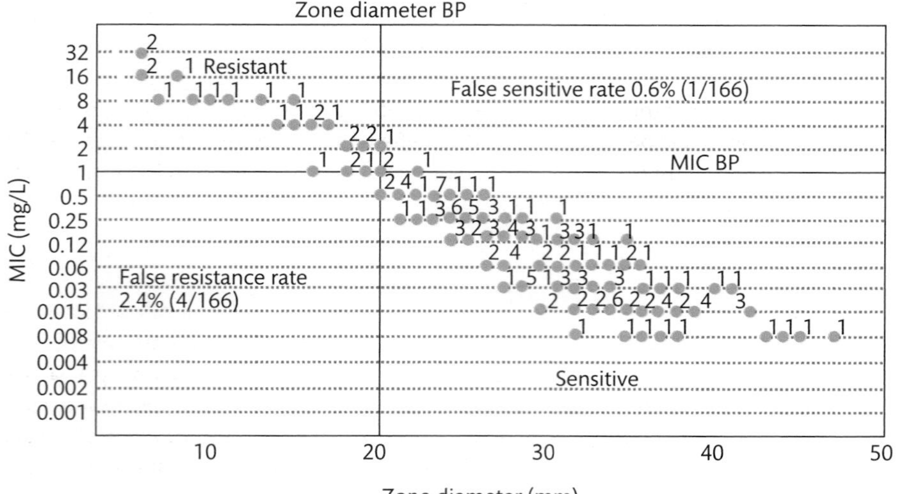

FIGURE 4.3
Example of error rate bounded analysis. (Courtesy of J. Andrews.)

breakpoint and applying a false susceptible and false resistant limit of ≤1 and 5%, respectively. This method of analysis has been used by the BSAC.

To convert zones of inhibition into susceptible, intermediate or resistant categories, thus inferring whether it is possible to treat an infection with a given antibiotic, reference is made to the above data in conjunction with pharmacokinetic (time course of antibiotics in serum, tissues and body fluids), pharmacodynamic (the relationship between drug concentration and antimicrobial effect), and clinical response data. All of the data together are analysed and a zone size calculated, which can be termed sensitive, intermediate or resistant. It is important to define what these terms mean and how they relate to the clinical management of the patient:

- Clinically resistant: level of antimicrobial susceptibility that results in a high likelihood of therapeutic failure.
- Clinically susceptible: level of antimicrobial susceptibility associated with a high likelihood of therapeutic success.
- Clinically intermediate: a level of antimicrobial susceptibility associated with uncertain therapeutic effect. It implies that an infection due to the isolate may be appropriately treated in body sites where the drugs are physically concentrated or when a high dosage of drug can be used; it also indicates a buffer zone that should prevent small, uncontrolled, technical factors from causing major discrepancies in interpretation.

It must be noted that national MIC breakpoints are based on the standard dosing used in that particular country and that dosage may vary between Europe, the USA and the rest of the world.

SELF-CHECK 4.3

How does the zone size relate to the effectiveness of the antibiotic being tested? What methods have been used to determine this?

Interpretation of disc susceptibility tests

In the laboratory setting, the zones of inhibition must be read for each antimicrobial susceptibility test set up (Figure 4.4). These zones are measured from the back of the plate where clear media are used or from the front if supplemented media are used. The zone diameter is measured (to the nearest millimetre) with a ruler, callipers or an automated zone reader, taking the zone edge, as judged by the naked eye, as the point of inhibition. Tiny colonies at the edge of the zone, films of growth due to the swarming of *Proteus* spp. and slight growth within sulphonamide or trimethoprim zones should be ignored. Colonies growing within the zone of inhibition should be subcultured and identified, and the test repeated if necessary. The biomedical scientist must confirm that the zone of inhibition for the control strain falls within acceptable ranges before interpreting the test.

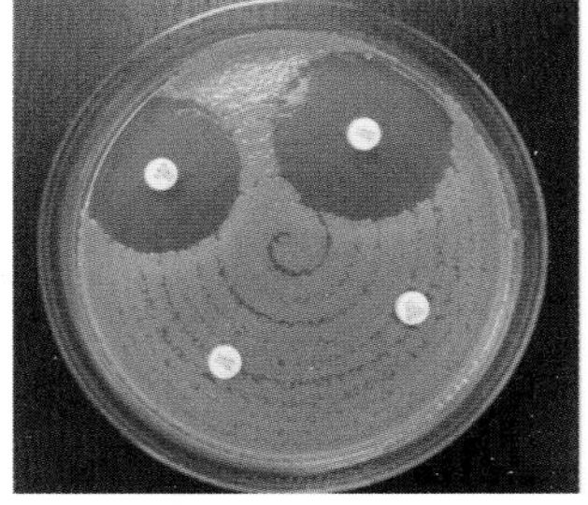

FIGURE 4.4
Showing growth of *E. coli* on Iso-Sensitest agar and zones of inhibition around antibiotic containing discs. (Courtesy of J. Collins.)

A template (Figure 4.5) can also be used for interpreting susceptibility and a program for preparing templates is available from the BSAC web site (http://www.bsac.org.uk). The susceptibility test plate is placed over the template and the zones of inhibition are examined in relation to the template zones. If the zone of inhibition of the test strain is within the area marked with an 'R' the organism is resistant. Antimicrobial susceptibility using discs is reported qualitatively, as sensitive, intermediate or resistant. For example if the zone of inhibition is equal to or larger than the marked area the organism is susceptible. The results would then be reported according to laboratory policy with reference to the hospital prescribing policy. It is important to remember that it is essential to have more than one antibiotic to which an organism is deemed

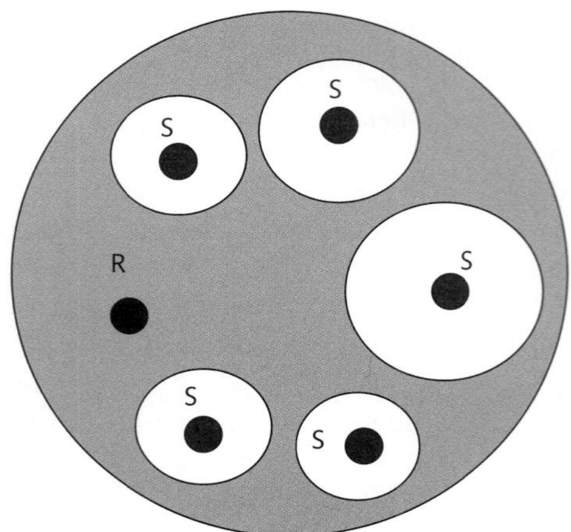

FIGURE 4.5
Template for interpreting disc susceptibility tests. R indicates the zone size which will be interpreted as resistant, and S is susceptible. (Courtesy of J. Andrews.)

susceptible. This is because the patient may have developed an allergic response or the drug may not be well tolerated. If the organism shows multidrug resistance then further antibiotics may need to be tested.

SELF-CHECK 4.4

What would be the consequences of mis-reporting the result of an antibiotic susceptibility test?

In an attempt to overcome some of the problems associated with reading antibiotic zone sizes, commercial companies have produced automated instruments. Such instruments allow zone sizes to be read rapidly and input into the host computer, and as a result transcription errors are removed. There are, however, problems with this instrumentation, in that colonies growing inside a susceptible zone may be missed and problems can be encountered with diffuse zone edges. It is believed the performance of these instruments will improve in the future with more powerful processing power and image recognition.

Direct susceptibility testing

As you have read earlier, disc susceptibility tests require a standardized inoculum in order to interpret the result correctly. However, there are some situations where it can be important to have a susceptibility test result at the same time as the culture is available. This, however, bypasses the standardized approach to susceptibility testing in order to achieve a potentially clinically useful result. Direct susceptibility tests are performed exactly as above but the sample is applied directly to the agar and antibiotic discs applied. This is often performed on positive blood cultures where a susceptibility result is often very important. Direct susceptibility tests are also performed on urine samples, which may have been highlighted as positive using an automated analyser or organisms have been seen using inverted microscopy. Some studies have shown only small variations between direct and conventional standardized tests. For urine samples, this will result in improved turnaround times and less labour as a suspension need not be made. However, it is essential that if the correct inoculum has not been achieved, then the test must be repeated - this is especially important for direct susceptibility results obtained from positive blood cultures.

SELF-CHECK 4.5

What are the potential problems in performing and reading direct susceptibility tests? Are there clinical benefits in performing them?

Double disc/combination disc techniques

Double disc tests have been used to detect inducible resistance in organisms that classically have inducible ***AmpC* β-lactamase** (e.g. *Enterobacter* spp. and *Serratia* spp.) or for detecting dissociated resistance in Gram-positive organisms (Figure 4.6). This is seen when testing erythromycin and clindamycin and occurs because erythromycin induces an enzyme in the test organism and as a result the organism is resistant to similar antibiotics (e.g. clindamycin). It must be remembered that the two antibiotic discs must be close to each other in order to demonstrate this particular type of resistance.

One of the main uses of combination discs is to confirm the presence of **extended spectrum β-lactamase** (ESBLs). These are enzymes produced by Gram-negative organisms which break down a range of β-lactam antibiotics, including penicillins and cephalosporins. Since these are such widely used antibiotics, it is imperative that organisms producing these enzymes are detected and their presence reported. Delayed recognition of ESBL strains has been associated with increased patient mortality.

Essentially, a plate is inoculated in the same manner as any other susceptibility test but a comparison is made between discs containing a cephalosporin, such as cefpodoxime, and the same disc containing clavulanic acid. The zone diameters are compared, and if the zone size is increased by >5 mm (or equal to) in the presence of clavulanic acid then this implies ESBL production. The advantage of combination discs is their low cost when compared to other methods such as the use of ESBL E-test strips.

Methods for detecting dissociated resistance and confirmation tests for ESBLs are available on the BSAC web site (http://www.bsac.org.uk).

AmpC

AmpC is an inducible enzyme occurring in clinically isolated Gram-negative strains such as *Enterobacter* spp. They hydrolyse broad and extended spectrum cephalosporins.

extended spectrum β-lactamase

Extended spectrum β-lactamases are enzymes that can be produced by bacteria, making them resistant to penicillins and cephalosporins such as cefuroxime, cefotaxime and ceftazidime. These antibiotics are some of the most important antibiotics available for the treatment of infectious disease.

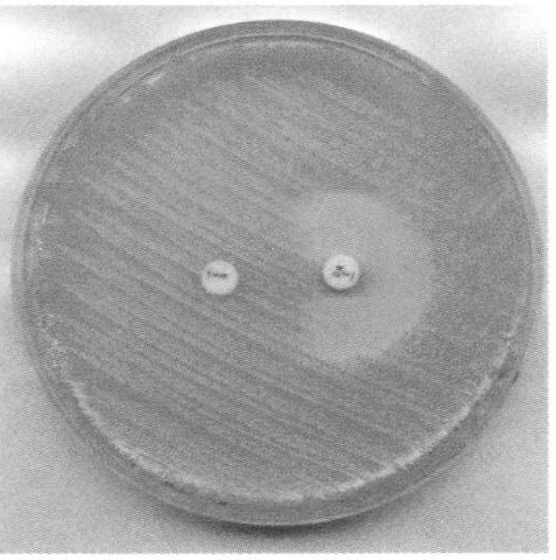

FIGURE 4.6
Dissociated resistance in *Staphylococcus aureus* detected using a double disc technique shown with erythromycin (left) and clindamycin (right) discs. (Courtesy of M. Ford.)

Gradient method of susceptibility testing

This is a method of susceptibility testing where an actual MIC as a numerical value is achieved rather than a zone size. The gradient method of testing is available commercially (E-test, bioMérieux, and MICE, Oxoid, Basingstoke, UK). For this method, antibiotic is applied as a gradient to an inert strip. When the strip is applied to the surface of an agar plate, previously inoculated with a lawn of the organism, there is instantaneous diffusion of the antibiotic into the medium and the gradient is maintained. After overnight incubation the ellipse is examined to determine the amount of antibiotic that has inhibited the growth of the organism (Figure 4.7). Gradient methods have mainly been used to confirm resistance rather than for primary susceptibility testing. This is because generally a whole plate is required for the test and it is very expensive when compared to conventional disc susceptibility tests.

FIGURE 4.7
E-test strip showing susceptibility of *S. pneumoniae* to erythromycin. (Courtesy of M. Ford.)

Gradient methods are not considered a 'gold standard' for evaluating the performance of other methods of testing. They should not be used to elucidate equivocal results when detecting methicillin resistance in staphylococci because they are subject to variation in the conditions of testing like any other microbiological method. A molecular method

mecA

mecA is a gene that codes for an altered penicillin binding protein (PBP2 or PBP2a). These proteins have a lower affinity for β-lactam antibiotics and confer resistance to the organism.

for ***mecA*** should be used to determine resistance, or for *S. aureus* a latex method for detecting PBP2 or PBP2a is available commercially (Mast, Merseyside, UK).

4.4 Broth methods

Broth MIC determinations are considered the 'gold standard' by the CLSI and this method of testing is preferred for determining organism susceptibility by the pharmaceutical industry as part of drug development programmes.

minimum bactericidal concentration (MBC)

Minimum bactericidal concentration is the lowest concentration of an antibiotic which is cidal to ≥99.9% of the test inoculum.

Broth MICs are undertaken infrequently in the UK and are generally only required when the **minimum bactericidal concentration (MBC)** of an organism is needed, for example as a possible aid in the treatment of endocarditis or for determining the susceptibility of slow-growing organisms. Mueller–Hinton broth (MHB) and Iso-Sensitest broth (ISTB) are available for undertaking broth susceptibility tests. For MHB the formula is the same as for the agar method (without the addition of agar) and with a Ca^{2+} and Mg^{2+} concentration of 20–25 mg/L and 12–25 mg/L, respectively. Broth methods have been adapted for use in automated systems that are available commercially. Some products use microtitre trays that require reconstitution of freeze-dried antibiotic with MHB. It must be remembered that if MHB is used, the CLSI MIC breakpoints must be applied to interpret susceptibility; those of BSAC cannot be used.

For broth MIC testing the inoculum should be 5 x 10^5 CFU/mL, except when testing organisms with extracellular β-lactamase where an inoculum of 10^6 CFU/spot is required. Methods for preparing inocula for MIC testing are freely available on the BSAC web site (http://www.bsac.org.uk) or in a variety of other publications.

Briefly, the organism for broth MIC testing is placed into broth and this is incubated overnight. This is then diluted to the required concentration of the organism. A rack of tubes containing sterile broth is prepared containing doubling dilutions of the test antibiotic. A known suspension of organism is placed into each tube, with an antibiotic-free tube as a control. A culture plate is inoculated with the organism suspension as a purity check. The tubes are then incubated overnight under the required conditions and examined for turbidity. After incubation it is essential to ensure that the test and control organism have grown on the antibiotic-free control plate, in the broth control tube and no contaminants are present. The MIC is determined as the lowest concentration that completely inhibits growth. This produces a numerical value which can be compared to published data to determine if the organism is sensitive, intermediate or resistant.

bacteristatic/bactericidal

Bacteristatic refers to an antibiotic which inhibits the growth of an organism but doesn't necessarily kill it. Bactericidal antibiotics are those which are likely to kill the organism.

For broth MICs a problem may be seen where one tube or well with no growth is observed but obvious growth is seen at higher and lower concentrations. The 'skipped' tube or well should be ignored and the concentration that finally inhibits growth should be recorded as the MIC. If more than one 'skipped' tube or well occurs the MIC should be repeated. Trailing endpoints may be seen with **bacteristatic** agents such as sulphonamide and trimethoprim. Trailing endpoints are generally not observed with **bactericidal** agents such as gentamicin. In these cases, a 90% decrease in growth as compared with the growth control should be recorded as the MIC. Partial inhibition may also be seen with growth of resistant variants or if there is contamination. Subculture of all tubes or wells is advised.

It must be remembered that the stability of the antibiotic must be assessed if prolonged incubation is undertaken. Iso-Sensitest broth (ISTB) is the medium used most frequently.

SELF-CHECK 4.6

What are the main differences between broth and agar susceptibility tests and the results obtained?

4.5 Automated susceptibility testing

Over the last few years, many routine laboratories have switched a large part of the susceptibility test workload onto automated platforms. This is in a bid to further improve standardization, as operator error in recording zone sizes is eliminated. In addition, these instruments can provide an identification which is very beneficial; as we have learned, some organisms can give large zone sizes but have mechanisms of resistance whereby the antibiotic may not work *in vivo*.

There are several instruments available, the most common being the bioMérieux VITEK and the Siemens Walkaway (Figure 4.8). For both of these instruments, interpretation of results is based on CLSI MIC breakpoints; however, in the future, it is likely that some standardization will occur and the instruments will be programed with EUCAST breakpoints. With these instruments, operator time is minimal and the main inoculation process can be easily deskilled.

The Siemens Walkaway instrument uses a broth dilution susceptibility test method where the components are dehydrated. The various antibiotics used are diluted in broth (Mueller-Hinton) with all of the required supplements necessary for optimal test performance. After the panels are inoculated with a standardized suspension, they are placed onto the instrument and incubated at 35°C for a minimum of 16 h before reading. The panels are read automatically by the instrument in comparison with growth controls contained within the panel. Growth in the panels results in turbidity and this is recorded by the instrument. The results are reported as susceptible, intermediate or resistant. The actual MIC is also recorded. Depending upon the panel type used, either an MIC or a breakpoint concentration is achieved. Numerous panel types are available for both Gram-negative and Gram-positive organisms. Figure 4.9 shows a typical antibiotic/identification panel used in the instrument.

Whilst the instruments provide excellent results and overcome the problems associated with reading disc susceptibility tests, there are some drawbacks. They cannot be used for performing direct susceptibility tests, for example on positive blood cultures. In addition, there are a number of organisms where these tests give poor results, for example on mucoid *Pseudomonas* strains. As a result, it is unlikely these instruments will totally replace conventional disc susceptibility tests.

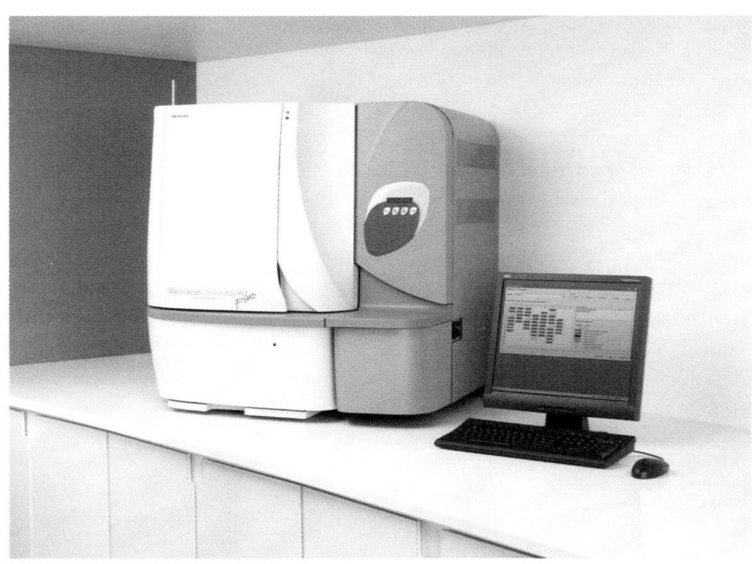

FIGURE 4.8
Siemens Walkaway automated susceptibility test instrumentation. (Courtesy of Siemens.)

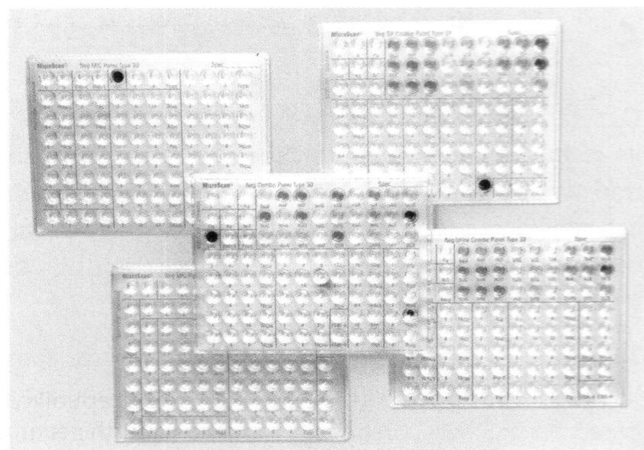

FIGURE 4.9
Microtitre panels containing dehydrated antibiotics used in the Siemens Walkaway system. (Courtesy of Siemens.)

4.6 Detection of mechanisms of resistance and the application of expert rules

Results of susceptibility testing are normally reported as sensitive, intermediate or resistant based on the results for individual antibiotics. Often only limited agents are tested and organisms are not identified to species level, which precludes the detection of some underlying mechanisms of resistance (e.g. *Amp*C β-lactamases).

In France and in some commercial systems (VITEK 2) a comprehensive panel of 16 to 20 antibiotics are tested in conjunction with an accurate identification of the test organism. This allows the full potential of interpretative reading to be exploited, including the application of expert rules which have implications for antibiotic therapy. A comprehensive publication on this subject is available on the BSAC website (http://www.bsac.org.uk).

It is also important for laboratories to recognize unusual or unexpected susceptibility patterns, for example Group A streptococci resistant to benzyl penicillin, vancomycin, teicoplanin or linezolid. Table 4.1 shows examples of exceptional resistance phenotypes that are worthy of further investigation. In these cases, if resistance is confirmed, organisms should be sent to a reference laboratory for further analysis and confirmation.

Natural (inherent/intrinsic) resistance

Resistance to antibiotics can be divided into that which is intrinsic/inherent resistance (naturally occurring) and that which is acquired. Naturally occurring resistance is shared by most or all of the species and is therefore considered a characteristic of that species. This type of resistance is less of a problem because it is well recognized during the early development of a drug. Of greater importance is acquired resistance, which implies that an originally susceptible organism has developed resistance to an agent to which it was originally susceptible.

It is important that laboratories are aware of natural (inherent/intrinsic) resistance of common pathogens before antibiotic susceptibility testing is undertaken. Examples of natural resistance are shown in Table 4.2; a more comprehensive list of resistance phenotypes of common pathogens can be found on the BSAC website (http://www.bsac.org.uk).

TABLE 4.1 **Examples of exceptional resistance phenotypes.**

Organism	Exceptional resistance/susceptibility
Enterobacteriaceae except *Proteus* spp.	Resistance to meropenem/imipenem
E. coli	Resistance to ertapenem, tigecycline
Klebsiella spp.	Susceptibility to ampicillin/amoxicillin
P. aeruginosa	Resistance to colistin
Pseudomonas spp.	Susceptibility to ampicillin/amoxicillin, cephalosporins other than ceftazidime of the fourth generation cephalosporins[1]
N. meningitidis	High level resistance to penicillin or resistance to ciprofloxacin
N. gonorrhoeae	Resistance to any third generation cephalosporin[2]
S. aureus	Resistance to vancomycin, teicoplanin, linezolid, quinupristin/dalfopristin, daptomycin, tigecycline
S. pneumoniae	Resistance to meropenem, vancomycin, teicoplanin, linezolid, quinupristin/dalfopristin
β-Haemolytic streptococci	Resistance to penicillin, vancomycin, teicoplanin, linezolid, quinupristin/dalfopristin, daptomycin, tigecycline
Enterococcus spp.	Resistance to linezolid, daptomycin, tigecycline Resistant to teicoplanin but not vancomycin
E. faecalis	Susceptible to quinupristin/dalfopristin
E. faecium	Resistant to quinupristin/dalfopristin

[1] Fourth generation cephalosporins: cefepime and cefpirome.
[2] Third generation cephalosporins: cefotaxime, cefpodoxime, ceftazidime and ceftriaxone.

TABLE 4.2 **Examples of natural (inherent) resistance in common clinical isolates.**

Organisms	Natural (inherent) resistance
All Enterobacteriaceae	Penicillin, glycopeptides, fusidic acid, macrolides, clindamycin, linezolid, daptomycin, quinupristin/dalfopristin, mupirocin
Klebsiellae	Ampicillin (*Klebsiellae* have chromosomal penicillinases)
Proteus mirabilis	Colistin, nitrofurantoin, tetracyclines
Salmonella species	Cefuroxime and aminoglycosides (active *in vitro*, not active *in vivo*)
All Gram-positive bacteria	Aztreonam, temocillin, colistin, nalidixic acid
Methicillin-resistant *Staphylococcus aureus*	All β-lactams
Moraxella catarrhalis	Trimethoprim
Listeria	Cefotaxime, cefpodoxime, ceftazidime and ceftriaxone, fluoroquinolones

SELF-CHECK 4.7

Why is it considered important that the identity of the organism is known at the same time as the susceptibility results are determined?

4.7 Additional tests

Antibiotic combination and synergy testing

Combinations of antibiotics have been used clinically to prevent the emergence of resistance when treating *Mycobacterium tuberculosis*, for the treatment of bacterial endocarditis and also in the treatment of patients with cystic fibrosis. *In vitro* methods that have been used to observe the effect of combining one or more antibiotics are the chequerboard titration, time kill curves and double disc techniques.

Chequerboard titrations

Chequerboard titrations have been used to calculate the fractional inhibitory concentration (FIC) using the following formula:

$$\text{FIC} = \frac{\text{MIC}_A \text{ (in combination)}}{\text{MIC}_A \text{ (alone)}} + \frac{\text{MIC}_B \text{ (in combination)}}{\text{MIC}_B \text{ (alone)}}$$

where A is one drug and B is the other.

Interpretation of the combination is as follows:

$$\text{Synergistic FIC} = \leq 0.5$$

(positive interaction where the combined effect is significantly greater than the expected result when drugs are tested alone);

$$\text{Additive FIC} = 1.0$$

(indifferent response);

$$\text{Antagonistic FIC} = \geq 2.0$$

(negative interaction where the combined effect is significantly less than their independent effect when tested separately).

The value of this test in the clinical setting has been debated. MacGowan *et al.* demonstrated synergy when erythromycin was tested in combination with itself and therefore doubted the value of the test. Others have recommend using synergy testing to aid in the treatment of patients with cystic fibrosis.

Time kill curves

These tests are mainly used as a research tool due to the labour intensity of performing the test. They have been used to compare the activity of drugs both alone and in combination, to measure the effect of increasing the concentration of drug on *in vitro* activity, and to detect the emergence of resistance.

The test to measure the effect of increasing the concentration of drug on *in vitro* activity has been used to determine if drugs have concentration or dose-independent activity (very little difference in the rate of bactericidal activity when the concentration exceeds four times the MIC, observed with the β-lactam antibiotics) or concentration or dose-dependent activity (number of organisms decrease more rapidly with each rising MIC concentration, seen with the quinolones and aminoglycosides). The *in vivo* pharmacodynamic parameters that correlate with clinical response in man and animals for agents that have concentration or dose-independent activity or concentration or dose-dependent activity respectively are as follows:

$$\frac{\text{Time}}{\text{MIC}}$$

where the concentration above the MIC should be 50–75% of the dose interval;

$$\frac{\text{AUC or } C_{max}}{\text{MIC}}$$

where:

a ratio of 30–60 is needed

AUC = Area under the curve

C_{max} = Peak concentration following standard dosing

4.8 Control of antimicrobial susceptibility testing

Control strains

Susceptible control strains have been recommended for monitoring test performance (not for interpreting susceptibility) and resistant strains to ensure that particular mechanisms of resistance can be detected. Ideally, appropriate controls should be included with every batch of tests performed in the laboratory on a daily basis as some antibiotic discs show reduced activity if not stored correctly. Control strains are available from the American Type Culture Collection (ATCC) and the National Collection of Type Cultures (NCTC) and limited controls are available commercially (several suppliers). Table 4.3 provides examples of susceptible and resistant control stains that can be used.

Maintenance of control strains

The method used for the maintenance of control strains should minimize the risk of mutation. The following procedure is suggested (http://www.bsac.org.uk). Briefly, store control strains at -70°C in glycerol broth or on beads in glycerol broth. Non-fastidious organisms may be stored at -20°C. Ideally, two vials of each control strain should be stored, one for an 'in-use' supply, the other for archiving. Each week, a bead or glycerol broth from the 'in-use' vial should be subcultured onto appropriate non-selective media and checked for purity. From this pure culture, prepare one subculture for each of the following 5 days. For fastidious organisms that will not survive on plates for 5 or 6 days, subculture the strain daily for no more than 6 days.

In the case of disc diffusion tests, acceptable ranges for control strains are published. For monitoring laboratory performance of disc diffusion a 20 point reading frame has been suggested (Figure 4.10) (http://www.bsac.org.uk). The problems that can occur with disc susceptibility testing are listed in Table 4.4.

TABLE 4.3 Control strains for antimicrobial susceptibility testing.

Organism	Strain		Characteristics
	Either	Or	
Escherichia coli	NCTC 12241 (ATCC 25922)	NCTC 10418	Susceptible
Escherichia coli	NCTC 11560		TEM-1 β-lactamase producer
Staphylococcus aureus	NCTC 12981 (ATCC 25923)	NCTC 6571	Susceptible
Staphylococcus aureus	NCTC 12493		*MecA*-positive, methicillin resistant
Pseudomonas aeruginosa	NCTC 12934 (ATCC 27853)	NCTC 10662	Susceptible
Enterococcus faecalis	NCTC 12697 (ATCC 29212)		Susceptible
Haemophilus influenzae	NCTC 11931		Susceptible
Haemophilus influenzae	NCTC 12699 (ATCC 49247)		Resistant to β-lactams (β-lactamase negative)
Streptococcus pneumoniae	NCTC 12977 (ATCC 49619)		Intermediate resistance to penicillin
Neisseria gonorrhoeae	NCTC 12700 (ATCC 49226)		Low-level resistant to penicillin
Pasteurella multocida	NCTC 8489		Susceptible
Bacteroides fragilis	NCTC 9343 (ATCC 25285)		Susceptible
Bacteroides thetaiotaomicron	ATCC 29741		Susceptible
Clostridium perfringens	NCTC 8359 (ATCC 12915)		Susceptible

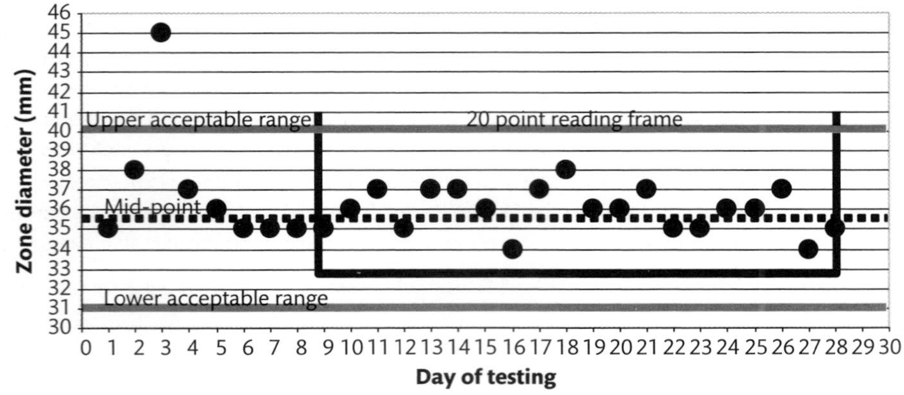

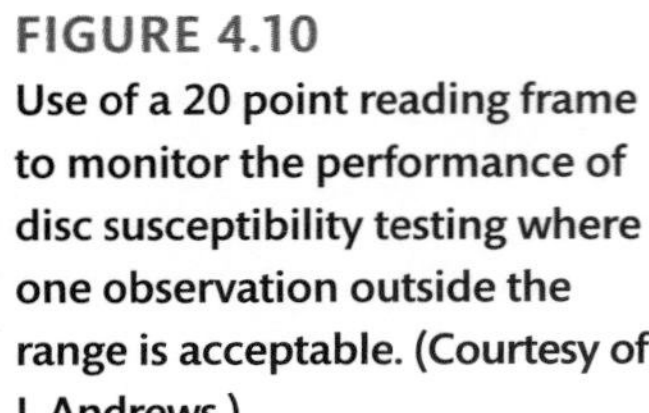

FIGURE 4.10
Use of a 20 point reading frame to monitor the performance of disc susceptibility testing where one observation outside the range is acceptable. (Courtesy of J. Andrews.)

TABLE 4.4 Potential sources of error for disc susceptibility testing.

Source	Error
Test conditions	Incorrect inoculum Uneven inoculum Excessive pre-incubation before discs applied Excessive pre-diffusion before plates incubated Incorrect incubation temperature Incorrect incubation atmosphere Incorrect incubation time Inadequate illumination of plates when measuring zones Incorrect reading of zone edges Incorrect zone criteria used to interpret susceptibility
Medium	Required susceptibility testing agar not used Not prepared as required by the manufacturer's instructions Batch to batch variation Antagonists present (e.g. with sulphonamides and trimethoprim) Incorrect pH Incorrect divalent cation concentration Incorrect depth of agar plates Agar plates not level Expiry date exceeded
Antimicrobial discs	Wrong agent or content used Labile agent possibly deteriorated Light sensitive agent left in light Incorrect storage leading to deterioration Disc containers opened before reaching room temperature Incorrect labelling of disc dispensers Expiry date exceeded
Control strains	Contamination Mutation Incorrect inoculum density Uneven inoculation Old culture used

4.9 Antibiotic assays

Antibiotic assays are performed for many reasons. New agents are assayed as part of drug development to establish their pharmacokinetic properties, initially in animals and later in man. Concentrations of a drug may also be measured as part of clinical trials to ascertain levels at various sites of infection. There has been a trend away from performing these assays in microbiology departments over the last few years and into biochemistry, where they are performed on modern fully automated analysers. However, many microbiology laboratories still perform these tests and the results are often interpreted by senior biomedical scientists or medical staff. Which department performs these tests is of little consequence as the tests need to be performed for a variety of reasons.

In a clinical setting, therapeutic drug monitoring is performed for the following reasons:

- To ensure that serum concentrations are high enough for clinical efficacy
- To ensure that toxicity is minimized for those agents with a narrow **therapeutic index**, e.g. the aminoglycosides
- To monitor patient compliance, e.g. in the treatment of tuberculosis.

therapeutic index
Therapeutic index is the ratio between the toxic dose and the therapeutic dose of a drug, used as a measure of the relative safety of the drug for a particular treatment. A drug with little difference between toxic and therapeutic doses is referred to as having a narrow therapeutic index.

Antibiotics that require monitoring primarily to minimize adverse effects are gentamicin, tobramycin, amikacin and vancomycin. In Tables 4.5 and 4.6 the normal levels for these agents is shown. For agents not commonly assayed, a service is generally provided by antimicrobial reference laboratories.

Assay methods - microbiological assays

Methods that have been used to measure concentrations of antibiotics are divided into microbiological and non-microbiological techniques.

Many of the original assay methods were developed by the pharmaceutical industry to monitor commercial products. However, microbiological assays have been used in clinical laboratories particularly with the introduction of potentially toxic agents such as the aminoglycosides. Although microbiological assays are relatively easy and cheap to perform, they are now used infrequently in clinical laboratories because patients are often prescribed more that one antibiotic, which affects the specificity of the assay and they have long 'turn around' times with results only being available after 18 to 24 h. Table 4.7 shows the antibiotic profile for a resistant *Klebsiella* spp. (NCTC 10896) that has been used to measure concentrations of gentamicin. A more detailed account of methods can be found in *Clinical Antimicrobial Assays* (see Further reading).

TABLE 4.5 Normal ranges for gentamicin, tobramycin, netilmicin and amikacin in patients following once or twice daily dosing.

Antibiotic	Once daily dosing predose level (mg/L)	Twice (BD) or three times (TDS) daily dosing		
		Site of infection	Predose level (mg/L)	Postdose level (mg/L)
Gentamicin Tobramycin Netilmicin	<1	Gram-negative pneumonia	<2	>7
		Infective endocarditis	<1	2–3
		Most other infections	<2	>5
Amikacin	<5		<10	>20

TABLE 4.6 Normal ranges for patients receiving vancomycin therapy.

Antibiotic	Concentration (mg/L)		
	Predose only	1 h postdose	2 h postdose
Vancomycin	10–20	20–40	18–26

TABLE 4.7 Antibiotic profile for resistant *Klebsiella* spp. NCTC 10896 that has been used to measure concentrations of gentamicin in clinical samples.

Antibiotic	Susceptibility	Antibiotic	Susceptibility
Penicillin	R	Gentamicin	S
Ampicillin	R	Neomycin	R
Cloxacillin	R	Trimethoprim	R
Carbenicillin	R	Chloramphenicol	R
Methicillin	R	Tetracycline	R
Fusidic acid	R	Streptomycin	R
Lincomycin	R	Co-trimoxazole	R
Erythromycin	R	Mecillinam	R
Rifampicin	R	Cefazolin	MIC 16 mg/L
Sulphadimadine	R	Cephradine	MIC 8 mg/L
Colistin	S	Cefoxitin	S
Tobramycin	S	Cefuroxime	S
Kanamycin	S	Carbapenems	S

R Resistant; S Susceptible.

Advantages of microbiological assays:

- Relatively cheap
- Relatively quick to develop
- Can be more sensitive than chemical methods
- Measure antimicrobial activity

Disadvantages of microbiological assays:

- Mixtures of antibiotics are difficult to measure
- Active metabolites may interfere with the assay
- Concomitant antibiotics may affect the result
- Often clinical details do not indicate if the patient is on more than one antibiotic
- Results are obtained after overnight incubation and therefore these assays are not suitable for urgent tests
- If the organism fails to grow there is a delay of 24 h before a result is available.

SELF-CHECK 4.8

Why are antibiotic assays performed only for a limited number of antibiotics?

Non-microbiological methods – enzyme-based assays

There are three broad classes of non-microbiological assay methods used to measure concentrations of antibiotics in clinical samples: enzyme-based assays, immunoassays and physicochemical assays, which includes chromatography.

Specific enzymes, which inactivate an antibiotic, have been used to assay specific antimicrobials. Examples of enzymes that have been used are chloramphenicol acetyltransferase (CAT) and aminoglycoside modifying enzymes; this method of testing involves the use of a radiolabelled cofactor. Enzyme-based assays were introduced to overcome the effect of concomitant antibiotics and to improve turnaround time, both disadvantages of microbiological assays. Although this method of testing gives results in approximately 1 h, the safety precautions associated with radioactive material (including the disposal of radioactive material) and the need for expensive scintillation counters have made this a costly procedure. In comparative studies it has been found that enzyme-based assays compared favourably with other methods of testing.

Advantages of non-microbiological methods:

- Assay not affected by concomitant antibiotics
- Results available in approximately 2 h
- Suitable for on-call purposes

Disadvantages of non-microbiological methods:

- Reagents not available as a kit, therefore preparation of enzymes and other reagents is undertaken by the laboratory
- Radiolabelled cofactor is expensive
- Involves the use of radioactive material, requiring special handling and disposal
- Expensive scintillation counter is needed.

Immunoassays

For immunoassays the antibiotic being assayed acts like an antigen and binds with a specific antidrug antibody. Most antibiotic assays are of the competitive binding type where a fixed concentration of tracer-labelled drug (the drug to be assayed labelled with a radioactive atom, fluorophore or an enzyme) competes with the drug to be assayed in the sample for a limited amount of antidrug antibody. The greater the amount of drug in the sample, the smaller the amount of tracer-labelled drug that can bind to the antidrug antibody. Examples of commercial assays are the fluorescence polarization immunoassay (FPIA, Abbott) and the kinetic interaction of microparticles in solution immunoassay (KIMS, Roche Diagnostics, Indianapolis, USA). The antibiotics that are generally assayed by immunoassay are the aminoglycosides and vancomycin.

Advantages of immunoassays:

- Rapid and easy
- Fast turnaround time
- Small samples can be processed and therefore they are applicable to paediatric samples
- Kits supplied commercially
- Continual upgrade of technology

- Can be assayed using clinical chemistry platforms and therefore results can be available 24 hours a day 7 days a week
- Perform well in National External Quality Assurance Schemes (NEQAS)

Disadvantages of immunoassays:

- Equipment may be expensive
- If the machine is purchased a laboratory could be committed to a system that is superseded

High-performance liquid chromatography

High-performance liquid chromatography (HPLC) is used when immunoassays are not available and microbiological assays are inappropriate. For HPLC, a form of liquid chromatography, the drug to be assayed is separated from other components in the sample and then quantified using a non-specific property such as ultra violet (UV) absorption.

In HPLC a mobile phase (liquid) is pumped through a column packed with a stationary phase on which separation takes place. The sample is introduced into the mobile phase by means of an injector and passes through the column where separation occurs by differential retention on the solid phase. After elution from the stationary phase, the mobile phase containing the separated solutes passes though a detector where the separated substances cause a response. The output from a detector is called a chromatogram. By comparing either the peak height or the area under the curve of a calibrator with those of the test, the concentration in the test sample can be calculated. Concentrations can be calculated using a calibration curve or a single-point calibration. For a single-point calibration the following formula is used:

$$C = C\,(R_{sample}/R_{calibrator})$$

where:

C = concentration of the calibrator

R_{sample} = peak height or peak area of the sample

$R_{calibrator}$ = peak height or peak area of the calibrator

HPLC is normally only undertaken by reference laboratories or research departments.

Advantages of HPLC:

- Very versatile and mixtures can be measured
- Large numbers of samples can be processed using autosamplers
- High precision coefficient of variation (CV) 5% or less
- Result within 2–3 h

Disadvantages of HPLC:

- Costly
- Operator dependent
- New assay development is lengthy

SELF-CHECK 4.9

Which method of antibiotic assay would be the best for routine monitoring of antibiotics in patients' serum? Give reasons for your answer.

CASE STUDY 4.1 Multidrug-resistant organisms

A 70-year-old male admitted via the Emergency Department with symptoms of septicaemia has a blood culture and urine sample collected. After overnight incubation the positive blood culture showed Gram-negative bacilli. The positive bottle was subcultured to purity plates and direct disc susceptibility tests performed. The urine sample was 'flagged' as positive on automated microscopy and the organism cultured onto CLED agar and a direct susceptibility test performed.

The next day the culture results yielded a 'coliform', which showed resistance to cefuroxime, ceftazidime, cefpodoxime, gentamicin and ciprofloxacin. The organism was susceptible to amikacin.

- What would you suspect from the susceptibility pattern of the organism?
- How would you confirm this and what other tests would you perform and why?

The organism was identified as *E. coli* and the ESBL combination discs gave the following results:

Cefpodoxime 10 μg	No zone
Cefpodoxime 10 μg + clavulanic acid	20 mm zone

The organism also showed susceptibility to both ertapenem and meropenem.

- What would be your conclusions and what antibiotics would you report as therapeutic options?

CHAPTER SUMMARY

After reading this chapter you should know and understand:

- Why antimicrobial susceptibility tests are performed and why some are required more urgently than others
- Why national guidelines are in place and the importance of them
- The methods involved in performing disc susceptibility tests and the pitfalls involved.
- How the zone size relates to the MIC of the organism
- Why and when automated methods can be used.
- Which organisms are naturally resistant to some antibiotics
- The controls required for antibiotic susceptibility testing
- The difference between bacteristatic and bactericidal agents
- Why and when combination therapy is used, common synergistic combinations and the relevance of the test
- Which antibiotics are required to be monitored in a patient's serum and the methods involved in determining the concentration
- The importance of reference laboratories

FURTHER READING

- Greenwood D, Finch R, Davey P, Wilcox M. *Antimicrobial Chemotherapy*, 5th edition. Oxford University Press, 2006.
- Andrews JM. BSAC standardized disc susceptibility testing method (version 6). *Journal of Antimicrobial Chemotherapy* 2007; **60**: 20–41.
- Louie L, Matsumura O, Choi E, Louie M, Simor AE. Evaluation of three rapid methods for detection of methicillin resistance in *Staphylococcus aureus*. *Journal of Clinical Microbiology* 2000; **38**: 2170–2173.
- The European Committee on Antimicrobial Susceptibility Testing (EUCAST). http://www.eucast.org.
- Livermore DM, Winstanley TG, Shannon KP. Interpretative reading: recognizing the unusual and inferring resistance mechanisms from resistance phenotypes. *Journal of Antimicrobial Chemotherapy* 2001; **48** (Suppl. 1): 87–102.
- Reeves DS, Wise R, Andrews J, White LO, eds. *Clinical Antimicrobial Assays*. New York: Oxford University Press, 1999.
- Louie L, Matsumura SO, Choi E *et al*. *Performance Standards for Antimicrobial Susceptibility Testing*; Fifteenth Informational Supplement, M100-S15, Vol. 25. Wayne, PA, USA: Clinical and Laboratory Standards Institute, 2005.

DISCUSSION QUESTIONS

4.1 Discuss the advantages and disadvantages of both automated and non-automated susceptibility tests. In conclusion, state which method you feel is best for routine laboratory testing.

4.2 Direct susceptibility tests are used widely despite the possibility of incorrect results being obtained. What reasons can you give to justify their continuation?

4.3 Do you think the antibiotic level in serum should be monitored for all antibiotics? Provide detailed reasons for your answer.

5

Blood cultures

Derek Law

This chapter will describe what a blood culture sample is, and when and why they are taken. It will cover in detail the procedures for taking and processing blood culture samples in the clinical laboratory, the organisms commonly found in blood cultures and the methods used to identify them.

Learning objectives

After studying this chapter you should confidently be able to:

- Understand how blood cultures are taken and processed
- Understand the principles of common blood culture systems
- Work safely with blood cultures
- Know which organisms are commonly isolated from blood cultures, their likely significance and their source
- Be aware of the usefulness of a blood culture Gram stain in providing a rapid diagnosis of septicaemia
- Understand culture methods for commonly encountered organisms and tests available for rapid identification
- Be aware of the problems of blood culture contamination
- Recognize samples requiring specialized testing, e.g. endocarditis, samples from children or immunocompromised patients

bacteraemia
Bacteraemia is the term given to the presence of bacteria in the bloodstream; bacteraemia may be transient, intermittent or continuous.

septicaemia
Septicaemia is the presence of bacteraemia with symptoms of systemic infection. Typical symptoms include pyrexia, chills, rigors, hypotension, shock and, in more severe cases, multiple organ failure.

5.1 Laboratory investigation of blood culture samples

Blood cultures are extremely important clinical samples, and are requested by clinicians when the signs and symptoms suggest the possibility of **bacteraemia** or **septicaemia,** a systemic (widespread) infection where organisms have disseminated from a localized site of infection.

When there is a strong suspicion of septicaemia the patient is usually started on empirical antimicrobial therapy aimed at several target organisms. If the patient's symptoms are not clear cut and evidence of an infection is inconclusive, the clinician must decide whether to start broad-spectrum antimicrobial therapy or wait until further information is available. The use of antibiotic therapy increases costs, may cause side effects and may complicate the achievement of an accurate diagnosis, although an appropriate antibiotic may improve the outcome of an infection. A positive blood culture may help the clinician to confirm an existing presumptive diagnosis and to adjust therapy to a more appropriate or narrow-spectrum antimicrobial. Blood cultures are important tools in the management of patients and prompt processing of positive bottles can help in this process. Early recognition of a positive blood culture is crucial in guiding antimicrobial therapy and influencing patient management. Following recognition of a positive sample, appropriate laboratory techniques must be used to allow prompt and accurate identification and susceptibility testing of the infecting organism.

Cross reference
Chapter 4, Section 4.1.

Key Point

It is essential the Gram stain result of a positive blood culture is telephoned as a matter of urgency to the clinician.

Because of the importance of obtaining rapid results with blood cultures we will see that this is one area of microbiology where automation is common place. In addition, considerable research has been done over the years to improve culture media and detection methods.

SELF-CHECK 5.1

Why would the patient be started on antibiotics before the sample in the bottle is even positive if septicaemia is suspected?

Blood culture methods

As blood cultures are important samples, which can provide vital information to clinicians on the condition of their patients, it is important that they are taken and processed correctly and promptly. The following section will deal with these aspects.

Procedures for sample taking, transport and processing

A blood culture involves taking a sample of blood from a patient and inoculating the blood into a blood culture bottle; the bottle is then transported to the laboratory for incubation. The procedure has three main stages: sample taking, transport and processing.

When taking blood cultures, blood is usually collected from a patient's vein via a syringe. This is known as venepuncture. The venepuncture site is first disinfected, usually with alcohol (70% isopropyl alcohol is preferred) followed by iodine to remove normal skin organisms. Blood is withdrawn into the syringe and then used to inoculate the blood culture bottles. Occasionally, blood may be taken through a catheter in the skin. Depending on the blood culture system used, there may be a cap that has to be removed to expose a septum; in some systems the septum is not sterile and must be first swabbed with alcohol.

Cross reference
Chapter 1, Section 1.3.

There are several factors to consider when taking blood cultures:

- They should be taken by experienced personnel, familiar with the blood culture system, and familiar with taking blood cultures.
- The person taking the blood culture should wash their hands with an antimicrobial agent and water (e.g. chlorhexidine gluconate) and wear gloves when the samples are being taken.
- Attention to **aseptic technique** is vital to avoid contamination. Contamination of blood cultures is discussed later in this chapter.
- If other samples of blood are required, e.g. serological (antibody) testing or biochemical or haematological analysis, the blood culture bottles should always be inoculated first. Many types of blood tube are non-sterile and inoculation of these tubes before the blood culture bottles may introduce contamination.
- Bottles should be inoculated with the recommended blood volume.

aseptic technique
Aseptic technique is the use of sterile equipment and procedures to avoid possible contamination.

Many blood culture systems comprise pairs of bottles - an aerobic bottle and an anaerobic bottle; it is usual to inoculate a single pair of bottles on each sampling. In some systems pairs of aerobic bottles are used and anaerobic bottles are added if infection with anaerobic organisms is suspected. Further pairs of bottles should be taken at intervals of several hours, as this both increases the yield of positive results and may allow better recognition of contamination. Two sets of bottles are usually adequate in most cases, although in cases of **endocarditis** (discussed in more detail later in this chapter) more blood culture sets should be taken. Special paediatric blood culture bottles are available, as often only a small amount of blood sample will be available.

endocarditis
Endocarditis is an infection of the heart valves.

Figure 5.1 shows a typical set of commercially available blood culture bottles. In this system, one of the bottles is aerobic and the other is anaerobic. Note the caps on the top of the bottles which cover the septum where the sample is inoculated. The barcode labels are used to identify the bottle. It is important that the barcode is not covered by any patient identification labels as the barcode must be accessible by the incubating instrument.

It is common practice to take blood cultures at the time of, or shortly after, a **pyrexial spike** as this is the time when bacteria are more likely to be found in the bloodstream. It is usual to inoculate 5–10 mL of blood into each of a pair of bottles at each sampling point, depending on the blood culture system used. The manufacturer's recommended blood volume should be used at all times. Lower volumes should be avoided as there is a relationship between the volume of blood cultured and positivity rate. However, larger volumes should also be avoided as it may interfere or overwhelm the neutralization capacity of the bottle. Instead, further sets of bottles should be inoculated.

pyrexial spike
A pyrexial spike is a transient increase in the patient's body temperature.

Once blood culture bottles have been inoculated they should be clearly labelled with full patient details; accompanying report forms should also include patient's symptoms and current and/or recent antibiotic therapy and be sent to the laboratory as soon as possible. It is important that full clinical details are provided as this gives information to laboratory staff on organisms likely to be encountered, further tests that may need to be carried out on positive bottles and whether bottles require additional procedures such as extended incubation or **terminal subculture**.

terminal subculture
Terminal subculture is where the blood culture bottle is inoculated onto culture media at the end of the incubation period regardless of whether it is positive or not.

Once samples are taken and labelled:

- They must be transported to the laboratory as quickly as possible.
- They should be placed in a clearly labelled incubator, provided and maintained by the microbiology laboratory.

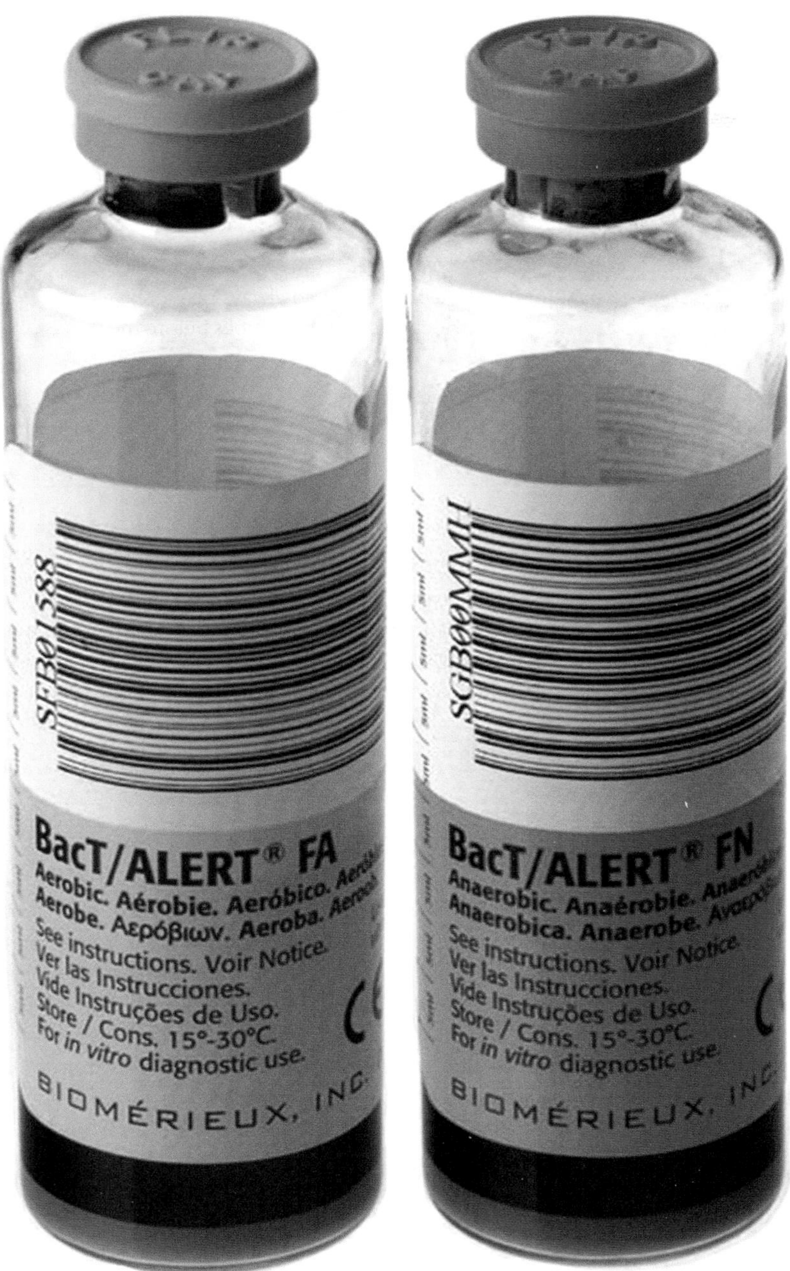

FIGURE 5.1
A pair of commercially available blood culture bottles. One bottle is an aerobic bottle (left) and the other is an anaerobic bottle (right). Note also the barcodes which help identify the bottle. (Figure kindly provided by R. Dopierala, bioMérieux.)

- On no account should blood culture bottles be refrigerated - it may adversely affect some organisms, delay organism growth and lengthen time for the bottle to become positive.
- If no incubator is available then they should be kept at ambient temperature.
- Once received in the laboratory, samples should be loaded onto any automated system as soon as possible to allow positive samples to be identified quickly.

Key Point

Blood cultures should be transported to the laboratory without delay and should be incubated as soon as possible.

HEALTH & SAFETY

Safety aspects of handling blood cultures

Blood cultures should be treated with care: breakages can occur and many systems use glass bottles which can cause cuts. Blood may harbour many dangerous pathogens including HIV, various types of hepatitis, and potentially haemorrhagic fever viruses, particularly in patients presenting with pyrexia of unknown origin (PUO). Any breakages should be handled with care using locally approved disinfectant and disposal methods, and the disinfectant used must have virucidal activity. Any blood stains on the outside of bottles should be wiped with an appropriate disinfectant, with the operator wearing suitable gloves.

Principles of common blood culture systems

The principle of all blood culture systems is to encourage rapid growth of any bacteria present in the blood sample inoculated into the blood culture bottle. Rapid growth is obtained by using highly nutritious media and incubation at an optimal temperature. Eventually, the number of organisms in the blood culture will reach a threshold where they can be detected either by some biochemical signal or by changes in turbidity.

Types of media used in blood culture systems

In all blood culture systems highly nutritious culture media are an absolute necessity. Many systems employ brain heart infusion broth or tryptic soya broths, which are able to support the rapid growth of a wide range of pathogenic bacteria including fastidious organisms. Equally importantly, the media must be capable of supporting growth from very small inocula. The amount of fermentable carbohydrate (usually glucose) is important, although high concentrations promote rapid growth and acidic conditions produced from glucose may cause autolysis of some organisms unless the medium is highly buffered. In anaerobic bottles the medium will often contain a reducing agent to promote anaerobiosis. The bottle will also be filled with an anaerobic gas mixture to further reduce oxygen tension and stimulate growth of anaerobic bacteria. Although fungi may grow in blood culture broths designed for bacteria, specialized fungal blood culture bottles are available with media optimized for fungal growth and detection. Similarly, for mycobacteria specialized blood culture bottles are available. In both cases prolonged incubation is the norm - up to 6 weeks for mycobacteria.

Key Point

Blood culture bottles contain highly nutritious culture media to promote rapid growth of pathogens.

SELF-CHECK 5.2

Why is it so important to use highly nutritious media in blood culture systems?

Traditional non-automated blood culture systems

Traditional blood culture systems use bottles filled with culture medium inoculated with a sample of the patient's blood (usually 5–10 mL in adults, 1–3 mL in children). Growth of organisms is identified by an increase in turbidity of the medium and/or haemolysis (breakdown of red blood cells (RBCs)). There are several drawbacks with this approach. Firstly, it takes several hours for RBCs to settle to the bottom of the bottle before turbidity associated with bacterial growth can be observed. Secondly, the degree of turbidity varies from organism to organism: coliforms and staphylococci grow rapidly and produce obvious changes in turbidity, whereas some organisms, such as streptococci and many fastidious organisms, produce less obvious turbidity and may be missed. Frequent checking of bottles in the first 48 h after receipt and **blind subculture** at 24–48 h and at the end of incubation is required. This frequent checking is a highly labour intensive process and every time the bottle is opened to sample it, there is a risk of contamination being introduced and also a potential infection risk to staff.

blind subculture
Blind subculture is subculture of the bottle regardless of the presence or absence of turbidity or indications of growth.

Because of the drawbacks of traditional systems, various improvements in methods were made. However, the biggest improvements in speed and sensitivity came in the 1980s with the development of automated continuous monitoring systems.

Improved blood culture systems

Traditional blood culture systems rely on bacterial growth and turbidity to indicate positivity. As we saw in the previous section, this can be problematic. Newer blood culture methods are able to detect bacterial growth earlier and this can reduce the time to positivity. Because **time to positivity** is critical in diagnosing septicaemia, most laboratories now use blood culture methods that can reduce this. Below is a description of some of the commonly used methods.

time to positivity
Time to positivity is the time taken from when a blood culture bottle is inoculated to when it is identified as being positive. Methods that reduce time to positivity are preferred.

Radiometric methods

The first automated systems used radiometric methods. These involved using culture media containing C^{14} **radiolabelled** glucose; metabolic activity by bacteria or fungi converted the labelled glucose to radiolabelled CO_2 which could be detected in the **headspace** of the bottle. This is achieved by sampling a small volume of the headspace gas and detecting radiolabelled CO_2 in a beta particle counter. Bottles are typically tested two to three times per day in the first 2–3 days and then once or twice daily for the remaining incubation period. Because bottles were not examined for turbidity they could be agitated during incubation, stimulating bacterial growth and shortening detection times. Although such methods were popular and widely used, issues with safety and disposal of radioisotopes led to the development of non-radioactive detection methods.

radiolabelled
Radiometric methods involve the use of radiolabelled substrates to detect bacterial growth.

headspace
Headspace is the gas in a blood culture bottle above the liquid.

Non-radioactive detection methods

Several commercial systems are now available that work by detection of CO_2 produced as a consequence of glucose metabolism during microbial growth. The production of CO_2 from glucose causes a change in fluorescence or reflectance in a sensor located in the bottom of the bottle. In both cases bottles are placed in a dedicated incubator/detector where they are agitated and the detector is continually monitored to assess the CO_2 content in the bottle. Agitation increases the growth rate of bacteria and reduces the time to positivity, especially in aerobic bottles. In traditional manual systems the lack of agitation may delay detection of a positive bottle. This is one of the main advantages of automated systems.

The bottles are routinely incubated in the incubator/detector for up to 7 days and are then discarded. Bottles can become positive at any time during the incubation period as monitoring is continuous and a signal alerts the user that a bottle is positive. Bottles can then be removed and processed as soon as they become positive. Many laboratories have blood cultures checked outside of normal working hours, allowing early detection and processing of positive bottles. The majority of blood cultures received in the laboratory are negative and reducing the amount of work required on these bottles is beneficial. Because automated methods do not require all bottles to be manually subcultured they produce savings in the time spent processing negative bottles and reduce the amount of culture media used.

Key Point

Although radiometric blood culture methods were commonly used these have largely been replaced by non-radiometric methods which detect CO_2 production by bacteria using sensors that respond to the concentration of CO_2 in the bottle. These methods avoid issues associated with use and disposal of radioisotopes.

Lysis centrifugation methods

Other systems have been developed such as the lysis centrifugation system. In this blood culture system, agents are added to the blood sample, prior to inoculation into a bottle, to lyse both red and white cells. The blood is centrifuged and the deposit containing organisms is harvested and used to inoculate various solid media. This allows a quantitative assessment of the number of bacteria per millilitre of blood. Such techniques are labour intensive but have found value where quantitative cultures are required and in mycobacterial and fungal blood cultures where they can speed up the time to detection for these organisms.

Cross reference
Chapter 9

Neutralization of antimicrobial agents and other inhibitory factors

Blood contains many antibacterial components including antibodies and complement. It may also contain a wide variety of antibiotics that the patient has received. These can reduce or inhibit the ability of bacteria to grow and to produce a positive signal either by turbidity or an automated signalling method. Most blood culture systems employ at least one method of inactivating some of these antibacterial systems; simple dilution by the addition of 5 mL of blood to 50–100 mL of medium can reduce the activity of many of these inhibitors. In other systems, special resins or charcoal are added to the medium. These bind many antibacterial agents and reduce their effect. Sodium polyethanol sulphonate (SPS) is added to media and is an effective neutralizer of innate bacterial inhibitory agents in blood. Some systems also contain para aminobenzoic acid (PABA) and thymidine to neutralize the effects of sulphonamides and trimethoprim respectively. In some cases commercially available **β-lactamase** can be added to bottles to neutralize β-lactam drugs.

β-lactamase

β-lactamases are enzymes that breakdown penicillin and related antibiotics (β-lactams) such as ampicillin and destroy their antibacterial activity.

Key Point

Traditional blood culture systems are labour intensive, requiring frequent examination of bottles and blind subcultures. Automated systems often rely on CO_2 production by bacteria as they grow - such systems offer increased positivity rates, shorter time to positivity and reduced labour costs.

5.2 Procedure for dealing with negative bottles

The majority of blood culture bottles received in the laboratory show no growth and should be discarded after 5–7 days' incubation. If a continuous monitoring system is used a terminal subculture need not be carried out. However, if a traditional manual blood culture system is used it is common for a blind subculture to be carried out. There are some exceptions where bottles may require longer incubation before being discarded, for example endocarditis (discussed below in this chapter) or in specific patients where there is a possibility of an infection with a fastidious or slow growing organism. For negative bottles a written report is issued to the ward at 48 h and at the end of incubation stating that the culture shows no growth.

5.3 Procedure for dealing with positive bottles

It is important for patient management that positive blood cultures are processed rapidly - in this section we will cover the procedures used to deal with positive blood cultures. There are three stages in this process, namely microscopy, culture and sensitivity tests, and finally identification of the pathogen.

The first stage in processing is to carry out a Gram stain on the blood culture

- The aim of a Gram stain of a positive blood culture bottle is to provide the clinician with a guide to the identity of a possible pathogen, allowing initiation of appropriate antibiotics. Up to 24 h may elapse before susceptibility test results and the organism's identity are known; initial therapy is therefore frequently guided by the Gram stain result.
- A presumptive identification of the bacteria in a positive blood culture is invaluable to clinicians in assessing the requirement for and type of antimicrobial therapy appropriate for a patient.

HEALTH & SAFETY

Safety aspects of processing positive blood cultures

In most cases positive blood cultures can be handled on the bench; it is good practice to wear gloves to protect against infection from spillages. Disinfectants with virucidal properties should be used to clean up blood culture spillages. If there is a suspicion that a Category 3 organism may be present (e.g. *Salmonella typhi*, *Brucella abortus*, or *Neisseria meningitidis*) all work must be performed in a Class I safety cabinet. *In some institutions all positive blood cultures are*

processed in a safety cabinet because of the potential risk of such organisms being present. **Removing fluid from blood culture bottles and preparing Gram stains can result in the generation of potentially infectious aerosols, and therefore use of a safety cabinet is mandatory. In such cases all materials for working with blood cultures (e.g. microscope slides, loops, hotplates) must be contained in the safety cabinet.**

A venting needle is used to aspirate liquid from a blood culture bottle. Used needles must not be resheathed but should be placed into suitable containers for incineration. Where organisms such as *Brucella* spp., *Francisella tularensis*, *Yersinia pestis* or *Bacillus anthracis* are suspected on the basis of clinical details and occupational information, it is both appropriate and prudent to send bottles direct to a laboratory better equipped for working with these pathogens.

Brucella **spp. are pathogens of farm animals and are most frequently isolated from vets or farmers. The other organisms are potentially organisms associated with bioterrorism threats.**

Microscopy of positive blood cultures

When a bottle becomes positive, either because of turbidity in a manual bottle or from a positive signal from an automated system, it is removed from the incubator and a sample of the fluid is taken for a Gram stain. Special venting needles can be used to aspirate a sample or in some systems the top can be removed and some fluid aspirated. Experience is required to ensure that the smears are neither too thick nor too thin. The slide should be fixed on a hotplate and flaming in a Bunsen burner should be carried out to completely kill all bacteria.

Key point

Gram staining is the first stage in processing positive blood culture bottles. The result of the Gram stain plays an important role in patient management and antibiotic therapy is often started on the basis of the Gram stain result.

Differentiation of bacterial types based on microscopy results

Gram staining of the slide is carried out as normal using neutral red as a counter stain, although some laboratories prefer to use dilute carbol fuchsin as the counter stain to aid visualization of small Gram-negative rods. In the majority of cases abundant bacteria are clearly visible and the type, arrangement and staining reaction should be recorded. This can provide important information on the identity of the organism; for example pairs of lanceolate Gram-positive diplococci may be *Streptococcus pneumoniae*, clusters of large Gram-positive cocci may be staphylococci.

In some cases the morphology and arrangement of bacteria may not be characteristic and it may not be possible to presumptively identify an organism in a positive bottle. This typically occurs with Gram-positive cocci occurring singly, in pairs and in small groups, which may be staphylococci or streptococci. Although a certain bacterial type may be immediately evident on a smear, for example Gram-positive cocci in clusters, further examination of the smear should be carried out to identify smaller numbers of other bacterial types. Mixed infections, although uncommon, do occur.

Microbiology medical staff should be informed of the Gram stain results and provisional identity so that advice on therapy can be provided to the ward clinicians. A written report should also be issued to the ward with the Gram stain result.

SELF-CHECK 5.3

Why is it important that the result of the Gram stain is reported to clinicians as quickly as possible?

Procedures in the event of negative microscopy

Occasionally, we come across situations where a blood culture bottle is positive but the Gram stain does not show any organisms even after careful examination. There may be several reasons for this:

- Bacteria may be present but too small or too poorly stained to see.
- In manual systems turbidity in bottles may be due to blood lipids or red blood cells that have not settled out.

In automated detection systems false positives do occur - most often due to metabolic activity of white blood cells in the bottle. Large numbers of white cells will often be seen on the Gram stain. Examine the signal from the bottle; Figure 5.2 shows typical traces from a negative blood culture bottle, a true positive bottle and a bottle that has become positive because of white blood cell metabolism. The traces of the true positive bottle and false positive bottle are clearly distinct and it is relatively easy to identify cases where a positive signal occurs because of white blood cell metabolism.

In all cases the procedures listed in the Method Box for organisms not showing on Gram stain of positive blood culture should be carried out.

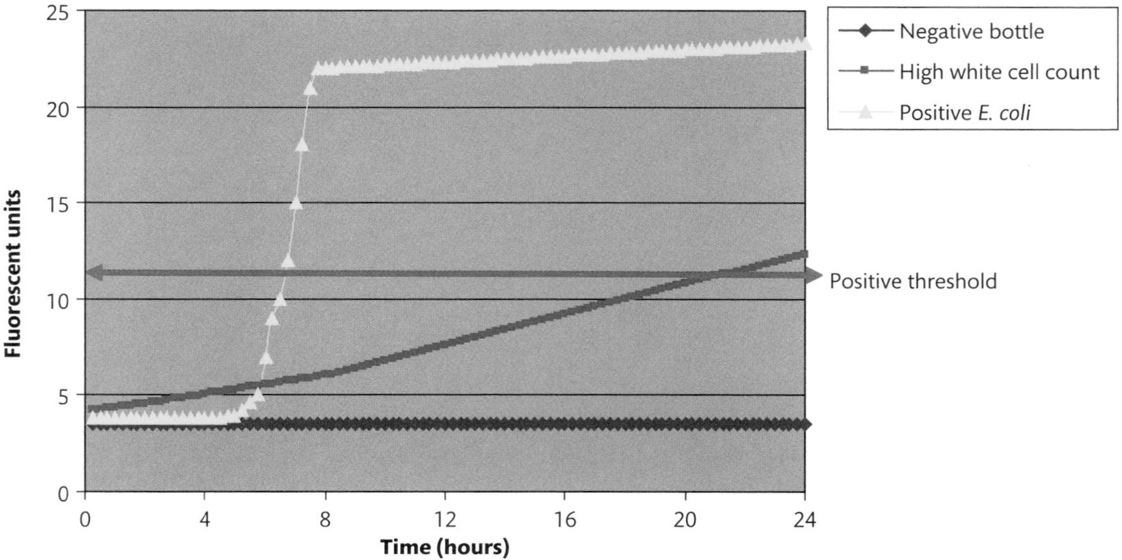

FIGURE 5.2

Typical growth indices obtained from a commercial blood culture system, highlighting the differences in profile between a true positive bottle and a false positive caused by a high white cell count. (Courtesy of D. Law.)

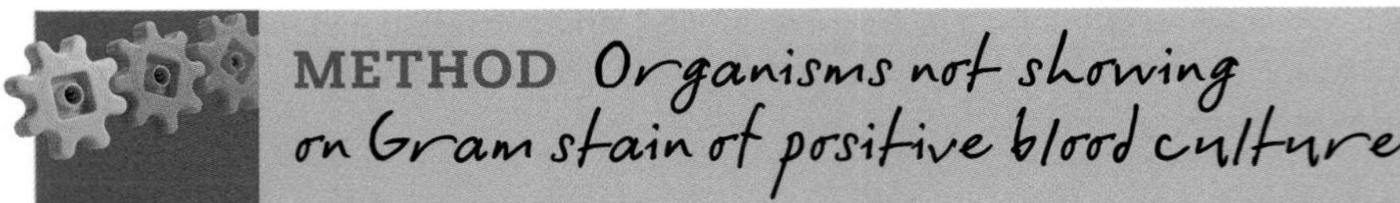

- **Repeat Gram stain - consider alternative counter stain, e.g. dilute carbol fuchsin.**
- **If still negative, consider alternative stains, e.g. acridine orange, or Giemsa.**
- **If no bacteria are seen on further smears, subculture the bottle onto blood agar, incubated aerobically and anaerobically, plus a chocolate agar plate incubated in CO_2 for 4 days. Plates should be examined daily to detect any slow-growing organisms.**
- **The blood culture bottle should always be reincubated.**

Key Point

Positive bottles with negative microscopy must be investigated further. A repeat Gram stain is the first step in this process. Positive bottles with negative microscopy should be cultured and the bottle reincubated. False-positive blood cultures do occur.

Culture regimes following microscopy

When a blood culture becomes positive, microscopy is the first stage of the process to identify and characterize the infecting organism. Following microscopy, the blood culture is cultured to isolate and identify the infecting organism. At the same time sensitivity tests to antibiotics should be carried out and identification tests set up if appropriate. The aim for the laboratory is to have a culture result, a presumptive identification of the pathogen and a sensitivity test result to appropriate antibiotics within 24 h of the bottle becoming positive. The efforts of the laboratory must be focused on achieving these aims; with the most common blood culture pathogens this is certainly possible.

Culture of positive bottles

When a blood culture bottle becomes positive, microscopy is carried out, and once the result of microscopy is known the bottle should be cultured. The results of microscopy dictate the type of culture plates set up for each positive blood culture bottle.

In all cases, a minimum of blood agar plates incubated aerobically and anaerobically and a chocolate agar plate incubated in CO_2 should be inoculated for each positive sample. The most common bacterial pathogens will grow on at least one of these media in 24 h with some requiring 48 h. In some cases, prolonged incubation may be required, particularly for anaerobic or fastidious organisms. If there is no growth on the plates at 24 h, the Gram stain should be repeated on the bottle to ensure that no errors have been made.

Cultures should be set up on positive bottles depending on Gram stain results and clinical details. Some suggested culture regimes for commonly encountered organisms are listed in Table 5.1. These can be modified based on local experience and local prevalence rates of pathogens.

TABLE 5.1 Culture media and identification tests required for organisms seen in positive blood culture bottles.

Gram stain result	Target organism	Media	Identification aids	Confirmatory tests
Gram-positive cocci in clusters	*S. aureus* Coagulase-negative *Staph.*	Blood agar in air Chocolate agar in CO_2 Blood agar anaerobically Susceptibility test in air + methicillin/cefoxitin at 30°C Chromogenic MRSA medium	DNase medium MRSA agar Slide coagulase	Slide and tube coagulase test Or latex agglutination test for *S. aureus* Commercial kits
Gram-positive cocci in pairs or chains	Streptococci *S. pneumoniae* *Enterococcus* spp.	Blood agar in air Chocolate agar in CO_2 Blood agar anaerobically Susceptibility test in CO_2	Islam plate if child or expectant mother Optochin discs on CO_2 or anaerobic plate High level gentamicin susceptibility	Catalase test Lancefield group Comercial identification kits
Gram-positive bacilli Coryneform	*Listeria* spp. *Corynebacterium* spp. *Propionobacterium* spp.	Blood agar in CO_2 Chocolate agar in CO_2 Blood agar anaerobically + metronidazole disc Susceptibility test in CO_2	Aerobic growth Anaerobic growth Selective *Listeria* medium Broth for tumbling motility	Commercial identification kits
Large, Gram-positive bacilli +/− spores	*Clostridium* spp. *Bacillus* spp.	Blood agar in air Chocolate agar in CO_2 FAA anaerobically + metronidazole disc Susceptibility test in O_2 and anaerobically	Anaerobic growth Aerobic growth	Commercial identification kits
Gram-negative cocci	*N. meningitidis* *N. gonorrhoeae*	Blood agar in air Chocolate agar in CO_2 Blood agar anaerobically Susceptibility tests in CO_2	Latex agglutination on blood culture broth	Initial confirmation with oxidase reaction and serological tests Confirm with serological tests
Gram-negative rod Coliform type morphology	Coliforms *Pseudomonas* spp. Non-fermenters, e.g. *Acinetobacter* spp.	Blood agar in air Chocolate agar in CO_2 Blood agar anaerobically + metronidazole disc CLED/McConkey Chromogenic coliform media Susceptibility test in air	Oxidase test Reaction on chromogenic medium Serology on possible *Salmonella* spp.	Confirm with biochemical tests, and serology where appropriate (*Salmonella*)
Gram-negative rod Small, poorly staining OR Growth in anaerobic bottle or appearance suggestive of anaerobes	Anaerobes *Haemophilus* spp. *Cardiobacterium* *Eikenella* *Kingella*	Blood agar in CO_2 Chocolate agar in CO_2 Blood agar anaerobically + metronidazole disc CLED/McConkey FAA anaerobically Susceptibility test in CO_2 Susceptibility test anaerobically	Growth in CO_2 Growth anaerobically	Susceptibility to metronidazole X and V factor dependence Commercial identification kits

If mixtures of organisms are seen on Gram stain then appropriate susceptibility tests should be carried out against all organisms expected to be isolated. Selective media should be set up to assist isolation of slow-growing organisms or those likely to be overgrown, for example in a mixture of streptococci and possible coliforms, neomycin or naladixic acid containing media can help isolate the *Streptococcus*. Or in the case of swarming *Proteus* and staphylococci the use of mannitol salt agar or McConkey's/CLED agar with appropriate antibiotic discs (gentamicin, naladixic acid or aztreonam) may help isolate the *Staphylococcus*.

Cross reference
Chapter 4.

Direct susceptibility tests

direct susceptibility test
A direct susceptibility test is a sensitivity test carried out directly on a sample rather than on an isolated bacterial culture. It is more difficult to standardize inocula on direct sensitivities and mixtures of organisms can be problematic.

It is important that direct susceptibility tests are carried out on all positive blood cultures, so that a provisional sensitivity result is available. **Direct susceptibility tests** should also be carried out on media and incubation conditions (CO_2/anaerobic) should be appropriate for the target organism. The antibiotics used in susceptibility tests should reflect those of the local antibiotic policies and local resistance patterns and must be continually re-evaluated.

It is important when carrying out a direct susceptibility test that the inoculum is acceptable. Whilst getting the correct inoculum is a matter of trial and error there are some guidelines. For Gram-negative organisms likely to be coliforms, a drop of blood culture fluid should be mixed with 5 mL of water and used to inoculate plates. For potential staphylococci three drops of blood culture should be mixed with 5 mL of water. For other organisms a single drop of fluid from the blood culture is used to inoculate the plate.

Disc susceptibility tests can be carried out on different media and incubated under different conditions depending on the Gram stain results and the likely identity of the organism. Disc tests also allow the identification of mixed cultures. Although direct susceptibility tests on blood cultures are important, susceptibility tests should always be repeated with a pure culture using the laboratory's standard methodology where inoculum and incubation time can be more carefully controlled.

If both bottles in a pair of blood cultures show the same types of organism on Gram stain then both bottles should be cultured, although a direct sensitivity test is often only carried out on one of the pair, unless the morphology in each bottle is distinct.

SELF-CHECK 5.4

Why is it important to perform direct susceptibility tests on positive blood cultures?

Additional tests

As well as carrying out direct susceptibility tests other media can be set up to aid in the identification and characterization of the organisms in the blood culture. Chromogenic media can be used which may help in identifying many common pathogens, for example MRSA, coliforms and yeasts, depending on the result of the Gram stain.

Cross reference
Chapter 3, Section 3.6.

There are some specific tests that can be carried out directly on the blood culture bottle to assist in identification of the organism.

Gram-positive cocci in clusters are often staphylococci but it is important to differentiate between *Staph. aureus* and coagulase-negative staphylococci, the latter frequently occurring as contaminants (this is discussed later in this chapter). Methods are available to differentiate *S. aureus* from coagulase negative staphylococci in blood cultures such as the heat-stable nuclease test, although this is not widely used. More commonly, DNase plates are set up with confirmation of identity by slide coagulase test on isolated colonies.

Key Point

The DNase test is commonly used to differentiate *S. aureus* from other staphylococci. This is typically carried out using DNase agar. *S. aureus* is DNase positive whereas most coagulase-negative staphylococci are DNase negative.

S. pneumoniae can be identified by carrying out a pneumococcal antigen test on a sample of the blood culture. This uses a commercially available latex reagent. Similar methods may also be available for identification of *N. meningitidis* in blood culture containing Gram-negative cocci.

Commercially available identification tests such as API strips and VITEK cards may be inoculated directly from blood culture bottles. However, blood and blood culture media components may interfere with the interpretation of certain tests and so results obtained in this way may produce incorrect identifications and often have to be repeated. For rapidly growing organisms such as *E. coli* and other coliforms, if plates are inoculated early in the day, there may be sufficient material growing on an aerobic blood agar plate 6–8 h later to allow an identification test to be set up and read the following day with the culture plates and susceptibility test. In all identification tests a non-selective culture plate must be set up to ensure that the culture is pure. However, most laboratories wait until the culture has grown sufficiently to provide single colonies to carry out identification tests or serological identification. Identification methods will vary from laboratory to laboratory but when commercial kits are used the manufacturer's instructions must be followed.

Although not widely available yet, it is likely that in the future commercially available **PCR** (polymerase chain reaction) tests will be used to identify key pathogens found in blood cultures.

PCR

PCR (polymerase chain reaction) is a technique involving amplification of specific DNA sequences which can be used to rapidly identify pathogens.

Key Point

It is imperative to set up sufficient cultures to allow a preliminary identification of the organism seen in the Gram stain. Direct sensitivity tests must also be set up. The antibiotics tested should reflect both the possible identity of the organism and local guidelines and local susceptibility patterns.

5.4 Bacterial pathogens

It is important for anyone dealing with blood cultures that there is an understanding of the common organisms isolated, the conditions associated with these organisms and possible sources of infection. We will now consider the organisms commonly found in blood cultures and discuss these aspects.

Pathogens likely to be a cause of septicaemia and bacteraemia

A search of the literature will show that a very wide range of bacteria can be isolated from blood cultures as a cause of septicaemia. Figure 5.3 shows the organisms encountered in UK blood cultures during 2002.

Although figures vary from one survey to another, most studies show that the majority of positive blood cultures yield one of the following pathogens: *S. aureus*, *E. coli* or other enterobacteria, *S. pneumoniae*, *Enterococcus* spp., *Pseudomonas aeruginosa* or *Candida albicans.*

Key Point

The incidence of various pathogens varies from hospital to hospital and reflects to some extent the medical/surgical specialities in a given institution.

S. aureus is one of the most frequent pathogens isolated from blood cultures; this organism enters the blood stream from an infected site in the body. Common sites of infection include abscesses in various tissues and organs, infected lesions in skin, for example cellulitis, and infected joints; occasionally pneumonia can be caused by *S. aureus* and septicaemia is often associated with this. The isolation of *S. aureus* from a blood culture is usually indicative of an infection at another body site and should prompt further clinical and microbiological investigation to identify the source.

pyelonephritis
Pyelonephritis is an infection of the kidney, which is a complication of urine infection.

E. coli is one of the other frequent causes of septicaemia. In most cases it is associated with an infection in the urinary tract, such as **pyelonephritis,** or in the elderly or immunocompromised

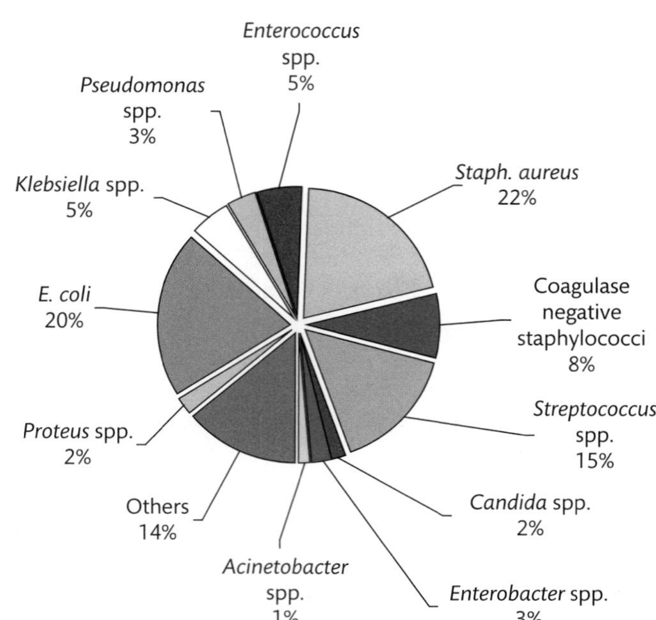

FIGURE 5.3
The relative frequency of significant bacterial pathogens from UK blood cultures in 2002. (Data taken from HPA bacteraemia report for 2002.)

a simple UTI may result in *E. coli* bacteraemia. Other sources of the organism include intestinal lesions and infections of the biliary tract. Similarly, other members of the Enterobacteriaceae also cause UTI and may be found in the blood as a result. *Salmonella enterica* may be found in cases of severe food poisoning or occasionally as a result of biliary sepsis. *S. typhi* and *S. paratyphi* cause enteric fever and rapid diagnosis and treatment of such infections is essential; any non-lactose fermenting, oxidase-negative colonies isolated from blood cultures should therefore be checked using antisera to see if they are *Salmonella*.

Key Point

Check all non-lactose fermenting, oxidase-negative, Gram-negative bacilli isolated from blood culture with antisera to exclude *Salmonella* spp.

Coagulase-negative staphylococci are frequently encountered in blood cultures but as we will see later in the chapter they are often contaminants, being derived from the skin during the taking of blood. They can be occasionally pathogenic, and cannot therefore be discarded without looking closely at the patient's symptoms; in particular they are often associated with infections of catheters and indwelling lines.

S. pneumoniae is often isolated from patients with chest infections (e.g. pneumonia) or occasionally meningitis. For some patients with pneumonia, blood culture may be the best means of microbiological diagnosis, as sputum production may be poor or the sample may be overgrown with normal flora.

Enterococci are frequently encountered as pathogens in blood cultures and have two principal sources: firstly, the urinary tract where they are a common cause of infection; and, secondly, the bowel is an important source. There is a strong association between *Streptococcus bovis* endocarditis and cancer of the bowel. Enterococcal septicaemia can be problematic to treat because of the limited range of active therapeutic agents.

β-haemolytic streptococci, principally those of Lancefield groups A, C and G, are encountered in blood, usually as a result of a skin infection (e.g. cellulitis and infected leg ulcers). *Streptococcus agalactiae* (group B *Streptococcus*) is most commonly encountered in cases of neonatal sepsis.

Pasteurella multocida may occur in blood cultures, often associated with an abscess or skin infection following an animal bite. It may also be associated with chest infections.

α-haemolytic streptococci are encountered most frequently in patients with endocarditis where the organisms are derived from the oropharynx; however, they do occur as contaminants in blood cultures, often through oral contamination during venepuncture.

Anaerobic Gram-negative rods, such as *Bacteroides* spp. and *Prevotella* spp., occur in blood cultures of patients with abdominal sepsis or following surgical procedures on the bowel, female reproductive system or oropharynx. Although prophylaxis with metronidazole has markedly reduced the incidence of anaerobic infections including septicaemia they may still occur.

N. meningitidis although rare is an extremely important pathogen, indicating septicaemia with the possible presence of meningitis. Suspicion of this organism in a blood culture should immediately be brought to the attention of a clinician; the clinical course of meningococcal septicaemia can be very severe with many complications and rapid, aggressive therapy is required. Biochemical tests are required to differentiate the organism from *N. gonorrhoeae* which may occasionally occur in blood cultures, often as a complication of a joint infection, and oral *Neisseria/Moraxella* which are often found as contaminants.

Candida albicans is recognized as the fourth or fifth most common blood culture isolate in some studies. *C. albicans* is most common in **immunocompromised** patients, patients with central venous catheters and in premature infants. Use of broad-spectrum antibiotics may predispose to *Candida* infection by removing bacterial flora and increasing colonization by yeasts. Cytotoxic therapy for cancer patients may damage the gut mucosa allowing *Candida* to move from the bowel into the blood stream. *Candida* infections carry a high mortality, partly due to the underlying status of the patients; therefore rapid diagnosis and speciation is important. Some species of *Candida* (e.g. *C. glabrata* and *C. krusei*) are often resistant to azole antifungals and their identification is required to ensure that correct therapy is given.

immunocompromised
Immunocompromised, or immunodeficient, is a state where the patient's immune function is absent or impaired so the ability to fight infection is impaired. This can be as a result of drug action (e.g. immunosuppressive drugs used for organ transplantation) or disease (e.g. HIV infection).

Key Point

When a fungal infection is suspected specialized fungal blood cultures should be taken.

Cross reference
Chapter 11.

SELF-CHECK 5.5

What are the commonest pathogens associated with septicaemia?

Polymicrobial bacteraemia

Although most cases of bacteraemia or septicaemia are caused by a single organism, occasional cases of **polymicrobial bacteraemia** occur. Many cases of polymicrobial bacteraemia are associated with serious underlying illness and frequently arise from the bowel. Organisms likely to be associated with such infections include enterobacteria, enterococci and anaerobes. The death rate from cases of polymicrobial bacteraemia is often higher than that from **monomicrobial bacteraemia**; this reflects both the nature of the underlying condition and the requirement for aggressive therapy against two different pathogens.

polymicrobial bacteraemia
Polymicrobial bacteraemia is bacteraemia where two or more significant organisms are isolated.

monomicrobial bacteraemia
Monomicrobial bacteraemia is bacteraemia where only one significant organism is isolated.

SELF-CHECK 5.6

Would it be easier to treat a polymicrobial or a monomicrobial bacteraemia? What are the reasons for your answer?

Mycobacteria

Mycobacterium tuberculosis may occur in blood during the primary phase of tuberculosis. However, other mycobacteria, such as *Mycobacterium avium-intracellulare*, may also occur in blood particularly in immunosuppressed patients, and especially those with AIDS. Mycobacteria are unlikely to be detected in conventional blood culture systems because the length of incubation of the blood culture bottle and solid media is too short.

Various techniques have been developed to help isolate mycobacteria from blood. The lysis centrifugation technique (described above) is useful; however, automated systems similar to those for conventional bacterial cultures have been developed. These incorporate a mycobacterial medium with a CO_2 sensor system; bottles are incubated for 3–4 weeks and checked for a positive signal on a daily basis. When such a bottle becomes positive, Gram stain and acid-alcohol fast bacilli (AAFB) stains are carried out on the material, as well as further subculture onto mycobacterial-specific media. PCR may be useful for rapid identification of an AAFB present in a blood culture bottle.

All work with mycobacterial blood cultures must be carried out in a Class II safety cabinet; such work is often carried out in specialized centres with dedicated automated systems for mycobacteria. It may be appropriate to send blood culture bottles for mycobacterial investigation to such centres for processing.

Cross reference
Chapter 9, Section 9.4.

Bacteraemia in children

Although most of the organisms described above can cause infection in children and babies there are certain pathogens that are more frequently encountered in children. In newborn infants bacteraemia is commonly caused by either *E. coli*, *S. agalactiae* or *Listeria monocytogenes* and often leads to the development of meningitis. Coagulase-negative staphylococci are also responsible for a significant proportion of cases of bacteraemia in children, often in association with intravenous catheters. Coagulase-negative staphylococci may also occur as contaminants and distinguishing true infection from contamination can be problematic (blood culture contamination is discussed later in this chapter). The problem is compounded by the small volumes of blood cultured from babies and the number of bottles taken. Infection with *Candida parapsilosis* is relatively common and is probably related to the use of intravenous lines, which are colonized with relative ease by this organism.

Key Point

It is essential to rapidly differentiate *S. agalactiae* from *Listeria monocytogenes* as soon as possible. This is because the patient is likely to have been started on a cephalosporin antibiotic and *Listeria* is resistant. In cases where *Listeria* has not been recognized early, neonatal deaths have occurred.

Until recently, *Haemophilus influenzae* type b (Hib) was a common cause of septicaemia in children, often associated with meningitis or epiglottitis. However, vaccination with the Hib vaccine has resulted in a marked decline in the incidence of Hib infections.

Bacteraemia in immunocompromised patients

Many hospitalized patients are immunocompromised, most frequently because of concurrent therapy for transplants, cancer, disease such as AIDS or congenital immunological defects. **Leucopenia** and **neutropenia** represent the greatest risk of infection and are related to the neutrophil count. Although pathogens found in immunocompetent patients also infect immunocompromised patients, many infections are caused by organisms of low virulence which are commonly members of the normal flora. Many of the patients have underlying illnesses, long hospital stays, indwelling venous catheters and have received broad-spectrum antibiotics. Infections caused by coagulase-negative staphylococci, *Listeria* spp., non-fermentative Gram-negative rods such as *Flavobacterium* spp. and fungi such as *Candida* spp. and *Aspergillus* spp. are common. Infections may also involve two or more organisms. It is sometimes difficult to determine whether an organism isolated from a blood culture is a pathogen or a contaminant

leucopenia
Leucopenia is a reduction in the total number of white blood cells.

neutropenia
Neutropenia is a reduction in the number of phagocytic white blood cells.

(methods to differentiate significant isolates from contaminants are described later in this chapter). Treatment of such infections is also problematic as the lack of defence mechanisms reduces the efficacy of many antimicrobial agents, and cidal agents or combinations should be employed wherever possible rather than static agents.

Key Point

Although many types of bacteria are isolated from blood cultures, the most common organisms are staphylococci, enterobacteria, streptococci and *Pseudomonas*. Polymicrobial infections account for over 10% of cases of septicaemia.

5.5 False-positive results

We have seen the importance that a blood culture has in patient management and how antibiotic therapy may be started or changed on the basis of a Gram stain of a positive bottle. However, not every organism detected in a blood culture is a pathogen and in many cases the organism is considered not significant - its occurrence in the blood culture may be due to contamination of the bottle. We will discuss the implications of contamination, ways to differentiate significant isolates from contaminants and ways to reduce contamination.

Blood culture contamination

Contaminated cultures are a source of frustration for clinicians and laboratory personnel alike. For the former it is not known whether treatment is warranted or not. In many cases, patients will be started on antibiotic therapy, which may then be stopped when the identity of the organism is determined. This results in increased drug costs, longer hospital stays, the possibility of side effects, and problems of achieving a diagnosis due to concurrent antibiotic therapy. For laboratory personnel processing of contaminated blood cultures takes up valuable time and resources.

Blood culture contamination rates vary widely from institution to institution with an average rate of around 2-3% of bottles. There is some suggestion that rates are increasing, possibly related to improved blood culture methods, greater use of indwelling catheters and changes in phlebotomy practices to minimize the risk of **needlestick injuries.**

needlestick injury
A needlestick injury occurs when a needle used to take a blood sample from a patient accidentally pierces the skin of a health-care worker. There is an infection risk associated with needlestick injuries.

SELF-CHECK 5.7

What are the possible consequences for a patient whose blood culture becomes positive, but the isolated organism is later found to be a contaminant?

Determining the significance of a possible contaminated culture

How can contamination be recognized and differentiated from truly significant bacteraemia? The most useful clue for determining whether an organism is a possible contaminant is to determine its identity. Certain organisms, such as *E. coli* and other enterobacteria, *S. aureus*,

S. pneumoniae, *P. aeruginosa* and *C. albicans*, are almost always genuine pathogens. On the other hand, coagulase-negative staphylococci, *Corynebacterium* spp., *Bacillus* spp. (other than *B. anthracis*) and *Propionobacterium acnes* are contaminants in many but not all cases. Of these organisms coagulase-negative staphylococci are most frequently encountered as a cause of blood culture contamination, representing over 70% of contaminants. However, coagulase-negative staphylococci are also increasingly implicated as a cause of true bacteraemia, especially in patients with prosthetic devices or central venous catheters. Some studies suggest that between 10 and 25% of coagulase-negative staphylococcal isolates may be significant.

Each case of possible contamination must be examined on its own merits with regard to the patient's clinical condition and other factors described below. Where there is doubt as to the significance of an organism further blood culture should be taken, and microbiological samples from other body sites should be considered to identify the source or focus of infection.

Factors to consider when determining the significance of an isolate

When we isolate an organism from a blood culture and its significance is unclear we have to look at other data to help us interpret the result. A combination of factors is used to determine the significance of an isolate.

Three factors can be considered which may help determine the significance of the isolate; these are:

1. The number of positive bottles out of the number taken.
2. The time taken for the bottles to become positive.
3. Clinical information.

Number of positive bottles

The first factor that helps differentiate contamination from true infection is to examine the number of positive sets of blood cultures from the total number of sets taken. Growth of a possible contaminant organism from only one of two sets can usually be viewed as contamination. In cases of genuine bacteraemia the same organism is usually isolated from multiple sets of blood culture bottles taken at different times, highlighting the value of taking multiple sets of bottles. The number of positive bottles within a set of bottles is not as useful because of the use of mixed bottle sets. For example the occurrence of growth in the aerobic bottle only may not necessarily indicate contamination, but may reflect poorer growth of the organism in the anaerobic bottle, which may take longer to become positive.

Time for the bottles to become positive

A further factor to consider is time to positivity: cultures that become positive in 3–5 days are more likely to represent contamination than cultures that become positive in the first 48 h. This may reflect the organism load injected into the bottle. In cases of bacteraemia or septicaemia the number of organisms in each millilitre of blood may be in excess of 100. Organisms may be multiplying rapidly, and consequently growth is rapid and time to positivity is short and may be measured in hours. In contrast, contamination in many cases is likely to be caused by injection of small numbers of organism into the bottle, growth is slow and time to positivity may stretch to several days.

Clinical information

Perhaps the most useful clue to assessing the significance of a blood culture isolate is the use of clinical information including: the presence of sepsis, hypothermia or pyrexia, high white cell counts and high or rising **C-reactive protein** levels.

C-reactive protein

C-reactive protein (CRP) is a protein found in blood; levels of CRP increase when there is inflammation or infection.

The presence of such symptoms, along with the isolation of a potential contaminant organism, warrants an assessment of both the patient and organism. In such situations further blood cultures may be taken, and therapy with an appropriate antibiotic may be started. If a coagulase-negative staphylococci or other potential contaminant is isolated from a blood culture but the patient is generally well, with little or no fever at the time the culture becomes positive, the isolate is frequently regarded as a contaminant.

Many patients have vascular catheters inserted into veins and blood cultures are frequently drawn from such catheters. A positive culture taken via a catheter may represent true bacteraemia, true contamination or catheter colonization; up to a quarter of catheters are colonized by coagulase-negative staphylococci. It is sometimes advocated that samples are taken through the skin at the same time as catheter blood samples are taken to allow an assessment of contamination.

If there is doubt about the significance of an organism repeat blood cultures are frequently taken; if the same organism is isolated from further bottles the likelihood that it is a contaminant is lower.

Key Point

Distinguishing whether an organism is a contaminant or pathogen is difficult. Factors to be considered include: the identity of the organism, the number of bottles positive, the time taken to become positive and the patient's symptoms. Further bottles may need to be taken to confirm or refute contamination.

Non-cultural markers of septicaemia

Although cultures remain the gold standard for diagnosis of septicaemia there is a need to identify other (often biochemical) markers, which may give an early indication of sepsis. These surrogate markers may allow early presumptive therapy, and continual monitoring of the marker may yield information regarding the effectiveness of antimicrobial therapy. They may also help differentiate contamination from true cases of septicaemia.

A wide range of markers have been examined for their association with sepsis. Temperature measurements, white cell counts, and C-reactive protein are all frequently used as indicators of sepsis. However, they are relatively non-specific and may be raised in other conditions. Other markers have been investigated; these include procalcitonin, macrophage migration inhibitory factor, soluble urokinase-type plasminogen activator receptor and soluble triggering factor expressed on myeloid cells. However, none has proven to be a specific marker of septicaemia in all patients. For some markers (e.g. procalcitonin) kits are available for the measurement of levels in blood.

Reducing the incidence of blood culture contamination

The most obvious way to reduce the problems caused by blood culture contamination for clinicians and laboratory staff is to reduce its incidence. The most common source of blood culture contamination is the skin of the patient at the site where the blood culture is taken. **Skin antisepsis** is therefore of crucial importance when taking blood cultures and inadequate skin preparation is the commonest cause of blood culture contamination.

skin antisepsis
Skin antisepsis is preparing the skin with disinfectant to remove bacteria before the blood is taken.

There are various skin disinfectant regimes in use but whichever is used the length of time that the disinfectant is applied to the skin is important.

The septa of blood culture bottles through which blood is inoculated although covered before use are often not sterile and these must be wiped, usually with alcohol, prior to inoculation.

Staff specifically trained to take blood cultures have been found to have significantly lower contamination rates than non-specialized staff. The use of dedicated teams to take blood cultures should be encouraged wherever possible, as reducing blood culture contamination can have a major impact on laboratory workload, reduce antibiotic usage and costs, and helps in clinical decision making.

Key Points

Contamination is a major problem but can be reduced by using trained, dedicated staff for taking blood cultures.

Differentiation of contamination from true bacteraemia can be difficult in some cases.

Various markers have been assessed to distinguish contamination from true septicaemia, but to date none is completely predictive.

SELF-CHECK 5.8

Would one contaminating organism cause a problem in a blood culture bottle?

5.6 Endocarditis

Principles of how bacteria cause endocarditis

Endocarditis is an infection of the endocardium, the inner lining of the heart and heart valves. It is a serious illness that is fatal if left untreated. Endocarditis arises when bacteria gain access to the blood stream through brushing teeth, surgical procedures, or in drug addicts through the use of dirty needles. In most cases the immune system rapidly removes these organisms so that they pose no risk to health. However, bacteria may occasionally settle on a heart valve and cause disease. This is most likely to occur if there has been pre-existing damage to the heart valve, following surgery, or it may be caused by rheumatic fever or congenital heart defects. It can also occur on normal valves with certain pathogens such as *S. aureus*. Damaged valves are more prone to infection for several reasons. For example, they cause defects in blood flow, resulting in stagnation of blood around some areas of the valve. This allows bacteria to adhere more readily to valve tissue. Damaged valves may also have fibrin deposits to which bacteria adhere. Once bacteria have settled on a valve they grow and over time small clumps of material develop. These clumps are known as vegetations. The vegetations contain organisms, fibrin deposits and blood clots. The vegetations cause valve dysfunction, resulting in abnormal blood flow, and can lead to heart failure.

Key Point

The aortic and mitral valves are the most commonly affected heart valves.

Figure 5.4 shows a heart that has been dissected to expose the mitral valve. In the normal heart this consists of smooth tissue but this heart, from a patient with endocarditis, shows numerous vegetations on the valve. These vegetations cause the valve to malfunction. Pieces of the vegetation can also break off and disseminate around the body.

Bacteria associated with endocarditis

The most common bacteria implicated in native valve endocarditis are oral streptococci, such as *Strep. mitis* and *Strep. oralis*; occasionally nutritionally dependent streptococci (*Abiotropha* spp.) are isolated from cases of endocarditis. *S. aureus* and *Enterococcus* spp. are also encountered with a reasonable frequency. Many other bacteria can cause endocarditis, for example the HACEK group of organisms (*Haemophilus* spp., *Aggregatibacter actinomycetemcomitans* (previously named *Actinobacillus*), *Cardiobacterium hominis*, *Eikenella corrodens* and *Kingella* species). The primary source of these organisms is the oropharynx.

Prosthetic valves as well as native heart valves may become infected to produce prosthetic valve endocarditis (PVE). The microbiology of this condition is different to that of native valve endocarditis. Two forms of PVE occur: early PVE and late PVE. Early PVE occurs within 2 months

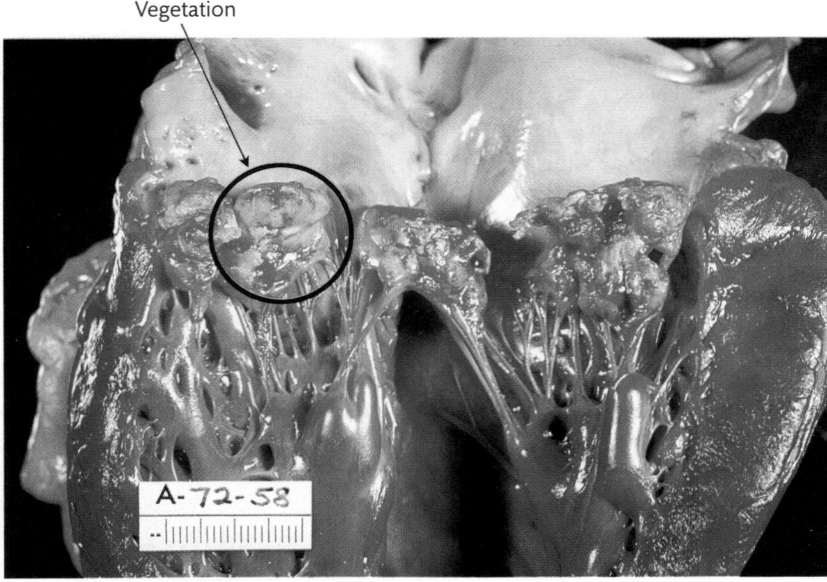

FIGURE 5.4
Dissection of the heart from a patient with endocarditis showing vegetations on the mitral valve. (Figure kindly provided by CDC/Dr Edwin J. Ewing Jr.)

of surgery and is likely to be caused by contamination of the valve during replacement surgery. Infection is typically caused by coagulase-negative staphylococci, *S. aureus*, fungi such as *Candida* spp. and *Aspergillus* spp., and *Corynebacterium* spp. Late PVE occurs a year or more after surgery. The organisms isolated from such cases are similar to those causing native valve endocarditis and the source of these bacteria is the same as those causing native valve endocarditis.

Endocarditis is a particular risk for intravenous drug abusers; the organisms causing endocarditis in such patients may be quite different from those causing endocarditis in non-drug abusers; commonly *S. aureus* or enterobacteria are involved.

SELF-CHECK 5.9

What is the primary source of the commonest organisms that cause endocarditis?

Signs and symptoms of endocarditis

Symptoms of endocarditis develop slowly; it is often termed infective endocarditis (IE). Symptoms include feeling generally unwell, tiredness and low grade pyrexia. Note that these symptoms are non-specific and are found in many illnesses. Heart murmurs caused by faulty blood flow through damaged valves can be heard. Frequently, small parts of the vegetation break off and get carried through the blood stream to other parts of the body causing nodules to develop in the eye or fingernails. Other symptoms include raised white cell count and enlarged spleen.

Diagnosis of endocarditis

The initial diagnosis of endocarditis is based on clinical signs. However, confirmation of the diagnosis is made in the laboratory by isolating and identifying the infecting organism. Prompt identification of the infecting organism and carrying out of antibiotic susceptibility tests are vital in order for the patient to receive the best possible therapy.

Numerous blood cultures are taken, often following a spike of temperature. Frequent sets are required to increase the likelihood of isolating the infecting organism as bacteraemia may be transient. Almost any organism can cause endocarditis; however, many of the causative organisms may be slow-growing and the general rule is to increase the incubation time for all blood culture bottles from endocarditis patients to up to 28 days. Inoculated plates (incubated anaerobically and in 5% CO_2) should also be incubated for at least 1 week and some recommend 3 weeks to improve detection of slow-growing organisms. Another organism encountered occasionally is *Coxiella burnetti*. Diagnosis of infection with this organism is based on clinical suspicion and positive serology.

Some laboratories carry out an intermediate and terminal blind subculture of endocarditis blood culture bottles, as some organisms may not always give a positive signal in automated systems.

In some cases diagnosis may be made after culturing a heart valve that has been removed from a patient.

SELF-CHECK 5.10

What symptoms would alert you to the possibility of endocarditis in a patient?

Identification and susceptibility testing

Once an organism has been isolated from a blood culture it should be identified to species level. Antibiotic susceptibility tests should be carried out; however, endocarditis is one of the few situations where disc or breakpoint susceptibility tests are often inadequate. To effectively eradicate bacteria from endocarditis vegetations requires that the antibiotics used are cidal rather than static. Minimum inhibitory concentration (MIC) tests should be carried out using either broth microdilution tests or E-test. The advantage of the former is that it allows bactericidal activity to be detected. In many cases of endocarditis a combination of drugs is often used to enhance cidal activity. Common regimes include the use of ampicillin and gentamicin for penicillin-sensitive streptococci or vancomycin and gentamicin in penicillin-allergic patients. Treatment is usually with intravenous antibiotics for the first 2 weeks of therapy and the patient may be switched to oral antibiotics if an improvement in condition is seen.

Cross reference
Chapter 4.

In some cases replacement of damaged valves is required once the valve has been sterilized by appropriate therapy. **Synergy testing** of various drug combinations may be warranted in some cases; however, such tests are difficult to carry out, standardize and interpret and are best left to specialized laboratories to carry out.

synergy tests
Synergy tests are tests carried out on combinations of drugs to see if their effects are potentiated when combined. In some cases combinations may antagonize each other which reduces the effect of the combination.

Patients who have had endocarditis have a risk of recurrence. Treated patients and those with damaged heart valves require antibiotic prophylaxis during various procedures such as surgery and dental extractions. The antibiotics used are dependent on the type of procedure and they remove bacteria entering the blood stream during such procedures.

Key Points

Blood cultures from endocarditis patients require extended incubation and blind sub-culture may be warranted.

Treatment of endocarditis requires prolonged doses of appropriate antibiotics often given in combination.

MIC tests are often used to guide antibiotic therapy rather than disc diffusion or breakpoint susceptibility methods.

CASE STUDY 5.1 Endocarditis

A 52-year-old man was admitted to the Emergency Department with a 2-week history of low grade pyrexia, tiredness and splinter haemorrhages. On examination he had an enlarged spleen, a CRP of 4.9 (normal range 0–1), an ESR of 49 mm/h (normal 1–15 mm/h), and a heart murmur. In the past he had surgery to correct a mitral valve defect. A diagnosis of subacute bacterial endocarditis was made and several sets of blood cultures were taken over a 24-h period.

The patient was started on a combination of ampicillin and gentamicin when all blood culture samples were taken, as streptococcal endocarditis was suspected. The aerobic bottle from the second set of blood cultures became positive 36 h after being taken. Gram stain revealed Gram-positive cocci in clusters. *S. epidermidis* was isolated from the bottle, and was regarded as a possible contaminant as no other bottles were positive at that time. After 72 h the first pair of bottles

became positive. Initial Gram stain revealed no organisms; the Gram stain was repeated, and counter stained with carbol fuchsin. Numerous small Gram-negative coccobacilli were seen. The bottles were plated onto blood agar and chocolate agar incubated in CO_2 and blood agar and fastidious anaerobic agar incubated anaerobically. Disc sensitivities were carried out anaerobically on blood agar and in CO_2 on sensitivity agar enriched with lysed blood and NAD. Examination of the plates at 24 h revealed no growth. At 48 h, however, small grey colonies were evident on the chocolate plate incubated in CO_2. At this time a further three bottles became positive and Gram stain with carbol fuchsin revealed small Gram-negative coccobacilli in each of these.

Identification was carried out using X and V discs and commercial kits; the organism was not dependent on X or V factors but there was a requirement for CO_2. Identification by a kit confirmed identity as *Haemophilus aphrophilus*. Susceptibility tests revealed sensitivity to ampicillin and gentamicin and this was confirmed by E-test MICs on sensitivity test media in CO_2. The patient was treated with 2 g amoxicillin IV 6-hourly plus 1 mg/kg gentamicin daily for 2 weeks. Amoxicillin was continued for a further 2 weeks and the patient made a complete recovery. The damaged valve was replaced with a prosthetic valve several months later.

- **Why was the initial Gram stain on the positive bottle negative?**
- **What precautions will the patient have to take when visiting the dentist?**

CHAPTER SUMMARY

After reading this chapter you should know and understand:

- Blood culture samples are one of the most important sample types processed in microbiology.
- Blood cultures must be taken using aseptic conditions to ensure no contaminating organisms are introduced into the sterile blood culture medium.
- One of the major symptoms of blood stream infection is pyrexia. Blood cultures are taken around the time of a pyrexial spike to ensure maximum numbers of organism are present for detection.
- Blood cultures are routinely processed in the laboratory using an automated method.
- The Gram stain is the first stage in identifying the organism likely to be causing the bloodstream infection. This allows the patient to be managed optimally, based on the type of organism seen.
- Positive blood cultures are inoculated onto a wide variety of different media. In addition direct susceptibility testing is essential.
- It can be difficult to assess the significance of an organism in a positive blood culture bottle. The organism may be a skin contaminant. The role of medical staff in determining the significance of the isolate is extremely important.
- Endocarditis is a serious clinical condition where bacteria cause vegetations on heart valves. Multiple blood cultures and serology tests may be required to diagnose this type of infection.

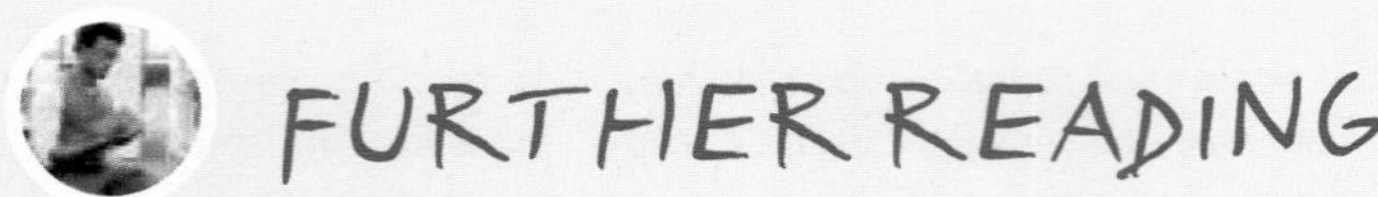

- **Gould FK, Elliott TSJ, Foweraker J *et al.* Guidelines for the prevention of endocarditis: report of the Working Party of the British Society for Antimicrobial Chemotherapy. *Journal of Antimicrobial Chemotherapy* 2006; 57:1035–1042.**

  Text covering all aspects of endocarditis and guidance for laboratories.

- **Elliott TSJ, Foweraker J, Gould FK *et al.* Guidelines for the antibiotic treatment of endocarditis in adults: report of the Working Party of the British Society for Antimicrobial Chemotherapy. *Journal of Antimicrobial Chemotherapy* 2004; 54: 971–981.**

  Text detailing the complex regimes which can be used for the treatment of endocarditis.

- **Hall KK, Lyman JA. Updated review of blood culture contamination. *Clinical Microbiology Reviews* 2006; 19: 788–802.**

- **Yagupsky P, Nolte FS. Quantitative aspects of septicemia. *Clinical Microbiology Reviews* 1990; 3: 269–279.**

- **Standards Unit, Evaluations and Standards Laboratory, Centre for Infections. *Investigation of Blood Cultures (for Organisms other than Mycobacterium Species)*, BSOP 37, 2005.**

  National SOP for the processing of blood cultures used in many laboratories throughout the world. Essential background information on clinical conditions and investigations.

DISCUSSION QUESTIONS

5.1 If a patient has suspected septicaemia the clinician will always treat before the blood culture becomes positive in the laboratory. How useful therefore are blood culture samples?

5.2 How useful is the Gram stain result on a positive blood culture?

5.3 What factors are involved in determining the significance of an isolate from a positive blood culture?

5.4 Is it important to have a full identification on every positive blood culture isolate?

5.5 Discuss the importance of ward procedures for taking blood cultures.

6

Investigation of urine samples

Clive Graham

Urinary tract infection (UTI) is considered to be the most common bacterial infection and urine the most common sample sent to the microbiology laboratory. In many laboratories hundreds of samples are received each day creating significant workload pressures, particularly when traditional labour-intensive methods are used.

The purpose of the laboratory is to provide a prompt result to the clinician but the laboratory needs to be mindful that therapy is often started empirically before the result is known. So perhaps rapid determination of the presence or absence of a significant infection followed by accurately assessing the significance and susceptibility of any growth should be regarded as its primary role.

Even then, these functions are confounded by the variability of samples received, both in quality and sample type, and the need for any laboratory method to cover all the eventualities. Often the methods used are designed simply to cover the majority of samples received.

Learning objectives

After studying this chapter you should confidently be able to:

- Describe the anatomy of the urinary tract
- List the common specimen types used to diagnose urinary tract infection
- List the common urinary tract pathogens
- Describe the routine tests performed on urine
- Describe which antibiotics are used to treat urinary tract infection

6.1 Overview of the urinary tract

In order to appreciate the clinical setting of urinary tract infections we must have an understanding of the normal anatomy of the urinary tract (Figure 6.1) and those mechanisms in place that help prevent infection from occurring.

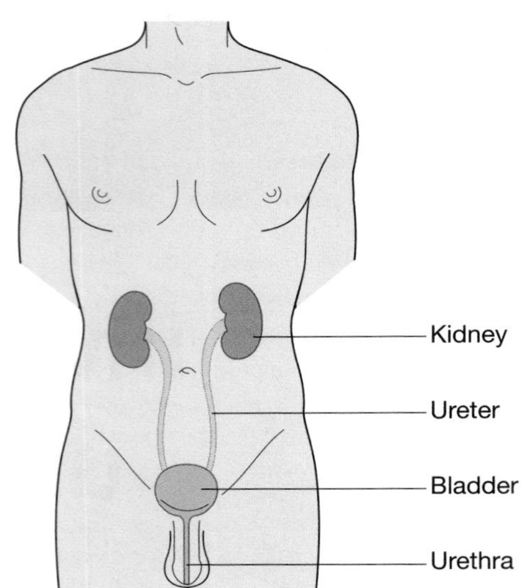

FIGURE 6.1
Anatomy of the urinary tract.

Urine is produced in the kidneys and passes through the ureter (a small tube) by a process known as peristalsis into the bladder. Urine collects in the bladder and is intermittently emptied via the urethra.

Ascending infection is the most common route of infection, that is to say microorganisms ascend the urethra into the bladder and may from there go on to infect the kidney. The organisms causing urinary tract infection are often from the gastrointestinal tract, for example *Escherichia coli.* Less commonly, bacteriuria results from haematogenous spread (via the bloodstream). Haematogenous infection is uncommon and will not be considered in detail further in this chapter other than to say that infection is usually caused by a limited number of organisms that include *Staphylococcus aureus, Salmonella* spp., *Mycobacterium tuberculosis* and *Candida* spp.

Normal flora of the urinary tract

Other than the distal urethra, the urinary tract is sterile.

Mechanisms designed to prevent infection

The bladder has a number of mechanisms designed to resist infection. The most basic of these is the physical washing out effect in the production of urine that dilutes any bacterial load and removes infecting organisms. This regular flushing helps prevent infection ascending either into the bladder or into the kidneys; furthermore a valve prevents reflux (backflow) of urine from the bladder into the ureter. One can imagine that the risk of infection would depend on the initial inoculum and multiplication rate of the organism in addition to any residual urine within the bladder, urine flow and frequency of voiding.

Obstruction to urine flow, for example through enlargement of the prostate gland in men, or if the bladder incompletely empties, for example through reflux of urine into the ureters (**vesicoureteric reflux**) due to inadequate development of the one-way valve between ureter and bladder, increase the risk of urinary tract infection developing.

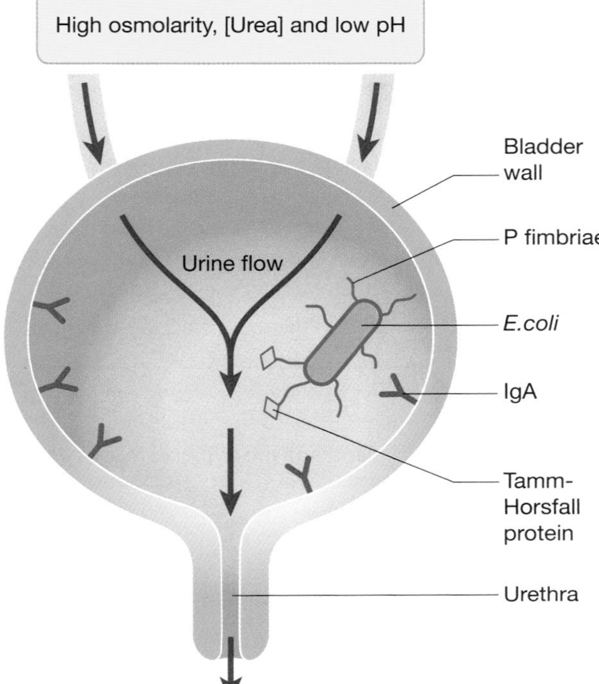

FIGURE 6.2
Diagram to show some of the bladder mechanisms which promote resistance to urinary tract infection. (Courtesy of M. Ford.)

Other physical properties of urine, such as extremes of pH, high osmolality, IgA and high **urea** concentration, tend to be protective (Figure 6.2). Historically urine has been used as an antiseptic. Furthermore, the cells lining the urethra, bladder and ureter resist colonization by bacteria. A type of antibody (immunoglobulin A) secreted by immune cells within the bladder wall is also found in the urethra and stops bacteria binding. Finally, Tamm-Horsfall protein, a mucoprotein shed by the kidney, has also been shown to bind and trap *E. coli*.

SELF-CHECK 6.1

What is the commonest cause of a urinary tract infection and what are the common host defence mechanisms?

Key Point

Urinary tract infection is much more common in females than in males.

Mechanism of infection in the catheterized patient

A **catheter** is a tube that is passed into the urethra and then pushed into the bladder; a small balloon on the end of the catheter is then inflated to prevent it coming out. The tube can be made out of a number of different products, e.g. latex, silicone, PVC or Teflon; some are coated with silver to reduce infection risk. Once inserted the catheter is connected to a collection bag. It is recommended that the system is kept closed, that is there are no breaks in the tubing to allow microorganisms to get into the inside (**intraluminal**) part of the tube. Catheters are required for a number of reasons; some are used for a short period of time, for example during an operation, whereas others are required for a prolonged period to deal with bladder or prostate problems. The insertion of the catheter is done aseptically to avoid introducing infection.

Cross reference
Chapter 1, Section 1.3.

It is now considered that intraluminal colonization by reflux of bacteria from a contaminated bag or from a break in a closed drainage system is a rare event, and the most common route of infection is colonization of the exterior of the catheter by bacteria from the outermost part of the urethra, either at the time of catheterization or later as a film of mucus develops between the catheter and urethra. The rate of acquisition of high-level **bacteriuria** is estimated to be 5% per day; therefore the key way to reduce risk is to reduce catheter use and duration. When infection does arise it is often due to antibiotic-resistant bacteria.

Key Point

Urinary catheters are widely used in hospitalized patients and this bypasses many of the patient's defences designed to prevent UTI.

Although infection may be introduced through catheterization there is also a risk of infection being introduced with other surgical procedures such as operations to look into the bladder or remove the prostate gland.

SELF-CHECK 6.2

What are the advantages and disadvantages of catheterizing a patient?

6.2 Urinary tract pathogens

E. coli is by far the most common cause of urinary tract infection. Infection within hospital is more likely to be due to other organisms, many of which are resistant to the antibiotics commonly used to treat infection in the community (Table 6.1). This is because patients admitted to hospital are often given antibiotics either to treat a suspected infection or to prevent

TABLE 6.1 Common bacterial causes of urinary tract infection in hospital and the community.

Organism	Community (%)	Hospital (%)
Escherichia coli	69.4	50.8
Proteus mirabilis	4.3	5.1
Klebsiella/Enterobacter spp.	4.7	7.3
Enterococcus spp.	5.5	11.9
Staphylococcus spp.	4.0	8.4
Pseudomonas aeruginosa	0	11.1
Others	12.1	5.4

infection, for example if they are having an operation. Such antibiotic use will select out skin and gut bacteria that are resistant to the antibiotic being used, and this may be a resistant strain of *E. coli* or another bacteria. These resistant strains in turn may cause infection in that patient or be passed on to another patient if infection control precautions are not followed.

The organisms causing community-acquired infection do vary, for example *Staphylococcus saprophyticus* is seen in sexually active young women but not in other groups. It also should be noted that antibiotic-resistant *E. coli* strains, in particular **extended-spectrum β-lactamase** producers (ESBLs), are increasingly being seen in patients who have had little or no contact with hospitals.

Cross reference
Chapter 4, Section 4.3.

Key Point

Commonly, the organisms causing UTI are from the patient's own gut flora.

Most community-acquired infections are in otherwise healthy individuals and are often termed 'uncomplicated' urinary tract infections. These infections respond well to antibiotics and recurrences are due to reinfections.

Complicated urinary tract infection describes infection in patients with a complicated urinary tract (e.g. abnormal because of structural or functional defects) or a patient generally more susceptible to infection, for example a patient on chemotherapy for cancer. These infections are due to a wider variety of bacteria, including other Gram-negative bacteria such as *Enterobacter* spp., *Pseudomonas* spp., *Providencia* spp. and *Morganella morganii*, and Gram-positive organisms such as enterococci, streptococci, staphylococci and yeasts, especially *Candida* spp.; *E. coli* is less common. For infections in catheterized patients, the longer the catheter has been left *in situ* the greater the proportion of isolates other than *E. coli*, although interestingly *Candida* spp. are seen more frequently in those with short-term catheters.

Urinary tract infection is not a common feature in immunocompromised hosts unlike the higher rates of infection seen in those with structural and functional renal tract abnormalities. However, immune defects may mean that infections that do occur are not adequately controlled and that treatment of infection in such patients requires the use of antibiotics that achieve good tissue levels.

Complicated urinary tract infection is often equated to hospital-acquired infection but it should be noted that with increasingly complicated cases being managed in the community this distinction is less clear than it was in the past. With complicated infections, response to antibiotics is less and relapse with the same organism is more likely to occur.

SELF-CHECK 6.3

What is meant by a complicated urinary tract infection and in what type of patients do these occur?

Virulence factors of microbes causing urinary tract infection

There are many individuals who develop urine infection who do not have identifiable risk factors. Many of these infections are due to *E. coli* and indeed certain (uropathogenic) strains of *E. coli* are more strongly associated with urinary tract infection. They have certain characteristics that facilitate their ability to move from their normal residence in the gut and to colonize and infect the urinary tract. An example of such a mechanism is P fimbriae – these are hair-like projections on the surface of the bacteria which can bind to cells found in the kidney. These strains of *E. coli* commonly cause acute pyelonephritis.

Epidemiology

UTI is most common during the first year of life and is more common in young girls, except in neonates where UTIs in males predominate. Lowest rates are seen in 11–15 year olds. The incidence of infection increases in young adult women where sexual activity is an important contributory factor; and increases again in old age. Indeed, around 20% of women develop a urinary tract infection during their lifetime; only after the age of 50 years is a similar incidence seen in males. Twenty-three percent of all health-care associated or **nosocomial infections** are urinary tract infections.

Infections of the lower urinary tract

Most cases (95%) of UTI are acute **cystitis**. It results from spread of organisms up the urethra and, like other sites of infection, *E. coli* is by far the most common cause. Symptoms of infection include pain when passing urine (**dysuria**), passing urine more frequently (often urgently) and pain in the lower abdomen (suprapubic pain).

urethral syndrome
Urethral syndrome occurs when the patient experiences symptoms of a UTI but significant bacteriuria is not detected.

The **urethral syndrome** is used to describe a patient who presents with the symptom of dysuria; this may (particularly in women) result from a urinary tract infection (due to *E. coli* or *S. saprophyticus*) sometimes with lower bacterial counts than anticipated with urinary tract infection. In both men and women, it may be part of an acute or chronic non-specific urethritis, classically due to *Neisseria gonorrhoeae* or *Chlamydia trachomatis* but *Mycoplasma* spp. may also be implicated.

There are other non-infective causes of the urethral syndrome. Prostatitis is inflammation of the prostate gland, a gland that in men surrounds the urethra just below the bladder. The causes of infection are usually sexually transmitted organisms such as *Chlamydia trachomatis* in young men, with more typical **uropathogens** causing infection in older men. It may follow instrumentation of the region by catheterization, etc. Inflammation may become chronic and in some of these cases there is no obvious earlier infection.

Acute epididymitis is inflammation of the epididymis, a tubing that connects the urethra to the testes. In older men this is usually due to typical UTI pathogens such as *E. coli* but in younger men it is more likely to be due to sexually transmitted bacteria.

Infections of the upper urinary tract

pyelonephritis
A urinary tract infection which has ascended to reach the kidney. This is a serious infection.

The hallmark symptoms of acute **pyelonephritis** are flank pain and fever; nausea, vomiting and rigors may also occur. In most cases infection is from ascending lower urinary tract infection, resulting in bacteraemia in a third of cases. The most common organism responsible is *E. coli*.

Chronic pyelonephritis is chronic inflammation of the kidney, often associated with chronic obstruction of the urinary tract or, in young children, from backflow of urine from the bladder into the ureter (vesicoureteric reflux).

SELF-CHECK 6.4

In general, which type of urinary infection is clinically the most important – upper or lower urinary tract infection?

Asymptomatic bacteriuria

This is commonly seen in elderly women and has no significant clinical consequence. Pregnant women should be screened for asymptomatic bacteriuria by culture as identification and treatment of asymptomatic infection reduces the risk of preterm birth. Treatment also reduces the risk of pyelonephritis developing (number needed to treat (NNT) = 7); recent evidence suggests it is thought that a single dose of antibiotic may be effective.

It is thought that the hormonal changes of pregnancy cause dilation of the ureters and increase the chance of ascending infection.

Key Point

Even though a urine sample may contain numerous pathogenic bacteria, there are occasions when they are not causing the patient harm. This is known as asymptomatic bacteriuria.

Asymptomatic pyuria

Asymptomatic pyuria is also found in patients with diabetes and the elderly (especially those with incontinence). In diabetics it is seen more frequently in those with **diabetic nephropathy**. In the elderly it can occur with or without bacteriuria and therefore in this patient group, pyuria should not be used to indicate whether a culture result is significant or not.

Common fungal pathogens

Fungal urinary tract infection is invariably due to *Candida* spp., usually *Candida albicans.* Most often infection is via the ascending route in a patient with a catheter *in situ* who is receiving antimicrobial chemotherapy. In asymptomatic patients, treatment may simply involve catheter removal (and stopping antibacterial therapy) with a repeat urine sample following catheter removal. For symptomatic patients or those with persistent infection, antifungal therapy is necessary. In women, vulvovaginal candidosis may give rise to positive cultures for *Candida* spp. in the urine; similarly, babies with 'nappy rash' may have urine contaminated with yeasts during collection especially if a bag or similar method is used for urine collection.

In neonates and immunocompromised patients, the presence of *Candida* spp. in the urine may indicate haematogenous spread of infection to the kidneys. Such infections have a significant morbidity and mortality associated with them and require prompt and aggressive antifungal treatment.

Atypical and fastidious organisms

Given that most urine cultures are done on a limited range of culture media, it should be noted that certain organisms that infrequently cause urinary tract infection may not be recovered on these media and additional culture methods may need to be considered in patients with persistent symptoms whose cultures are negative. For example *Haemophilus influenzae* (a cause of chest infections) is occasionally seen in children.

Mycobacteria

Mycobacterium tuberculosis infection of the renal tract may result in sterile pyuria and should always be considered in such patients. Infection of the kidney is due to blood-borne spread from a primary site elsewhere, invariably the lungs. The initial infection in the renal cortex, spreads to the papillae, caseates and then discharges into the urine. Excretion of organisms is intermittent, hence the importance of collecting repeat samples on three consecutive days (early morning or first void). Such specimens need to be processed under Containment Level 3. As urine may contain *Mycobacterium smegmatis*, a normal commensal of the genital tract, staining of urine for mycobacteria is not recommended and the urine sample should be directly inoculated onto culture media.

Cross reference
Chapter 9, Section 9.4.

Parasites

Parasitic infections of the urinary tract are rare in the UK but may be seen in those returning from abroad. *Schistosoma haematobium* (Figure 6.3) is the parasite that classically infects the urinary tract, causing haematuria and bladder irritation. A minimal volume of 10 mL should be examined as soon as possible after collection; if this is not possible the sample should be refrigerated. *S. haematobium* has a circadian rhythm; egg excretion peaks and is least variable from 10.00 a.m. to 2 p.m. Exercise before egg collection is unnecessary. Antigen detection methods have also been reported for urine (and serum) but are not widely available.

Other schistosomes and other parasites, such as *Entamoeba histolytica* and *Echinococcus granulosa*, can also rarely affect the urinary tract.

Finally, the Amazonian Indians may be infected by a bloodsucking catfish known locally as *candirù* but with the Latin name *Vandellia cirrhosa*. This fish is said to be attracted to urine and capable of swimming up the urethra into the bladder.

SELF-CHECK 6.5

Why is it important for not just one sample to be sent to the laboratory to detect the presence of mycobacteria?

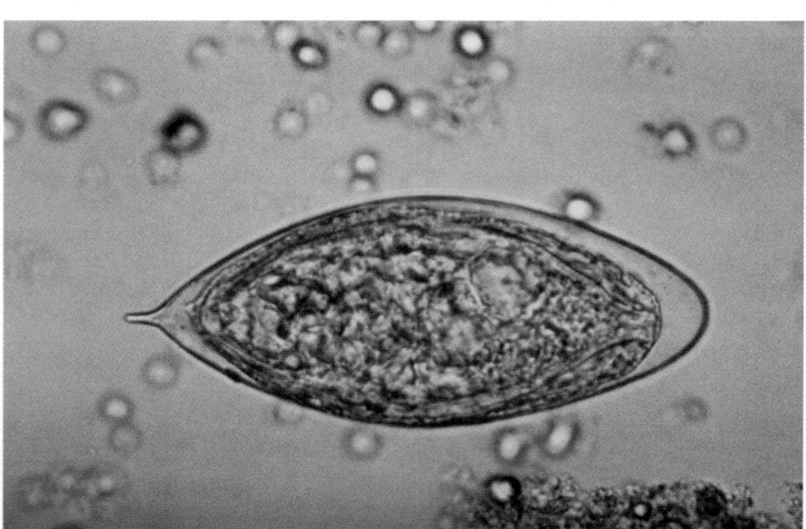

FIGURE 6.3
Image of a *Schistosoma haematobium* trematode parasite. Note the egg's posteriorly protruding, terminal spine, which is highly characteristic of the parasite. (Courtesy of CDC.)

6.3 Urinary samples

Cross reference
Chapter 1, Section 1.3.

Specimen collection

As is true for many aspects of microbiology, the quality of bacteriology results is only as good as the quality of the specimen. It is recognized that the distal urethra is colonized with bacteria, although it should be noted that this often involves skin commensal organisms such as anaerobes or *Lactobacillus* spp. which would rarely be considered significant in a urine culture. They may also have faecal organisms such as *E. coli* (which obviously could be deemed relevant) if the sample is not collected properly. There are a number of urine specimen types; these are listed below.

Midstream urine (MSU)

Patients should be given clear instructions on collection of urine. The first urine passed washes away organisms from the distal urethra and this should be discarded into a toilet or other receptacle. The midstream urine sample can then be collected into a suitable sterile container.

This process requires:

- Good control of micturition
- Appropriate hand-eye co-ordination
- Sufficient mobility to locate a sterile container in the urine stream

It is obviously not suited to the very young, the elderly or the infirm.

Some collection receptacles are associated with lower contamination rates.

Clean catch urine

This is the equivalent of midstream urine in young children where the volume passed can be relatively small and the sample is literally caught into a sterile container by the parent or staff member when the child happens to pass urine. It obviously takes patience and is not the method preferred by parents.

'Pad' urine

This is when an absorbent pad is located inside the nappy and checked every 10 minutes until wet. It is used in babies where obtaining a clean catch sample may be difficult. It is the preferred method of parents but does suffer the same problem with contamination that bag urines have; furthermore, if the pad becomes soiled it needs to be discarded. Pad urine samples may result in lower white cell counts on urine microscopy.

'Bag' urine

This is similar to pad urine but a sterile bag is placed in order to collect urine; such samples run a high risk of perineal contamination.

'In-out catheter urine'

This is the method of urine collection frequently used in children in North America. It has the advantage of being very rapid, not having to wait for urine to be passed spontaneously, but is relatively more traumatic and runs a risk of introducing infection into the bladder.

Catheter stream urine (CSU)

These are samples taken from patients with indwelling catheters, which may be either short term (for example used postoperatively) or long term where there is an underlying urological or other abnormality that requires the presence of a catheter. It should be noted that catheter-associated bacteriuria is common and in itself not significant; samples should only be taken if there is concern about a patient having an infection.

Samples need to be taken through a port or similar sampling point as breaking of a closed urinary catheter system runs the risk of introducing infection. An aseptic technique should be employed.

Suprapubic aspirate

These are samples taken from neonates and younger children only when it has been impossible to decide (based on pad or clean catch urine) if a child has a urine infection or not. The sample is taken by pushing a needle through the skin and aspirating urine directly from the bladder. Although there is a risk of contaminating the sample with skin organisms, the skin is cleaned thoroughly before the sample is taken and therefore any culture result must be viewed as significant.

Other samples from the urinary tract are sometimes received in the laboratory, in particular when that laboratory deals with samples from urology patients. Often these are from the upper renal tract, which should be sterile. Examples include those from nephrostomy - a tube that goes directly into the kidney.

SELF-CHECK 6.6

Why is it important not to collect the first few millilitres of a urine sample when collecting MSU?

6.4 Transport of specimens to the laboratory

Cross reference
Chapter 1, Section 1.3.

Urine is a culture medium and delays in processing will allow bacteria to replicate, especially at higher temperatures, and this may result in either erroneous significant cultures being reported or mixed growths, where it is not possible to pick out the organism responsible for the patient's infection. When delay is anticipated, for example out of routine laboratory hours, the following methods should be considered to prevent deterioration in the sample:

- Refrigeration at 4°C: generally a maximum period of 24 h is recommended to preserve cellular elements. In terms of bacterial growth, surface viable counts of simulated specimens are said to remain constant for 72 h and Kass recommended 48 h (see Further reading, below).

- Dip-slide technique: this involves commercially available agar-coated slides which are dipped into a freshly voided urine sample, placed back in their own containers, and then transported to the laboratory. Unfortunately, there is no record of the cellular or other components of the urine but conversion charts allow the quantification of culture results.
- Boric acid: boric acid is a preservative. It is designed to maintain bacterial and cellular components of urine where refrigeration is not practicable. The main disadvantage of this method is that bacterial counts may be reduced, and this is particularly important if counts are less than 10^5/mL to begin with (high inoculums are not affected) or if an inadequate volume of urine is dispersed into the container, as the concentration of boric acid would be higher than anticipated. Due to improvements in sample transport this preservative is now used very rarely.

6.5 Initial processing of samples

A clear, colourless urine sample is unlikely to grow significant numbers of bacteria but a cloudy sample can result from crystals, bacteria, red blood cells or white blood cells. Tests performed at the bedside or in the outpatient clinic can be used to help decide which samples should be sent for culture or if empirical therapy is appropriate. Catheters are foreign bodies and may elicit a raised white cell count or the patient may be infected with a nitrate-negative organism and therefore use of the dipstick tests described below should be avoided in this patient group. Reagent strip testing is often used; although a wide range of biochemical tests can be performed, two tests are particularly helpful in diagnosing infection.

Leucocyte esterase

Leucocyte esterase is an enzyme found in white blood cells called neutrophils; these are usually found in raised amounts in the urine of patients with infection. False-negative reactions occur with vitamin C, boric acid, glycosuria, high concentrations of protein and certain antibiotics such as cefalexin, gentamicin and nitrofurantoin; false-positive reactions occur with another antibiotic, clavulanic acid, contaminated samples or in febrile children.

Nitrate reductase (Greiss test)

Nitrate reductase is an enzyme found in many, but not all, bacteria associated with urinary tract infection. Of particular note, *S. saprophyticus* does not produce this enzyme.

CASE STUDY 6.1 UTI in a child

A 6-month-old boy is taken by his mother to the Emergency Department with fever. Although the doctor suspects a viral infection a urine sample is collected by the nursing staff. The urine sample is tested by dipstick and found to be positive for leucocytes but negative for nitrates.

- **Should a urine sample have been tested by the GP?**
- **What is the significance of a positive dipstick in a child?**
- **What further tests should be carried out on the urine?**
- **What are the implications if this child has a UTI?**

Urine microscopy

Although a number of methods may be used for urine microscopy including Gram stain and phase contrast microscopy, one of the most widely accepted is the microtitre tray. This is known as the Rant-Shepherd method and is described in the Method Box. Interestingly, the first paper that described this method was refused publication as the editors thought that it would never gain wide acceptance.

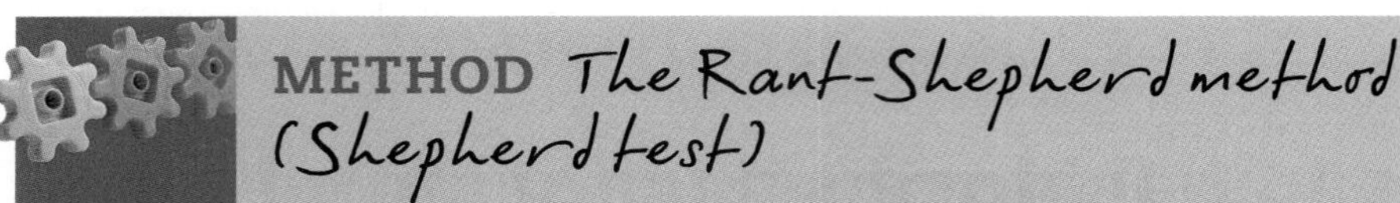

- **A urine specimen container is inverted and 80 μL of urine is transferred by pipette into a microtitre tray, and is allowed to settle for 10 min. This allows cells, casts and crystals to sink to the well floor. The settled material is then observed using an inverted microscope.**
- **A cell count is performed but to determine the count per cubic millimetre of urine one has to know the field of view of the microscope and the fluid depth (which in turn depends on the radius of the microtitre well).**
- **Often such calculations have been done for the particular equipment in the laboratory to give a standard factor. The viewer then just needs to count the red and white cells seen on microscopy and multiply this by the standard factor to obtain the count per cubic millimetre.**

Eukaryotic cell types in urine

Certain cell types are routinely reported when seen on microscopy.

- Leukocytes (white blood cells): in uncentrifuged urine, 10 leukocytes/μL is abnormal. Lower counts may be seen if there is a delay in microscopy as the hypotonicity of urine may cause cell disruption; furthermore if the peripheral cell count is low then the urine count may not be elevated. If no organisms are cultured the patient is said to have '**sterile pyuria**'; there are a number of potential reasons for this (Table 6.2).

sterile pyuria
Sterile pyuria is where the urine contains numerous white cells but no evidence of infection can be found.

TABLE 6.2 **Causes of sterile pyuria.**

Causes
Chlamydia trachomatis and other sexually transmitted infections
Prostatitis
Neoplasia of the renal tract
Renal calculi
Catheter or other foreign body
Renal tuberculosis
Fever in children (regardless of cause)
Antimicrobial chemotherapy (resulting in negative cultures)

- Erythrocytes (red blood cells): again any number of erythrocytes in urine is potentially abnormal although 'haematuria' may also be seen with menstruation. Haematuria may occur with infection but also occurs with renal calculi, malignancy and **glomerulonephritis.** Some automated methods reveal that a few red cells are seen in 'normal' individuals; indeed a cut off is required when an automated microscopy method is used. In healthy volunteers, 95% had less than 8 erythrocytes/μL. In such individuals all red cells seen are dysmorphic (squashed appearance) suggesting a glomerular origin. Dysmorphic red cells are best detected by a special type of microscopy known as phase contrast.
- Epithelial cells: presence of epithelial cells indicates the urine has come into contact with the skin and may be contaminated.

glomerulonephritis
Glomerulonephritis is inflammation of the glomeruli, or small blood vessels of the kidney.

Significance of casts in urine wet films

Casts are proteinaceous imprints of the fine tubular network present in the kidneys; they may occur in a number of different renal pathologies. Although a number of cast types can be in urine, the most commonly seen are:

- Hyaline casts: these are made by aggregation of the Tamm-Horsfall protein and are seen in 'stressed individuals'. Epithelial cells are sometimes attracted to their surface (epithelial casts). Epithelial casts are said to occur with damage to the tubular epithelium and their morphology varies over time.
- Granular casts: these can be split into different types depending on how they are formed; four types have been described but only the first two will be considered:
  - Type 1 is associated with white blood cells casts and with degeneration of these casts. It results from a process of active inflammation within the kidney, white cells entering the tubular lumen, becoming bound in to the cast matrix and degenerating. In most instances this occurs with active tubular or interstitial injury and inflammation.
  - Type 2 is the most common type of granular cast. It consists of hyaline casts with granules (thought to be lysosomes, from associated epithelial cells) scattered throughout the matrix.

Key Point

Casts are tubular structures containing aggregates of particulate matter that form in the distal nephron. These can break off and pass into the urine and are seen on microscopy of the sample.

SELF-CHECK 6.7

Why is it important to carry out microscopy on a urine sample and report the result back to a clinician?

Types of pathological crystals found in urine

Crystalluria is a common finding and most cases are due to the precipitation of crystals of calcium oxalate, uric acid, triple phosphate, calcium phosphate and amorphous phosphates or urates. Only in a minority of cases is it associated with pathological conditions such as urine stone formation or acute uric acid nephropathy. Crystalluria may also be associated with various drugs. Formation of urine stones has been linked to certain bacterial infections, especially those caused by urea splitting bacteria such as *Proteus* spp. Identification of urine crystals is not usually done in the microbiology laboratory.

Parasitic infection and detection

A method for examining urine for parasites is given in the Method Box.

METHOD Examination of urine specimens for parasites: centrifugation method (*Schistosoma haematobium*, Figure 6.3)

- **Ensure specimen is in a conical-bottom container.**
- **Centrifuge at 500 g for 2 min.**
- **Decant and discard the supernatant, then mix the sediment.**
- **Using the whole deposit, place 1 or 2 drops onto a clean microscope slide and apply coverslips. Examine the entire area of each slide under low-power magnification.**
- **A filter method (with filter paper, polycarbonate or polyamide) may also be used, and is often employed in epidemiological studies.**
- **It should be noted that egg excretion varies and although most patients with overt haematuria will have detectable eggs this may not be true for patients with more subtle urinary tract symptoms.**

Automated techniques available for processing urine samples in the laboratory and Point of Care Testing (POCT)

It should be noted that routine microscopy and culture is still recommended for diagnosing infection in children less than 3 years of age and pregnant women.

A high proportion of urine samples generally grow insignificant amounts of bacteria or are culture negative. A procedure that would eliminate the need to culture those specimens that are unlikely to grow significant bacteria would be advantageous both in the laboratory, saving time and resources, and at the point of care where, potentially, antibiotic use could be reduced. Another advantage is that many of these methods are objective rather than subjective and therefore may be less prone to error or bias. Automated methods are suited to high-volume work but obviously the method employed will vary depending on cost and anticipated throughput.

Automated dipstick method

This has the advantage over manual reading of dipsticks because the result is not subject to interpretation but is read in a standard way, after the appropriate time interval, etc. Although mainly using leucocyte esterase and nitrite detection, other indices can also be measured. Automated systems include Clinitek (Siemens, Erlangen, Germany) and Urisys 1100 (Roche, Basel, Switzerland).

These methods continue to have the drawback of hand-held dipsticks, namely low specificity and positive predictive value. They are, however, cost effective if negative samples are not cultured.

Automated microscopy

Automated microscopy methods, such as Yellow Iris™ or R/S 2000, were designed to speed up standard urine microscopy methods. The latest marketed version, iQ200 (Iris Diagnostics, Chatsworth, California, USA) uses digital imaging and auto-particle recognition software to recognize 12 urine particles (erythrocytes, leucocytes, leucocyte clumps, hyaline casts, unclassified casts, squamous epithelial cells, non-squamous epithelial cells, bacteria, yeast, crystals, mucus and sperm) and quantify results (Figure 6.4). The advantages over flow cytometry is that images that will be familiar to microbiologists can be viewed on demand rather than the scattergrams produced by flow cytometry.

Flow cytometry

Flow cytometry is widely used in non-microbiology laboratories, for example it is used in haematology laboratories for detection of the different cell types in blood. An early type, UF-50, was used to detect red blood cells, white blood cells, epithelial cells, casts and bacteria in urine; detection of the latter was not optimal (false negatives were a problem) and various modifications have been made. The latest version, UF-1000i (Sysmex Europe GmbH, Hamburg, Germany), is said to be more sensitive.

To detect cells in urine, 200 µL of urine is diluted with a citrate buffer solution plus cationic surfactant. This mixture is supplemented with 40 µL of fluorescent dye that stains nuclei acid within the bacteria; the final mixture has a dye concentration of 1 ppm. This is then hydrodynamically focused and passed through a sheath flow illuminated with a laser beam. The forward scatter light intensity and lateral fluorescent intensity of each cellular component of the urine is converted into electrical signals by a photomultiplier and these are measured for each sample. Urine bacterial analysis (UBA) can distinguish bacteria from other cellular and non-cellular components of urine (e.g. erythrocytes, leucocytes, epithelial cells, fungi and crystals).

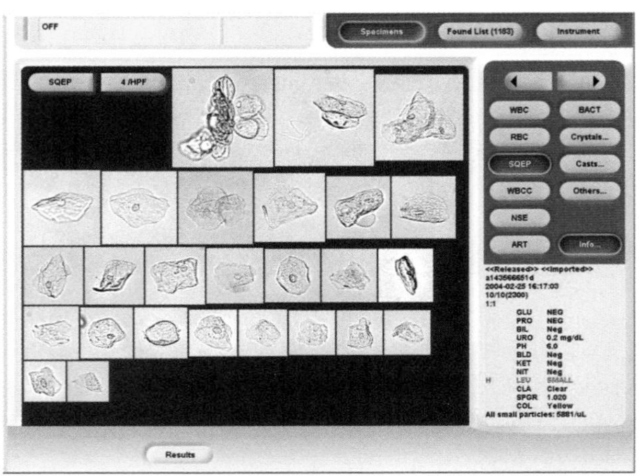

FIGURE 6.4
Screen image from an iQ200 Sprint Urine Analyser showing the presence of epithelial cells in the sample. (Courtesy of Instrumentation Laboratories.)

This technology has been assessed for linearity (increasing bacterial suspensions give higher measurements); the carryover rate is estimated to be 0.029%. Variation (between and within runs) is less than with traditional culture methods especially with lower bacterial counts and correlation with culture results is excellent. The false-negative rate is low - 0.7% in one study.

Other automated methods for detecting bacteriuria

Other automated methods, such as bioluminescence and filtration, are available but are not widely used in the UK.

SELF-CHECK 6.8

What are the advantages of automated methods for the analysis of urine samples?

6.6 Culture

Significant culture results

A significant culture result is taken as greater than 100,000 (10^5) colony forming units (CFU)/mL of urine. Kass established a figure of equal to or greater than 10^5 organisms/mL of midstream urine as being able to distinguish women with acute pyelonephritis. Even in his original studies he noted that 6% of asymptomatic women had this degree of bacteriuria.

The significance of the above value has been brought into question, with much lower counts (10^2 coliforms per mL) being deemed significant in symptomatic women. However, in the laboratory it is often difficult to be certain that a specimen has been collected properly and the use of such low counts could lead to the over-reporting of 'significant' results and, potentially, over-use of antibiotics, especially if one considers that most urine samples are subject to a delay before processing that may result in bacterial replication. For this reason, many laboratories use a cut off of a pure growth of 10^4 organisms per mL; indeed 90–95% of patients with acute pyelonephritis have bacteria at this concentration. For antibiotic studies, bacteriuria of $\geq 10^3$ CFU/mL of urine is used; this value has a specificity of 90% and sensitivity of 80% for the diagnosis of acute cystitis.

SELF-CHECK 6.9

Why is the number of bacteria isolated in urine samples considered very important?

Procedures for quantifying urine cultures

As indicated above, it is important to quantify the growth of bacteria. There are a number of methods one can use to do this; two are given in the Method Box. Other quantification methods are available, such as those used in dipslides (above) and the paper foot method (see Leigh and Williams, in the References list).

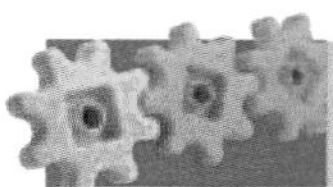

METHOD Quantification of urine cultures

- The *calibrated loop technique* is based on the fact that using a calibrated loop a fixed volume (usually 1 μL or 10 μL) of urine is delivered to the agar plate.
- The urine is then streaked out so that individual colonies may be more easily counted. These can then be multiplied up to see if the colony count is significant: for example using a 1-μL loop, 1 colony would equate to 1000 CFU/mL of urine; using a 10-μL loop, 1 colony would equate to 100 CFU/mL and would be more appropriate, for example, when dealing with a suprapubic aspirate.
- The *multipoint inoculator method* uses a multipoint inoculator (0.3–2 μL depending on the pin size) to transfer urine from a tray onto agar for culture; the urine may require re-refrigeration until the inoculator is full.
- This method can be done manually or automated and is limited in its ability to detect growths of less than 10^5/mL. Modification of this method can be used to incorporate sensitivity testing and identification.

Common isolation media

Most bacteria and fungi associated with urinary tract infection can be isolated on standard agar. Often cysteine lactose electrolyte deficient (CLED) agar is used as a single plate, with more than one urine sample applied so that costs are contained. This agar gives a yellow colour with lactose fermenting colonies such as *E. coli* (Figure 6.5) and a blue colour with non-lactose fermenting colonies such as *Proteus* spp. It supports the growth of *Streptococcus pyogenes* and other fastidious organisms that do not require blood and it also inhibits the swarming of *Proteus* spp. Andrade indicator (fuchsin) may be incorporated into the medium to enhance colonial appearance. The colour changes seen depend on the pH of the media, ranging from bright red with a pH of 6.0 to deep blue with a pH of 7.4 (Table 6.3). CLED

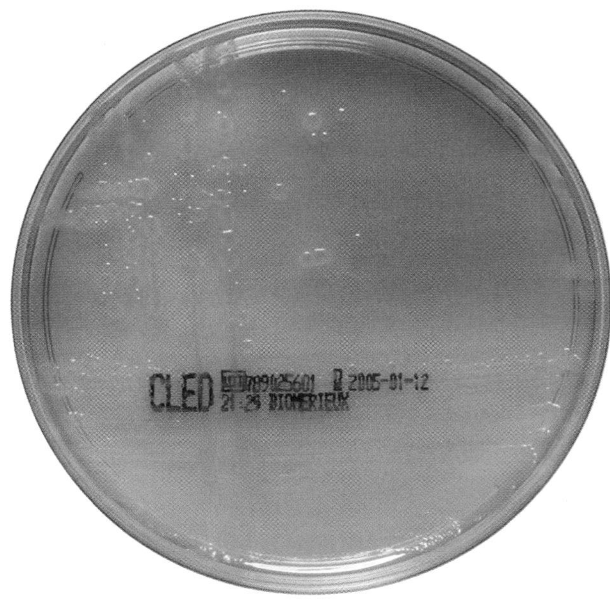

FIGURE 6.5
Growth of *E. coli* on CLED agar. (Courtesy of bioMérieux.)

TABLE 6.3 Colony morphology on CLED agar with and without Andrade indicator.

Organism	CLED agar	CLED with Andrade indicator
Escherichia coli (lactose fermenting strains)	Yellow, opaque colonies with a slightly deeper coloured centre (1.25 mm diameter)	Bright pink, semi-translucent colonies with a surrounding pink halo
Proteus mirabilis	Translucent blue colonies around 1 mm	Blue-green translucent colonies
Klebsiella spp.	Mucoid, variable colour, yellow to whitish-blue	Grey-green mucoid colonies
Enterococcus faecalis	Yellow colonies, 0.5 mm in diameter	
Staphylococcus aureus	Deep yellow colonies about 0.75 mm diameter	Smooth, entire, opaque bright golden yellow colonies
Coagulase-negative staphylococci	Pale yellow or white colonies	Smooth, entire, opaque white colonies
Salmonella spp.	Flat, blue colonies	
Pseudomonas aeruginosa	Green colonies, matt surface and rough periphery	
Corynebacteria	Very small grey colonies	

incorporating Andrade indicator should not be incubated for more than 24 h as lactose fermenters will cause pinkish coloration of the media, masking any non-fermenters present.

Differential agars are available for common urinary tract isolates; these give different colours depending on the organism present. Chromogenic agars contain substrates that when hydrolysed by bacterial (or in some cases fungal) enzymes produce coloured compounds. Several different media are available commercially; there are a number of studies comparing detection and identification rates and all are superior to CLED in detecting mixed culture results. As well as better detection of mixed cultures, they may also reduce inoculation and work up time by laboratory personnel.

Cross reference
Chapter 3.

Fastidious bacteria such as *Streptococcus pneumoniae* may cause infection in children and these organisms would not be isolated if a selective agar is used alone. In highly specialized situations, depending on the clinical context, cultures for *Mycobacterium tuberculosis* or *Mycoplasma* spp. or *Ureaplasma* spp. may be performed.

SELF-CHECK 6.10

What culture medium would you consider to be the gold standard for isolation of urinary pathogens?

Confirmatory tests

Experienced biomedical scientists can often identify organisms from primary culture plates. Organisms such as *Pseudomonas aeruginosa* have certain susceptibility profiles and may produce a pigment (pyocyanin) which produces a distinctive green discoloration to the agar.

Most presumptive identification needs to be confirmed; in most instances this is through biochemical tests. Historically, bacterial species used to be grouped together, for example '$>10^5$ coliform isolated'; however, it is now recognized that identification of the different bacteria responsible for urinary tract infection is essential for the accurate monitoring of susceptibility data and accurate determination of susceptibility using BSAC/CLSI or other methods.

Cross references
Chapter 2 and Chapter 4, Section 4.3.

Contaminated urine samples

Bacterial contamination of urine may be due to poor specimen collection, prolonged transport or both. It is important as contaminated samples may need to be repeated, it may result in the wrong diagnosis (and inappropriate antibiotic use) or delay appropriate treatment of an infection, and it increases costs. Despite awareness of contamination as a problem, there has been little improvement in contamination rates over time, with recent studies having a median contamination rate of 15%.

Contamination is seen more commonly in women than in men and interestingly does not appear to be reduced where written collection instructions or collection kits are used; however, this may reflect the quality of instructions, the abilities of the instructor, the receptiveness of the patient and where the instruction is given. Comparisons with other institutions can be used to assess the quality of specimen collection and processing.

Mixed bacterial growth may indicate contamination of a specimen during collection although in certain clinical situations, for example in catheterized patients, such culture results may be significant; indeed mixed bacteraemia can occur in patients who have mixed urine cultures. Table 6.4 gives a guide to further processing of such mixed growths.

SELF-CHECK 6.11

How does the laboratory determine if the sample is contaminated?

TABLE 6.4 **Significance and reporting of urine culture results.**

No. of isolates		Specimen type	Microscopy result	Laboratory interpretation	Susceptibility testing
1	$\geq 10^4$/ml	All	Any	Probable UTI	Yes – consider comment if no WBC present
2	1 predominant at $\geq 10^5$/ml	All	> 10 WBC / µl present	Possible UTI	Yes
2	Both organisms $\geq 10^4$/ml	All	> 10 WBC / µl present	Possible UTI	Yes
>2	No predominant organism	All except Suprapublic aspirate	Any	Mixed growth	No
1 or 2	Any growth	Suprapubic aspirate	Any	Probable UTI	Yes
>2	Any growth	Suprapubic aspirate	Any	Possible UTI	Combined sensitivity

Cross reference
Chapter 4.

Antimicrobial susceptibility testing

As well as determining if any bacteria isolated from the urinary tract are significant, the laboratory also needs to report what antibiotics the organism appears susceptible to. Certain antibiotics are often used for urinary tract infection and these antibiotics would be tested in all cases but not necessarily reported to the requesting doctor.
Factors influencing the choice of antibiotics reported include:

- Susceptibility profile of the organism
- Pregnancy
- If the patient is in hospital or in the community
- If the antibiotic is available in oral or IV form
- Renal function
- Broad or narrow spectrum
- Antibiotic costs

Microbiologists aim to promote the use of the most narrow spectrum, cost effective antibiotic therapy that has the least risk of adverse effects to the patient. Reporting of long lists of agents to which an organism is susceptible to may promote injudicious use of antibiotics.

Correct identification of organisms is necessary as certain antibiotics should not be used for certain coliforms that are naturally resistant to aminopenicillins, for example see Table 6.5. These antibiotic-organism combinations should not be used; similar rules apply to Gram-positive organisms; for example enterococci should be regarded as having intermediate susceptibility to trimethoprim.

Antifungal susceptibility tests are not routinely performed; indeed most *Candida* other than *Candida albicans* do not need to be identified to species level.

TABLE 6.5 Intrinsic resistance (R) in Enterobacteriaceae.

Organism	Amoxicillin	Co-amoxiclav	Piperacillin	Cefazolin	Cefoxitin	Cefuroxime	Aminoglycosides	Tetracycline/ tigecycine	Colistin	Nitrofurantoin
Citrobacter freundii	R	R		R	R					
Enterobacter cloacae	R	R		R	R					
Klebsiella spp.	R		R							
Morganella morganii	R	R		R		R		R	R	R
Proteus mirabilis				R		R		R	R	R
Proteus vulgaris	R			R		R		R	R	R
Serratia marcescens	R	R		R	R	R	*		R	

* All *Serratia marcescens* produce a chromosomal AAC(6′)-Ic enzyme that may moderate the activity of all aminoglycosides except streptomycin and gentamicin.

CASE STUDY 6.2 Antimicrobial prescribing policy

An elderly lady is admitted to hospital with suspected urinary tract infection. A sample is sent to the laboratory and reports a significant growth of *E. coli* with a long list of different antibiotic susceptibilities.

The clinician caring for the patient reads the reports and decides to use IV cefuroxime to treat the patient. As a consequence the patient develops *Clostridium difficile* infection.

If a narrower range of antibiotics had been suggested then a more narrow-spectrum agent, such as trimethoprim, may have been used which would be less likely to result in this complication.

- **What are the disadvantages to using intravenous antibiotics?**
- **Which antibiotics are associated with *Clostridium difficile* infection?**
- **Is trimethoprim resistance common in urine isolates in your hospital? Why is this?**

Direct susceptibility testing

Cross reference
Chapter 4, Section 4.3.

Susceptibility testing could be done at the time the urine is inoculated onto the culture plates or after culture results are complete, when only those urines harbouring significant growths would be inoculated. Most laboratories have systems in place where those urines likely to produce significant culture results (based on microscopy) are inoculated directly onto susceptibility test plates so that final culture results are available within 24 h of urine receipt; furthermore if further sensitivity work is required this is not delayed.

In terms of accuracy, direct susceptibility testing produces few errors with overall proportions of very major and major errors being low and comparable to the reproducibility of disc testing within the same laboratory.

Key Point

It is important to use the most appropriate narrow-spectrum antibiotic to treat a urinary tract infection. Broad-spectrum antibiotics may promote resistance and wipe out the patient's normal flora.

Automated identification and susceptibility

Many laboratories are using automated systems for identification and susceptibility testing. As resistance mechanisms are becoming more complex with one or more mechanism present in a particular bacteria such systems may more accurately identify them. Such systems can either be rule based, for example Microscan (Siemens, USA), Phoenix (Becton Dickenson, USA) and VITEK (bioMérieux, France), or pattern based, for example VITEK 2 (bioMérieux, France).

Cross references
Chapter 3 and Chapter 4.

Hazardous organisms

Urine rarely contains hazardous bacteria but it should be noted that in cases of typhoid and paratyphoid *Salmonella* spp. may be excreted in the urine during the first week of the illness.

When such infections are being considered urine samples should be processed under Containment Level 3.

6.7 Urinary antibiotics

Effective urinary concentrations are essential for the effective treatment of urinary tact infections. For certain infections such as pyelonephritis, adequate interstitial fluid concentrations are also required. Infection confined to the bladder in an individual with an anatomically and functionally normal system requires a shorter course of therapy than 'complicated cases' with known or suspected abnormalities.

Susceptibility data need to be interpreted with caution as many infections are treated in the community without resorting to bacterial culture; therefore samples received in the laboratory may be pre-selected and have higher resistance rates than those from patients in the community.

Key points relating to those antibiotics used to treat urinary tract infection are given below.

Trimethoprim

It is recommended that trimethoprim is not used in the first trimester of pregnancy as its antifolate activity could potentially be linked to neural tube defects (spina bifida); it is automatically suppressed in antenatal clinics in some laboratories. The evidence for this approach is not conclusive and trimethoprim is unlikely to cause harm if the patient has adequate folate reserves. Some authorities suggest that trimethoprim should not be used for enterococcal urinary tract infections - this is based on the fact that there is some reversal of the antibacterial activity of trimethoprim by thiamine and 'folates' and documented treatment failures in the literature. The European Committee on Antimicrobial Susceptibility Testing (EUCAST) recommend that enterococci are reported as of intermediate susceptibility.

Cefalexin

Cefalexin is safe to use in all stages of pregnancy. Enterococci are intrinsically resistant.

Amoxicillin

Amoxicillin achieves high concentration in the urine but its use has decreased as many strains of *E. coli* are now resistant.

CASE STUDY 6.3 Multidrug resistance

Eleven days following a hip replacement, 75-year-old Mr Smith developed fever and loin pain. A catheter urine sample grew >10^5 *E. coli*, resistant to amoxicillin, cefalexin, trimethoprim and ciprofloxacin but sensitive to nitrofurantoin.

- **Are there any other sensitivity tests you would like to perform?**
- **How can such infections be avoided?**

Co-amoxiclav

Co-amoxiclav is a combination of amoxicillin with the β-lactamase inhibitor clavulanic acid. It is too broad spectrum to be routinely used for urinary tract infection.

Nitrofurantoin

Nitrofurantoin is rapidly absorbed but produces low plasma levels and therefore despite a wide spectrum of activity is only used for treatment of urinary tract infections. It should be avoided in mild renal impairment because of inadequate urine concentrations. It is also contraindicated in children less than 3 months of age as it may cause neonatal haemolysis due to glucose-6-phosphate dehydrogenase (G6PD) deficiency, and is also contraindicated in pregnancy near to term and in the breast feeding mother.

Ciprofloxacin

Ciprofloxacin is a fluoroquinolone with high tissue concentration so, unlike many of the drugs listed above, it may be suitable for the treatment of complicated urinary tract infection. It should be avoided in pregnancy because of the risk of **arthropathy** and should be only used with caution in children.

arthropathy
Arthropathy is a disease or abnormality of a joint.

Gentamicin

Gentamicin is not absorbed in the gastrointestinal tract and must be given systemically, that is used intravenously or by intramuscular injection. It is nephrotoxic and is usually given in combination with a second agent.

Broader-spectrum agents - aztreonam, cefuroxime, carbapenems - may also be tested in resistant isolates and the laboratory needs to have a system in place for detection of extended spectrum β-lactamases. Less commonly used oral agents include fosfomycin and pivmecillinam.

Many antibiotics used to treat urinary tract infection are concentrated in the urine and because of this breakpoints that are used to determine whether an isolate is resistant or susceptible are higher for urinary isolates than for the same organism isolated from another body site.

SELF-CHECK 6.12

For a routine urinary tract infection would it be better to treat with oral or IV antibiotics? What are your reasons and which antibiotics would you suggest?

Treatment failure

Treatment failure is more likely to occur in the very young and the very old because they are more likely to have a functional or anatomical defect. This may warrant further investigation as only treatment of this defect may prevent further infections from occurring. It may also occur with empirical treatment if there is resistance to the initial agent used and therefore urine culture is mandatory in such patients.

> Key Point
>
> **Most urinary tract infections are easily treated and result in few complications. However, the organisms can get into the bloodstream and cause life-threatening septicaemia.**

Detection of antimicrobial substances

Some laboratories use methods to detect antimicrobial substances in urine; this usually consists of a susceptible organism to which you can simply look for inhibition of growth or as part of a microtitre plate assay based on a colorimetric substrate, *p*-nitrophenyl-β-D-glucopyranoside, in combination with a *Bacillus subtilis* strain. It may be useful in explaining cases of sterile pyuria, but as this should be recognizable by the clinician it is no longer performed in many laboratories. Interestingly, boric acid can interfere with this assay.

After reading this chapter you should know and understand:

- The anatomy of the urinary tract is sterile and is difficult to infect in normal healthy individuals
- Urinary tract infections are more common in females than in males
- Uropathogenic strains have specific properties which allow them to colonize the urinary tract and cause infection
- Which antibiotics accumulate in urine and are therefore commonly used to treat such infections
- The common culture media used, e.g. CLED and chromogenic agars; chromogenic media enable a better presumptive identification of common infecting strains
- The significance of isolating organisms from urinary catheters can be difficult as the catheter can be 'colonized' and there is difficulty in managing such patients
- The reason for performing microscopy, either automated or manual, and how this can affect the significance of the organism isolated
- The significance of the various cell types found in urine and the significance of these when detected on microscopy.

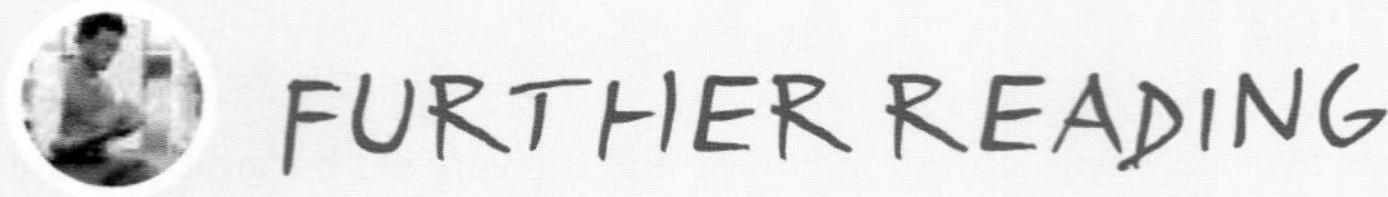

- **Bekeris LG, Jones BA, Walsh MK, Wagar EA. Urine culture contamination. A College of American Pathologists Q-probes study of 127 laboratories. *Archives of Pathology Laboratory Medicine* 2008; 132: 913–917.**

Paper investigating why urine contamination is high in some institutions and how it can be reduced.

- Kass EH. **Bacteriuria and the diagnosis of urinary tract infection.** *Archives of Internal Medicine* 1957; **100**: 709–714.

  One of the original articles on which the quantitative examination of urine is based and the rationale for it.

- NICE. Antenatal Care - Routine Care for the Healthy Pregnant Women. **NICE Clinical Guideline March 2008. http://www.nice.org.uk/nicemedia/pdf/CG062 NICEguideline.pdf.**

  Report providing guidance for the clinical management of pregnant women, with reference to the problems associated with asymptomatic bacteruria.

- Rubin RH, Shapiro ED, Andriole VT, Dais RJ, Stamm WE. **Evaluation of new anti-infective drugs for the treatment of urinary tract infection.** *Clinical Infectious Diseases* 1992; **15**: S216–S227.

  Guidelines which discuss acute uncomplicated cystitis, acute uncomplicated pyelonephritis, and complicated infections of the urinary tract.

- Shepherd M. **A revision of the microtitre tray method for urine microscopy.** *PHLS Microbiology Digest* 1997; **14**: 236–237.

  Text discussing methods for manual urine microscopy.

DISCUSSION QUESTIONS

6.1 What are the common causes of urinary tract infection? Why can the causes of UTI be different?

(a) In the community?

(b) In hospital?

6.2 What is meant by the term significant bacteriuria?

6.3 What methods can be used to screen patients for urinary tract infection?

6.4 Which antibiotics are commonly used to treat urinary tract infection?

6.5 A high proportion of urine cultures are negative - why is this?

6.6 What are the advantages and disadvantages of the different methods of urine culture?

7

Investigation of wound, tissue and genital samples

Steve Davis and Mark Tovey

The skin and mucosa form an extremely effective physical barrier to the outside environment as part of the host's immune defence system. If this integrity is breached in any way, entry of bacteria can occur, which could potentially lead to infection. The majority of organisms associated with skin, mucosal and soft tissue infections are **commensals**; these may act as opportunistic pathogens once this defence has been overcome. This may occur in a number of ways: as a result of surgery or other medical intervention, trauma, burns, inflammation or due to an underlying medical condition. Genital infections can also be a result of opportunism as well as sexual transmission.

The most common presentations of skin and tissue infections are:

- Cellulitis
- Folliculitis
- Boils and carbuncles
- Abscess formation
- Impetigo
- Ulceration
- Erysipelas
- Discharging mucosal surfaces

Wound, genital, eye, ear, nose and throat swabs are some of the most commonly received samples in the clinical laboratory. This chapter describes the laboratory investigation of these and other specimens, and examines both commonly isolated and other significant pathogens.

Swab or tissue specimens can essentially be taken from any part of the body. This chapter therefore covers an extremely broad proportion of microbiology specimens; it is not conceivable to cover every possible infection of every possible site. We have deliberately separated skin and tissue infection from that of the genitalia. These infections are very

much site-specific and can be found towards the end of the chapter. Before investigating types of infections it is important to briefly appreciate the anatomy of the skin.

Learning objectives

After studying this chapter you should be able to:

- Describe the ways in which the skin and mucosa act as a defence mechanism and how this barrier may be breached
- Describe the different types of infection that occur in skin and soft tissue and on the genitals
- List the most common pathogens and their virulence mechanisms, including toxins produced
- Outline how these toxins are detected and the role of the relevant reference laboratory
- Describe the media and incubation conditions used to isolate these organisms

7.1 The skin and mucosa

Overview of the skin

The skin is the largest organ of the body. It consists of an external cellular layer, the epidermis, and a deep connective tissue layer called the dermis (Figure 7.1). The skin is critical in the body's defences against injury and dehydration, as well as against foreign bodies. It is equally important in sensation (e.g. pain) and the regulation of temperature.

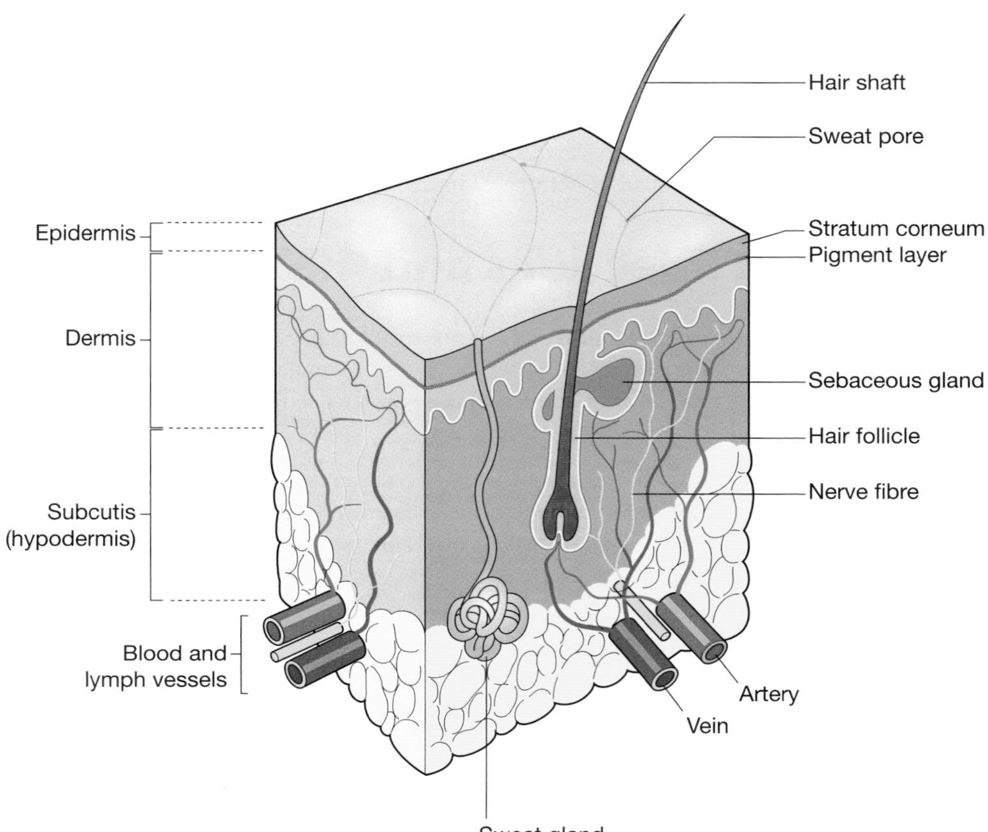

FIGURE 7.1
Diagram of the skin.

Below the dermis is the **superficial fascia** (subcutis) consisting of looser connective tissue and fat. It is here you would find cutaneous nerves, blood and lymphatic vessels and sweat glands. Finally, the deep fascia is a denser, more organized section of connective tissue that gives rise to structures such as muscle.

The skin acts as a primary defence system against unwanted bacteria or viruses. Physical trauma, such as burns, injury or surgery can breach the integrity of the skin and may create an infection opportunity. Mucosal membranes, including eyes, nostrils and ears, are often easier to breach as they usually do not have skin as the primary defence mechanism.

Key Point

The skin is an excellent barrier to microorganisms and provided it remains intact it will act as a 'brick wall' to prevent infection.

Normal flora of the skin

commensal organisms
Defined as those that live on a surface to the benefit of both host and organism. They constitute our 'normal flora'.

Few areas of the body are sterile. The skin is exposed to the environment and so is colonized with **commensal organisms.** These are often transient - they may be temporarily removed, by washing for example, but will usually return by natural means. Common skin commensals include coagulase-negative staphylococci (CNS), viridans-type streptococci, *Corynebacterium* spp. and small numbers of *Staphylococcus aureus* and *Candida* spp.

pathogenic organisms
Although possibly part of our normal flora, their main role is to survive and multiply, often at the expense of the host. Such organisms possess more mechanisms by which to initiate infection.

Under normal circumstances we live in close harmony with these organisms, but as soon as the skin is damaged or our immune system is compromised in some way, the risk of infection increases. The normal flora may also be disrupted by hospitalization or by taking courses of antibiotics. Since most commensals and most **pathogenic organisms** are relatively sensitive to treatment, antimicrobials often select resistant bacteria by removing bacteria susceptible to the antibiotic. This is especially so in the hospital setting, which may lead to hospital-acquired infection, for example methicillin-resistant *S. aureus* (MRSA).

7.2 Skin infections and their causes

Infections of the skin and/or subcutaneous tissue may be caused by a number of organisms. Traditionally, swabs are taken of suspected body sites, although tissue or biopsy specimens may often result in more efficient recovery of these bacteria. It is important that all organisms isolated from a particular swab (e.g. a wound swab) are assessed for clinical significance. Skin flora are often isolated along with pathogenic bacteria, and the qualified biomedical scientist is expected to know what should be investigated as a likely cause of infection and what should be regarded as insignificant.

Skin infections

The following are a few examples of different forms of skin infection.

Cellulitis

Cellulitis is a spreading skin infection often involving the superficial fascia, but may be caused by damage to venous or lymphatic systems, resulting in swelling (oedema), redness (erythema), tenderness and pain. Common causes of infection include *S. aureus* and β-haemolytic streptococci (β-HS). These are shown side by side in Figure 7.2a and b.

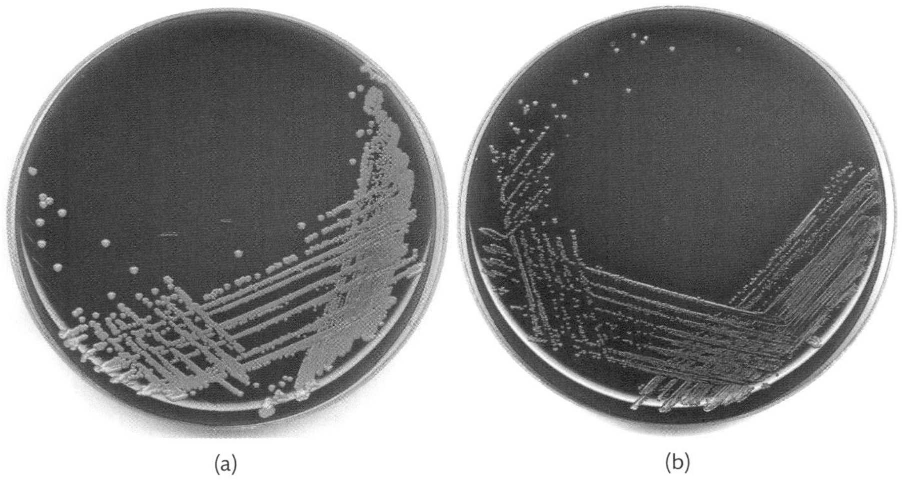

FIGURE 7.2
(a) *Staphylococcus aureus* and (b) β-haemolytic *Streptococcus* on Columbia blood agar.

Cellulitis due to wounds may also be infected by *Bacteroides* spp. and/or anaerobic cocci. Due primarily to the size of the wound, burns may result in severe skin infection involving one or more of the following: *S. aureus*, β-HS, *Pseudomonas aeruginosa*, Enterobacteriaceae, *Candida*, *Acinetobacter* and *Bacillus* spp. (especially *Bacillus cereus*).

Bites are a common focus of infection; many bacteria have been implicated in human or animal bites, where organisms of an oral origin infect the wound (e.g. *S. aureus*, viridans streptococci and anaerobes). Similarly, the **HACEK** group of organisms is often implicated in human bites. Recognition of these organisms can be difficult due to their colonial similarity to other organisms often present in the sample. Their identification is also troublesome, mainly due to their fastidious nature and infrequent detection. *Pasteurella multocida*, *Capnocytophaga* spp (dog) and *Streptobacillus* spp (rat) have been specifically implicated in animal bites. Insect bites may also result in bacterial wound infection, but these are usually secondary infections with *S. aureus* and β-HS.

HACEK
A group of organisms, namely *Haemophilus* spp, *Aggregatibacter actinomycetemcomitans* (previously named *Actinobacillus*), *Cardiobacterium hominis*, *Eikenella corrodens* and *Kingella* spp. whose initial letters form the abbreviation. They are CO_2 dependent and initial infections can lead to complications such as endocarditis.

Folliculitis

This occurs when a hair follicle becomes infected and inflamed. An erythematous pustule forms around the hair, often caused by *S. aureus*. Outbreaks of *P. aeruginosa* folliculitis have been traced back to swimming pools, while *Candida* species have been implicated in patients taking long-term antibiotics.

Boils and carbuncles

Boils are a localized area of inflammation caused by an infection under the skin, in a gland, or in a hair shaft, whereas carbuncles are deep subcutaneous abscesses, larger than a boil, usually with one or more openings draining pus onto the skin. *S. aureus* is the most common pathogen.

Abscesses

Abscesses are formed when pus starts to accumulate in the tissues. These may be superficial or of a deep-seated nature, associated with a particular organ. Essentially, any organism isolated may be of significance; however, common causes include *S. aureus* and anaerobes. For example:

- Brain abscesses may be caused by anaerobes, '*Streptococcus anginosus*' group, *S. aureus*, *Streptococcus pneumoniae*, β-HS and, more rarely, *Haemophilus* and *Nocardia* species.

- Breast abscesses are most commonly caused by *S. aureus*, anaerobes, *Pseudomonas* and *Proteus* species and often result in nipple discharge, oedema and erythema.
- Dental abscesses are occasionally caused by the normal oral flora (e.g. viridans streptococci) but facultative or strict anaerobes are more common causes. '*S. anginosus*' group and occasionally *Actinomyces* species may also be responsible.
- Liver abscesses may be bacterial or amoebic in origin. They are often life threatening and require urgent drainage and/or antibiotic treatment. They are commonly caused by enteric organisms, e.g. Enterobacteriaceae, anaerobes, *P. aeruginosa* and *Enterococcus* spp.

Impetigo

Impetigo is a superficial skin infection that results in swelling and pustules, often forming crusty sores. They are common around the mouth and are often associated with *S. aureus* or β-haemolytic streptococci.

Ulceration

Ulcers often form when circulation is compromised. This may be due to some form of vascular disease, neuropathic changes or as a result of too much pressure on already weakened skin (e.g. bedsores in hospitalized older patients). Chronic leg ulceration due to poor venous circulation is most commonly encountered. Ulcers are often longstanding and seldom sterile. It is thought that the presence of some organisms may be beneficial as they produce proteinases that may help break down necrotic tissue, but too many organisms may produce excessive proteinases and hinder the healing process, as any new tissue is also broken down. Organisms implicated in ulcer infections traditionally include *S. aureus*, β-HS, anaerobes and *P. aeruginosa*. Water-related infections brought about by trauma during swimming or boating have given rise to rarer causes such as *Aeromonas* and non-cholera *Vibrio* species (e.g. *Vibrio parahaemolyticus* and *Vibrio vulnificus*).

Erysipelas

This is a superficial skin infection affecting mainly the dermis and the upper reaches of subcutaneous tissue. It largely involves the lymphatics within these regions, and results in marked swelling, often with small vesicles and a definite border. It is usually caused by β-HS Lancefield Group A, although Group G *Streptococcus* and *S. aureus* have also been implicated.

SELF-CHECK 7.1

Some organisms have been mentioned a lot in the above list of common skin infections, e.g. *S. aureus*. Why do you think this is so?

Erysipeloid

Confusingly, the genus *Erysipelothrix* occasionally causes erysipeloid, and not erysipelas, in humans. The organism is a short, slim, Gram-positive non-sporing rod and *Erysipelothrix rhusiopathiae* is the only species known to cause erysipeloid. In 1886, however, Loffler had established this genus as the cause of 'swine erysipelas' and subsequently *Erysipelothrix* species have been proven to be of major economic importance, as they cause septicaemia and arthritis in swine, calves, lambs, turkeys and other farm animals.

Necrotizing fasciitis

Necrotizing fasciitis is a rare condition affecting the lower reaches of epidermis and subcutaneous tissue. Toxin production results in extensive skin and muscle destruction and necrosis, often as a complication of trauma or surgery. It is often caused by Group A streptococci, *V. vulnificus* and *Clostridium perfringens*, the so-called flesh-eating bacteria, but also heavy mixed populations, including several different anaerobes, can give similar effects.

Gangrene

Gangrene refers to the decay and death (necrosis) of tissue brought about by poor blood supply. It is commonly associated with trauma (e.g. gunshot wounds) and a common causative organism is *C. perfringens*. This bacterium and its spores are naturally found in the human and animal gut and in soil. After trauma, the oxygen potential at this site drops due to interruption to the circulation. If spores of *C. perfringens* contaminate this site they are able to become vegetative bacteria in the oxygen-free environment. *C. perfringens* then begins to produce toxin and initiates gangrene. The organism also produces gas as a by product, causing the skin to 'bubble up', sometimes accompanied by a 'creaking' sound under the skin (**crepitus**). This collectively is known as **gas gangrene**. If treated early enough, the outcome is usually favourable. However, where treatment is delayed (e.g. in war situations) severely affected tissue and/or limbs may need to be amputated to prevent death.

SELF-CHECK 7.2

C. perfringens is a strict anaerobe. Why would this organism be involved in gangrene?

Key Point

Remember how important clinical details are - you may look for different organisms if the patient has been bitten, or has been swimming abroad, etc. You may not receive such details and the biomedical scientist must consider every possibility when reading culture plates!

7.3 Mucosal infections

The mucosa or mucous membranes line various parts of the body that are either exposed to the environment or internal organs. They are primarily involved in either absorption or secretion, and include the eyes, ears, nostrils and throat. Generally the sites are protected by the bacterial flora present at the site; the exception being the eyes, which are usually bacteria free due to the production of **lysosyme** and local IgA.

Eye infections

A wide range of microorganisms are capable of causing eye infections. Although it may be common to isolate contaminating skin flora from eye swabs, every organism should be assessed for significance before being ruled out as a causative agent.

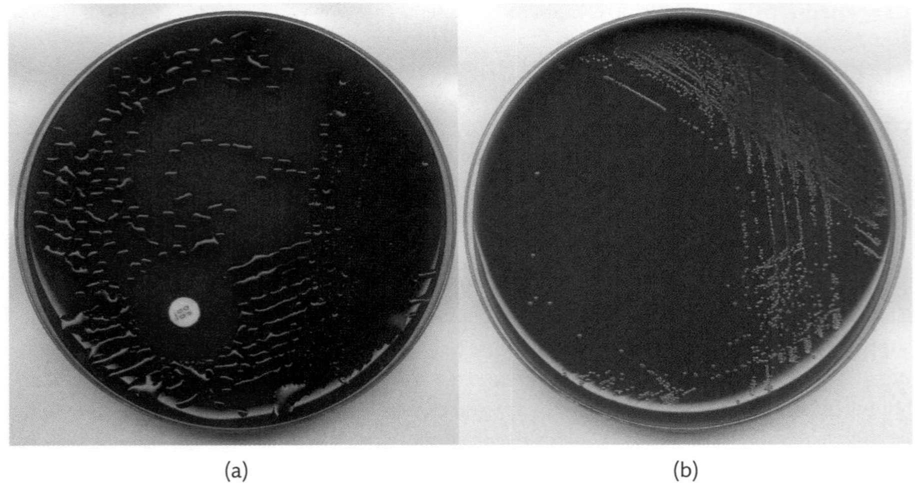

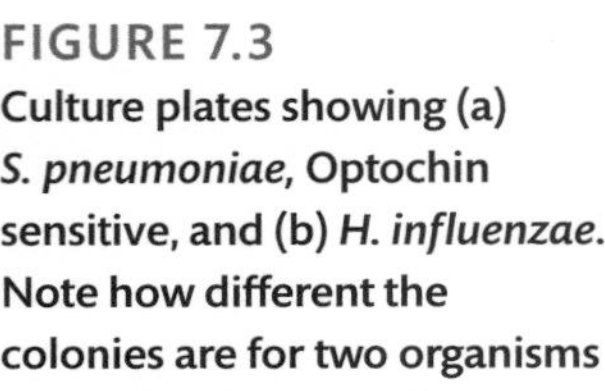

FIGURE 7.3
Culture plates showing (a) *S. pneumoniae*, Optochin sensitive, and (b) *H. influenzae*. Note how different the colonies are for two organisms that infect the same site.

Conjunctivitis - inflammation of the conjunctiva - is the most common ocular infection encountered and often results in red or 'sticky' eyes. Common causes include: *S. aureus*, *S. pneumoniae* and *Haemophilus influenzae*. Examples are shown in Figure 7.3a and 7.3b. Other causes include β-HS, *Moraxella catarrhalis*, *Neisseria gonorrhoeae*, *P. aeruginosa* and the Enterobacteriaceae.

Ophthalmia neonatorum - conjunctivitis of the newborn - is primarily caused by vertical transmission of *N. gonorrhoeae* or *Chlamydia trachomatis* from the mother at birth. If left untreated infection can result in blindness. In addition to the above causes, other pathogens include Group B streptococci. Adult conjunctivitis caused by *N. gonorrhoeae* is usually a result of concomitant genital infection, although it is also a risk to laboratory staff as it may be laboratory acquired.

***Chlamydia* and viral conjunctivitis** - trachoma can be caused by serotypes A–C of *C. trachomatis* whilst inclusion conjunctivitis may be caused by serotypes D–K. The former serotypes are more associated with rural communities in under-developed areas, whilst the latter are more likely to be urban and associated with sexual transmission. Viral causes of conjunctivitis are predominantly adenoviruses.

Keratitis - inflammation of the cornea - may be caused by the protozoan *Acanthamoeba* species. They may be detected in corneal scrapings and are often associated with incorrect use and storage of contact lenses.

Orbital cellulitis - infection of orbital tissue usually as a consequence of trauma, either accidental or through surgical procedure and may lead to blindness or thrombosis. Tissue aspirates tend to be more appropriate specimens than eye swabs. Typically, *S. aureus*, β-HS and anaerobes are implicated, whilst *H. influenzae* may still be isolated from children, although the use of the Hib vaccine has led to a fall in the prevalence of the type b strain.

Ear infections

Ear infections may be subdivided into otitis externa (external ear infection) and otitis media (middle ear infection).

Otitis externa may be caused by a similar range of organisms to those affecting the skin and soft tissues, with *P. aeruginosa* and *S. aureus* predominating. Group A streptococci, anaerobes, Enterobacteriaceae and fungi (e.g. *Aspergillus* species) are also common.

Otitis media is diagnosed in symptomatic patients presenting with signs of acute illness along with the presence of fluid in the middle ear. Extreme untreated cases have led to deafness. It often occurs when organisms from the upper respiratory tract migrate along the Eustachian tube to reside in the middle ear. The main offenders are *S. pneumoniae*, *M. catarrhalis* and *H. influenzae*. Other causes include *S. aureus*, β-HS and Gram-negative bacilli.

Nasal swabs

Although between 15 and 40% of us carry *S. aureus* in our nostrils, nasal carriage markedly increases the risk of infection at other body sites, for example post surgery. It may also be associated with recurrent skin infections. Depending on the individual's circumstances, eradication of the organism from the nostrils may be required. Currently mupirocin nasal drops are extremely successful at eradicating carriage, but occasionally the use of systemic or other topical antibiotics is required.

Nasal swabs may also be employed to screen for certain 'target' organisms, especially MRSA screening on new hospital admissions and Group A streptococci on burns patients.

Throat infections

Sore throats are often of viral origin, are mild and often get better without the need for specific treatment. This is one area where general practitioners were criticized by the government, as they were accused of helping cause the bacterial resistance problems by giving patients unnecessary antibiotics.

Severe sore throat or tonsillitis, however, will normally require appropriate antimicrobial therapy.

Pharyngitis – the most common cause of pharyngitis is the Group A *Streptococcus*. It can be spread by casual, close or direct contact with infected individuals through respiratory droplets, and several outbreaks have been noted. The incubation period is between 2 and 5 days, and symptoms include sore throat, fever, headache, pyrexia, and occasionally nausea and vomiting. Group C and G streptococci can also cause pharyngitis.

Diphtheria – an acute upper respiratory tract disease caused by *Corynebacterium diphtheriae*. The organism is capable of producing a toxin (see later) which can damage the epithelium of the pharynx and form a pseudomembrane. Death can ultimately occur due to obstruction of the airways. Screening throat swabs for *C. diphtheriae* is recommended in patients with the following risk factors:

1. Membranous pharyngitis/tonsillitis
2. Patients who have travelled abroad recently, particularly to Russia, former Soviet states, Eastern Europe, Africa, South America and South East Asia
3. Contacts of the above

Epiglottitis – an infection mostly associated with young children and the organism *H. influenzae*. The introduction of the Hib vaccine in 1992 has led to a fall in the number of cases; however, the organism should always be considered, especially in those who have not been vaccinated. In uncomplicated sore throats, the organism is not routinely investigated. However, this highlights the importance of clinical details and should epiglottitis be suggested, additional culture plates need to be inoculated.

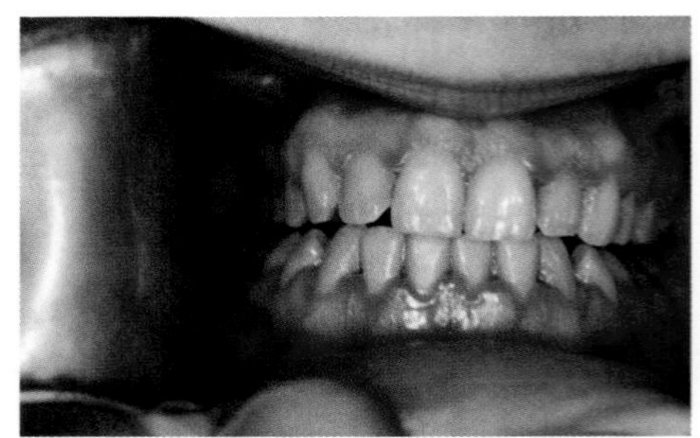

FIGURE 7.4
Image of a patient with Vincent's angina. Note the pronounced gum disease which if untreated is likely to lead to ulceration. (Courtesy of CDC/Minnesota Department of Health, R.N. Barr Library; Librarians Melissa Rethlefsen and Marie Jones.)

Vincent's angina - an infection associated with poor oral hygiene or systemic disease and characterized by pharyngeal or gum ulceration. Also known as trench mouth as many WWI soldiers probably suffered from the disease due to very poor oral hygene (Figure 7.4). Laboratory detection is by direct microscopy and is confirmed by the presence of white cells, cigar-shaped fusiforms (*Fusobacterium* spp.) and spirochaetes (*Borrelia vincentii*).

Fusobacterium necrophorum - this anaerobic organism is now recognized as a cause of persistent sore throat, particularly in young adults, and may occasionally lead to **Lemierre's disease,** if undiagnosed. However, routine investigation is difficult as differentiation from anaerobic flora usually present in the mouth is troublesome and time-consuming.

Lemierre's disease

A disease mainly caused by *Fusobacterium necrophorum* but other fusobacteria can be involved. It mainly affects young adults. Lemierre's syndrome follows a streptococcal throat infection which has led to a peritonsillar abscess. *Fusobacterium necrophorum* can flourish in this abscess and can penetrate into the neighbouring jugular vein. Bacteria can then disseminate into the blood stream. It is associated with a high mortality rate.

- Less common causes of pharyngitis include: *Arcanobacterium haemolyticum*, non-toxigenic *C. diphtheriae*, *N. gonorrhoeae*, *S. aureus* and fungal infections.
- Throat swabs may also be taken to check for carriage of *Neisseria meningitidis* from contacts of meningococcal meningitis.

SELF-CHECK 7.3

Why do you think that GPs don't treat many throat infections with antibiotics? What would be the problem if they did?

7.4 Organisms, virulence factors and toxin production

In this section we will briefly look at some of the virulence factors and toxins produced by common pathogens discussed in this chapter. We will also consider the growing role reference laboratories play in the investigation of some organisms.

Staphylococcus aureus

As we have seen, *S. aureus* is capable of causing a wide range of infections. Its high level of invasiveness and pathogenicity is predominantly due to the ability to produce a wide range of extracellular enzymes and toxins:

- Coagulase - clots plasma, assists in spread throughout tissues and can interrupt phagocytosis
- Lipase - digests fat

CASE STUDY 7.1 An outbreak of S. aureus

In 2007, an outbreak occurred in South Yorkshire of boils and skin infections in children. It initially centred around one school. It was noted that *S. aureus* with a distinctive antibiotic profile was isolated from several of the children. When the isolates were forwarded to the reference laboratory it was found that most were producing the Panton-Valentine leukocidin toxin. Although this was not a particularly aggressive strain, this community-acquired outbreak of methicillin sensitive *S. aureus* (CA-MSSA) spread rapidly throughout the school and affected a large number of children. The symptomatic children and their close family plus the majority of pupils in the school were screened for nasal carriage of *S. aureus*. Fortunately, the organism had a characteristic antibiotic profile (resistant to penicillin, gentamicin, tobramycin and trimethoprim but sensitive to most other Gram-positive antibiotics, including methicillin and erythromycin). After receiving the initial results, it was possible to predict which families were carrying the toxigenic strain in their nostrils and hence needed to have mupirocin treatment. Several months after the onset of this outbreak, sporadic cases still occurred, indicating the difficulty in eradicating community-based outbreaks.

- Haemolysin - destroys red blood cells
- Hyaluronidase - assists in spread of the organism in tissues by breaking down hyaluronic acid, a key component of connective tissue
- Staphylokinase - helps induce fibrinolysis (affects clotting)
- Protein A - affects complement activation
- Toxic shock syndrome toxin - causes rash, shock and shedding of the skin
- Enterotoxins - associated with food poisoning
- Exfoliative toxin - causes skin loss
- Leukocidin - destroys white blood cells
- Panton-Valentine leukocidin (PVL)—a particularly aggressive toxin produced by *S. aureus* that is capable of forming pores in cell membranes and has been associated with necrotizing pneumonia.

Significant isolates, as well as organisms isolated from outbreaks, are often sent to reference laboratories to monitor local and national prevalence of particular strains.

Group A streptococci (*Streptococcus pyogenes*)

Infections caused by this pathogen range from sore throats to necrotizing fasciitis. The organism exists as Gram-positive cocci, typically in chains. As with *S. aureus*, a wide range of enzymes and toxins are produced. For example:

- Streptolysins - both types are responsible for digestion of blood and cause haemolysis of red blood cells, as seen on blood-containing agar plates. Streptolysin O is inactivated by oxygen, is antigenic and induces production of antistreptolysin O (ASO) antibody, whereas streptolysin S is oxygen stable and non-antigenic.
- Hyaluronidase - assists in spread of the organism in tissues by breaking down hyaluronic acid, a key component of connective tissue.
- Leukocidin - as previous.
- Streptokinase - helps induce fibrinolysis (affects clotting).

- M proteins - virulence factors associated with colonization and resistance to phagocytosis. More than 50 have been identified in different strains of Group A streptococci according to antigenic structure. R and T proteins are not involved in virulence, but play a key role in epidemiological typing.

adhesins
Adhesins appear to mediate bacterial adherence to host epithelial cells, thereby assisting in pathogenesis.

- Lipoteichoic acid - one of several **adhesins** that lends itself to colonization and ultimately infection at a given site, e.g. the pharynx.
- Streptococcal pyrogenic exotoxin (SPE) - formerly known as erythrogenic toxin, it is responsible for the rash seen with scarlet fever and occurs when the toxin disseminates in the blood. It is also associated with a streptococcal form of toxic shock syndrome, as well as necrotizing fasciitis.

SELF-CHECK 7.4

Just look how many toxins Group A haemolytic streptococci produce! Think about the damage done to the host upon an infection with this organism. Why is it so significant?

Group A streptococci isolated from systemic infections (from positive blood cultures) are sent to reference laboratories. In this way surveillance of infection and changes in trends are analysed both locally and nationally. From an epidemiological standpoint, this keeps the laboratory aware of circulating strains and may highlight a previously unrecognized outbreak.

Pseudomonas aeruginosa

opportunistic pathogens
Opportunistic pathogens exploit weaknesses in host defences to create infections.

biofilms
Biofilms are complex masses of mono- or polymicrobial communities that are capable of secreting extracellular substances, producing a complex and sometimes highly organized glycocalyx structure within which they are embedded. This creates a protective layer that the organisms use to survive hostile environmental conditions or, in clinical situations, guard against host defence mechanisms, such as lymphocytes. They also prevent antibiotics from reaching their target sites, the most obvious example being the mucoid *P. aeruginosa* strains often found in the lungs of cystic fibrosis patients and the difficulty encountered in trying to eradicate the organism. In the microbiological diagnosis of infection, this biofilm may have to be disrupted in order to culture the responsible organisms.

P. aeruginosa is a Gram-negative rod, found in the human intestinal tract, water, soil and sewage and moist environments, such as sinks. It is an **opportunistic pathogen** of humans and infections are often hospital-acquired. It rarely affects healthy tissue, but can produce damage in tissue that is in poor condition or compromised in some way.

Virulence factors include:

- Fimbrae - are used to adhere to epithelial cells of the host, especially those of the upper respiratory tract of hospitalized patients.
- Alginate - production of this slimy substance, consisting of polysaccharide, also assists with respiratory tract adhesion. It enables the organism to form the basis (matrix) of the **biofilm** that *Pseudomonas* species create to survive within their environment.
- Elastase - is capable of cleaving collagen, immunoglobulin and complement. It can interfere with the epithelial lining of the respiratory tract and affects ciliary operation.
- Cytotoxin - appears to be toxic to most eukaryotic cells.
- Phospholipase and lecithinase - act against lipid and lecithin within cells.
- Pyocyanin - the blue pigment produced by many *P. aeruginosa* which is a redox-active phenazine compound that kills both mammalian and bacterial cells alike by the generation of reactive oxygen intermediates. It has been linked to reduced bacterial clearance from the lungs of immunocompromised patients.

Clostridium perfringens

Clostridium perfringens is an anaerobic Gram-positive rod that resides in the human gut and the environment. *Clostridium* spp. are capable of producing spores which allow the organism to survive in a dormant state until favourable conditions return.

Virulence factors include:

- α-Toxin (phospholipase C) is one of the toxins produced in cases of gas gangrene, resulting in rapid deterioration of tissue carbohydrate and gas production in necrotic tissue, especially muscles. Alpha toxin is the main cause of cellular breakdown and toxin spread, and is produced by *C. perfringens* type A1.
- Enterotoxin is produced by *C. perfringens* type A2 and causes self-limiting food-poisoning within 8–12 h of eating contaminated meat.
- β-Toxin causes a severe jejunitis known as **pigbel**. It is produced by *C. perfringens* type C, usually as a result of eating undercooked pig meat. Symptoms include abdominal pain, dysentery and vomiting. It is rare in the developed world, but has been recorded in China, Bangladesh, Papua New Guinea and parts of East Africa.

Corynebacterium diphtheriae

The organism exists as a facultative Gram-positive rod, typically seen in palisades of 'Chinese letter' formation on Gram stain. Whilst uncommon in the UK, laboratories need to be aware that diphtheria is still present in several countries and consequently the need to maintain the capability of detecting *C. diphtheriae* remains, in terms of both experience and relevant media. There are four biotypes of *C. diphtheriae*, namely *C. diphtheriae* var. *gravis*, *intermedius*, *belfanti* and *mitis*. As mentioned previously, both toxigenic and non-toxigenic strains can be found and the toxin is bacteriophage induced. Classical symptoms of diphtheria are presented only when toxin is produced and as a rule var. *gravis* is more pathogenic as it is known to produce the most exotoxin.

The pathogenesis of nasal, nasopharyngeal and tonsillar diphtheria involves the production of this powerful and rapidly fatal toxin, which then disseminates into the bloodstream through the damaged mucous membranes of the upper respiratory tract. The patient requires urgent antitoxin to prevent potentially fatal cardiac and neural complications. In addition, an acute inflammatory response leads to the formation of the classical grey-yellow pseudomembrane at the site of infection. This can obstruct the larynx, block the passage of air and may cause death by asphyxiation.

All isolates of *C. diphtheriae* should be sent to a reference laboratory for confirmation and toxin testing. This service is available 365 days a year in the UK and is traditionally performed using the ELEK plate (Figure 7.5) although molecular detection is currently being investigated.

Whilst many strains of *C. diphtheriae* prove to be non-toxigenic, increased travel to Eastern Europe and recent influxes of migrant workers from these regions increase the chances of isolating toxigenic forms. Some strains of *Corynebacterium ulcerans* and *Corynebacterium pseudotuberculosis* are also known to produce diphtheria toxin.

Mycobacteria

Mycobacteria are unusual causes of skin infections but should be considered for certain case histories. For example, **fish tank granuloma** is particularly associated with keeping or handling tropical fish and may also be found in swimming pools. The condition is caused by *Mycobacterium marinum* giving rise to a red, scaly rash, often on the hands and arms.

Mycobacterium ulcerans should be investigated if the tropical '**Buruli ulcer**' is suspected. The organism is believed to enter the skin through cuts and abrasions. An ulcer develops,

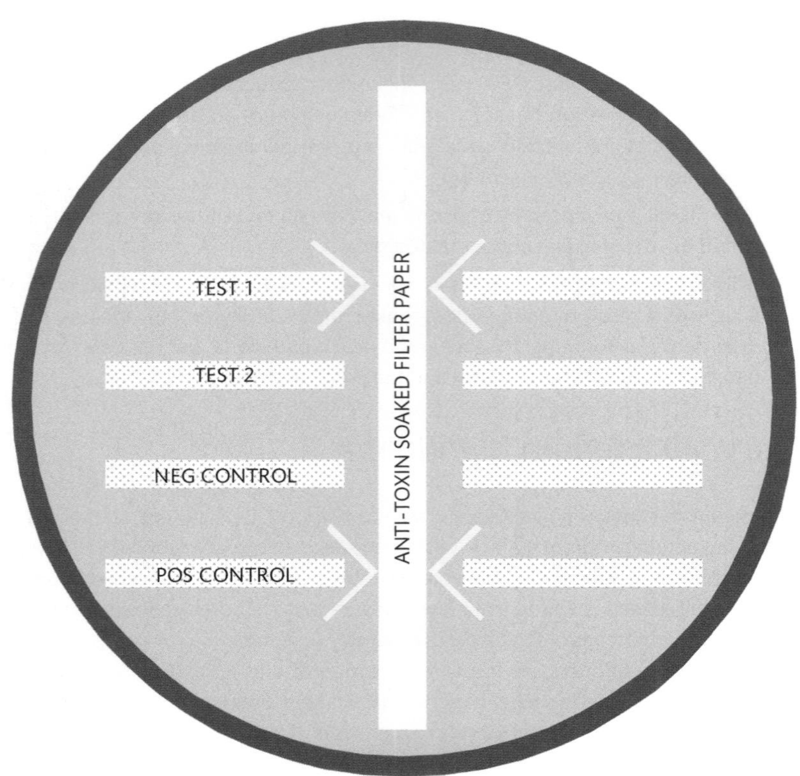

FIGURE 7.5
Image of an ELEK plate. Test strain 1 is toxigenic and is producing precipitation between the strain and the diphtheria antitoxin soaked filter paper (similar to the positive control). Test strain 2 is toxin negative (no precipitation).

followed by necrosis and ultimately skin and bone destruction due to toxic events. Patients often undergo extensive surgery to remove the lesion.

7.5 Sample processing

Swab or tissue specimen may be collected from virtually any part of the body so it is virtually impossible to describe how each specimen is processed individually. However, it is important when processing these specimens that the scientist has some knowledge of the likely pathogens that may be encountered from these samples. This will help you decide which media should be selected.

Microscopy techniques

For most wound and mucosal swabs, the use of microscopy is unnecessary due to the large number of commensal organisms often present, resulting in the possibility of misleading information. Gram stains of pus samples or sterile fluids are less likely to be contaminated with skin flora and can give meaningful results.

Cross reference
Chapter 9.

The presence of white blood cells along with organisms may be suggestive of an inflammatory response to the organism in question. Reporting this finding to a clinical microbiologist may give some immediate bearing on the treatment and management of the patient. Infections of this nature are rarely polymicrobial; hence you would expect to find only one type of organism

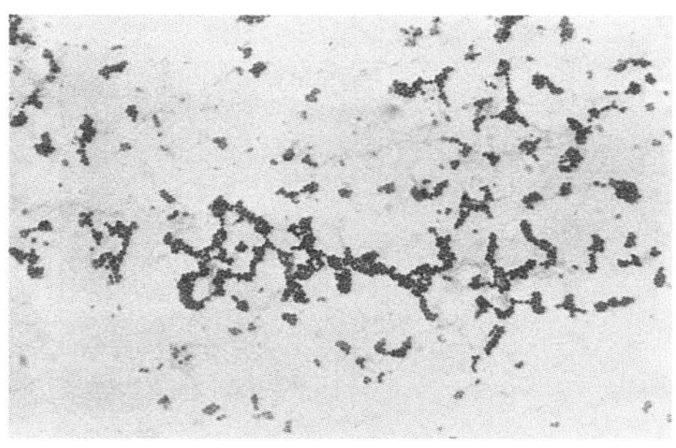

FIGURE 7.6
Photograph of Gram stain of staphylococci. Remember that whilst this is suggestive of a pathogenic *S. aureus* infection, you can't tell from the Gram stain which staphylococcal species it is. (Courtesy of CDC/Dr Richard Facklam.)

present, for example Gram-positive cocci (Figure 7.6). Biomedical scientists should be wary of mixtures of different bacteria in normally sterile samples, as this is often the type of picture you may see from specimens such as drain fluids contaminated with gastrointestinal flora.

Gram stains from samples of connective tissue may also be of clinical benefit, but it is important to remember these specimens may be easily contaminated by normal skin flora and so results may be misleading. Tissues such as heart valves should be sterile and so should always have a Gram stain performed on them, due to the urgency of the request.

SELF-CHECK 7.5

A Gram stain of a wound swab can give misleading information back to the clinicians. Why could this be?

Tissue samples from abscesses, particularly facial and dental abscesses, should be examined for 'sulphur granules' which may demonstrate the presence of *Actinomyces* species. The granules are clumps of organisms and should be crushed between two slides and Gram stained. *Actinomyces* species are branching Gram-positive bacilli.

As previously described, Vincent's stain is required to diagnose Vincent's angina and AFB staining techniques (either Auramine or Ziehl–Neelsen) are required to demonstrate presence of *Mycobacterium* species.

Cross reference
Chapter 9, Section 9.4.

Key Point

Sterile fluids, pus and tissue specimens should have microscopy performed and the presence of bacteria should be reported to clinicians. Transport swab specimens are less appropriate for Gram staining, because both the swabbed site is more likely to contain normal flora and the transport media used (usually charcoal) may conceal the presence of organisms.

Common isolation media

Most of the bacteria and fungi associated with skin and soft tissue infections can be isolated using routine cultural techniques. Whilst the precise selection of media used may vary between laboratories nationally, they will all tend to reflect the organisms that are likely to be encountered. Broadly speaking, selective and non-selective media are used in combination. Specific selective plates, however, may be employed for particular target organisms.

Cross reference
Chapter 3: Culture media.

In general, the following types of media are routinely employed:

- Blood agar
- Chocolate agar
- McConkey or CLED agar
- Specific anaerobic agar with metronidazole disc added
- Selective media relevant to isolation of specific pathogens, e.g. aztreonam agar for *S. aureus*.

Blood and chocolate plates should be incubated in an environment enriched with 5–10% CO_2. Selective media containing aztreonam or nalidixic acid are often used. These media suppress the growth of most colonizing Gram-negative organisms, allowing for the selection of Gram-positive organisms, such as staphylococci and streptococci.

Normally sterile fluid specimens that may yield fastidious organisms, such as *Haemophilus influenzae*, should be cultured to chocolate agar. These samples include ear and eye swabs, as well as knee aspirates.

Specimens that may yield *Candida* spp. or other fungi, such as *Aspergillus* spp., should be cultured to Sabouraud dextrose agar. These samples include ear swabs, mouth swabs, where oral thrush is suspected, and swabs from moist or sweaty sites, for example feet.

Culturing for anaerobic bacteria requires an anaerobic environment. This is usually achieved by employing an anaerobic incubator or jars. Metronidazole discs (5 μg) are usually added to the primary inoculum, and the presence of a zone of inhibition after incubation indicates the presence of anaerobes, although growth will often require prolonged incubation.

Key Point

By definition, anaerobic bacteria will not grow in the presence of oxygen. Fortunately, in the UK, the vast majority are still sensitive to metronidazole and hence metronidazole serves both a diagnostic and therapeutic role. NB Exceptions include *Actinomyces* and *Propionibacterium* species which are intrinsically resistant to metronidazole.

Throat swabs are routinely inoculated onto blood agar plates and are often incubated anaerobically to isolate any β-haemolytic streptococci that may be present. If diphtheria is a possibility, the specimen should also be inoculated onto a Hoyle's plate (in CO_2) which contains potassium tellurite and is selective for *C. diphtheriae*.

SELF-CHECK 7.6

Why are the different media you use to isolate pathogens incubated under different atmospheric conditions?

7.6 Orthopaedic samples

Orthopaedic tissue samples are given a special mention due to the unique nature of the specimens. The costs to both the NHS and the patient of infected **prostheses** are huge. Consequently, antibiotic cocktails are incorporated into the orthopaedic cement in an attempt to prevent infection. Hip replacements were pioneered by Sir John Charnley in the early 1960s and since then joint replacement (**arthroplasty**) has become commonplace. Arthroplasty is most commonly used for patients with osteoarthritis or inflammatory abnormality such as rheumatoid arthritis, with hip and knee replacements predominating. Modern surgical and anaesthetic techniques, appropriate patient selection, modified prosthesis design, prophylactic antibiotics, laminar airflow in operating theatres and good post-operative care have reduced infection rates substantially, but there is still a finite risk associated with each procedure.

Infecting organisms may gain entry during primary implantation surgery or through the **haematogenous** (bloodstream) route and may establish either acute or chronic infection. The presence of a foreign body (the prosthetic device) in the joint creates a focus for invasion and means fewer organisms are required to establish infection.

The most common organism to cause acute infections is *S. aureus* and in chronic infections either *S. aureus* or coagulase-negative staphylococci (CNS) predominate. Many other organisms, including other skin flora, streptococci, coliforms, enterococci and rarely anaerobes, mycobacteria or fungi, may also be implicated. This means that any organism cultured from a sample associated with a prosthetic joint or other orthopaedic device could be significant. Consequently, multiple samples should be taken including several tissue samples, bone, fluid samples and swabs of the 'infected' site.

Once infection is established around a prosthetic joint, organisms may form a biofilm. Similarly, 'persisters' within the biofilm are very difficult to treat and removal of the prosthesis is often necessary.

SELF-CHECK 7.7

Why are organisms in a biofilm so difficult to eradicate?

Orthopaedic sample processing

The likely presence of antibiotics, either administered directly to the patient or in the cement used to support the replacement joint, indicates that both direct *and* broth enrichment culture is important. The obvious disadvantage of broth enrichment is the increased possibility of contamination. Additionally, as many chronic infections are due to skin flora, any organism cultured may be relevant, making differentiating infection from contamination difficult, especially when CNS are isolated. To aid this process, at least four to five tissue samples per site should be taken and processed individually. Under such circumstances the specificity of isolating CNS from two or more samples has been shown to be 97%.

Exclusion of contaminants during operative and laboratory processing is critical and strict aseptic technique and the use of a Class II cabinet is advocated. Subculture is normally onto standard media, although some workers recommend the use of chocolate agar, as this may help detect small-colony variants of staphylococci that may be isolated from deep samples as a consequence of antimicrobial therapy. Vancomycin, gentamicin and tobramycin are the antibiotics that are generally incorporated into orthopaedic cement.

SELF-CHECK 7.8

Why would you need to use enrichment broth for culture orthopaedic samples when a direct culture is performed?

7.7 Genital tract and associated specimens

As with the skin, it is important to be able to distinguish between commensal and pathogenic organisms of the genital tract. Genital commensals can vary both by age and between the sexes, hence each patient's specimen needs to considered individually and the microbiology assessed accordingly.

Normal flora of the genital tract

Due to the external nature of the penis, male genital commensals are similar to that of the skin. They consist of CNS, viridans-type streptococci, *Corynebacterium* species and small numbers of *Staphylococcus aureus* and *Candida* species. Internally, however, the male genital tract is largely sterile; this is maintained by prostate secretions and the mechanical action of urination.

Normal flora of the female genital tract is more variable and is strongly related to hormonal physiology and development. At birth, maternal hormones result in elevated levels of vaginal glycogen. Lactobacilli survive well in this environment, hence the initial flora of healthy neonates should reflect this, along with the presence of diphtheroids, staphylococci, streptococci and *E. coli*.

The bacterial flora of babies soon change and in female babies levels of glycogen soon fall and the normal flora in pre-pubescent girls may largely consist of coliforms and enterococci. The vaginal environment changes again during puberty; circulating oestrogens result in the elevation of glycogen levels and re-colonization by lactobacilli. Finally, hormonal changes at menopause lead to re-establishment of a flora consisting of an almost pre-pubescent state.

These are only general female changes that may occur with age. It is important to remember that the vaginal flora are constantly under other pressures, such as antibiotic therapy, pregnancy and during menstruation. Organisms that may be harmlessly present in small numbers but can be implicated in disease if their numbers multiply include yeasts, anaerobes, *Gardnerella vaginalis* and coliforms.

Specimen quality and type

Due to the nature of the sample, appropriate specimens are often difficult to obtain. Samples may be poorly collected, and incorrect samples are often received. Avoiding faecal contamination when collecting genital specimens can also improve the chances of recovering microbes causing infection. The types of specimen you may encounter include: high vaginal swab (HVS), vaginal discharge, endocervical swab, vulval swab, labial swab, penile swab, genital ulcer swab, urethral discharge/swab, semen and screening swabs for gonorrhoea. Transport media play an important role in the support of delicate pathogens, such as *Neisseria gonorrhoeae* and *Trichomonas vaginalis*, although both survive for a very limited time in this media. Viral transport media contain antimicrobials to help prevent bacterial contamination of cell lines and hence are inappropriate for bacterial culture.

Aspirates may be collected from **Bartholin's gland**, **pouch of Douglas** fluid or fallopian tubes. Intrauterine contraceptive devices (IUCDs) or products of conception (POC) may also be submitted to the laboratory for investigation.

SELF-CHECK 7.9

Why can it be so difficult to isolate an organism as important as *N. gonorrhoeae* from a genital sample?

Sexually transmitted infections (STIs)

There are a number of organisms that are capable of sexual transmission and result in a variety of clinical syndromes. It is advisable to screen for multiple infections once an initial STI is clinically diagnosed. Screening and contact tracing, performed by dedicated genitourinary medicine (GUM) departments, have a vital role in the control of gonorrhoea, syphilis, Chlamydia and human immunodeficiency virus (HIV) infection (not discussed here).

Gonorrhoea is a form of urethritis involving the mucous membranes of the urethra and caused by the organism *N. gonorrhoeae*. It commonly results in a purulent urethral discharge, usually more apparent in the male than the female. Additionally, men may also experience pain in passing urine (dysuria). The most common complications are epididymitis (inflammation of the epididymis), prostatitis (invasion of the prostate) and orchitis (testicular infection) which can result in sterility. Rare complications include gonococcal cellulitis, urethral abscess and penile lymphangitis. Urethral discharge is the preferred specimen to collect from symptomatic males; urethral swabs are commonly collected from asymptomatic individuals.

Infection is confined to mucosal surfaces lined with columnar epithelia, that is the urethra, cervix, pharynx, rectum and conjunctiva. Vaginal squamous epithelial cells may not be infected by *N. gonorrhoeae*; hence endocervical or urethral swabs are preferred to HVS if gonorrhoea is suspected. Although endocervical infection is often associated with discharge, itching (pruritis) and dysuria, approximately 50% of infections in women are asymptomatic. Ascending spread from the cervix, through the uterus to the fallopian tubes can cause **salpingitis**, or to the ovaries, resulting in **ovaritis**. **Pelvic inflammatory disease** (PID) is the most important complication and can result in infertility.

Rectal infection (proctitis) can occur in women with endocervical infection. This is often caused by autoinoculation and is usually asymptomatic. Proctitis in men who have sex with men is common through anal intercourse and can be symptomatic. Treatment of partners as well as the individual is essential to avoid reinfection.

Disseminated gonococcal infection (DGI) is a rare complication of gonorrhoea and results in spread to other body sites. Symptoms include joint and/or muscle pain (arthralgia/myalgia), polyarthritis, pustular rashes, septic arthritis, endocarditis and meningitis (very rare). Blood culture and joint fluids are appropriate samples for diagnosing certain forms of DGI. Eye infections due to *N. gonorrhoeae* are discussed in Section 7.3.

Trichomoniasis, an infection caused by the protozoan *Trichomonas vaginalis* (TV), is predominantly spread by sexual contact. Patients may be asymptomatic for years, but for many patients infection causes symptoms including vaginal discharge, itching and dysuria. Low birth weight and premature delivery can result from infection during pregnancy. There is also evidence that *Trichomonas* infection can predispose the person to an increased susceptibility to HIV infection after intercourse with an HIV-infected partner. The life cycle of the organism is shown in Figure 7.7.

Microscopy is generally the preferred method of diagnosis, although culture is still regarded as the 'gold standard', but this tends to be slower and more labour intensive. Consequently, wet preparation or acridine orange staining is employed by most UK laboratories, although

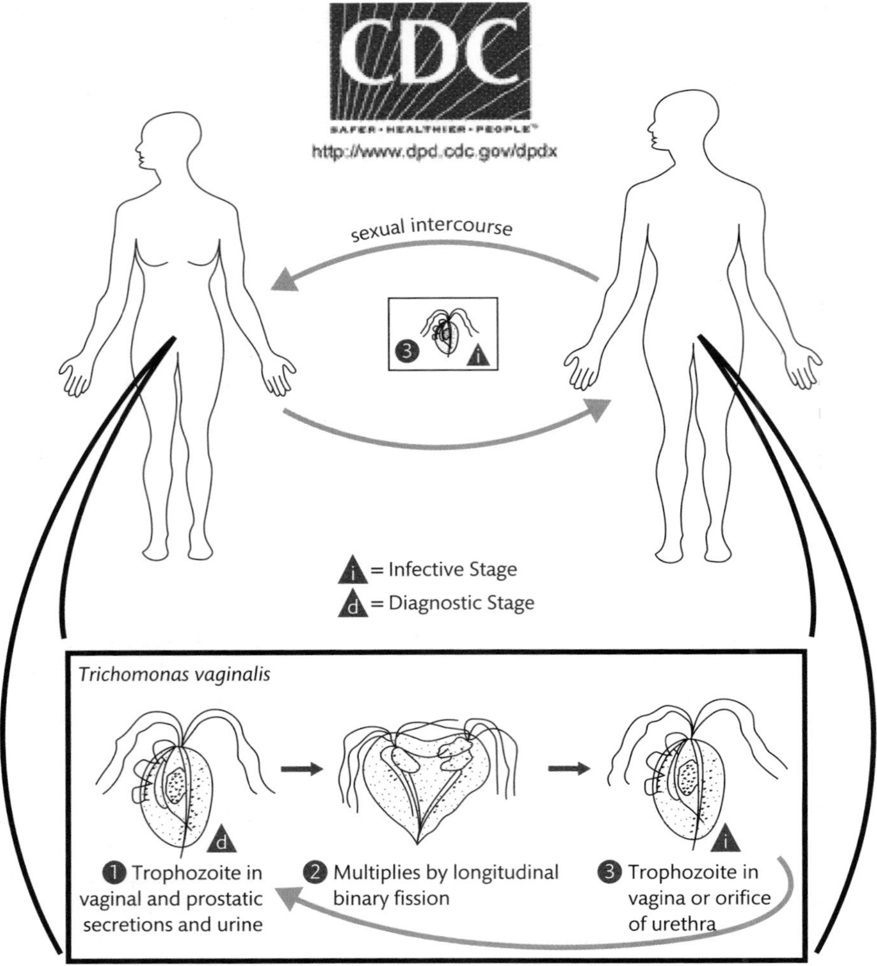

FIGURE 7.7
Life cycle of *Trichomonas vaginalis*. (Courtesy of CDC/ Alexander J. da Silva, PhD/ Melanie Moser.)

the sensitivity of both methods is questionable. As conventional bacterial transport swabs do not offer optimal support for TV over longer periods of time, employing acridine orange may theoretically detect non-viable TV which may be overlooked in wet preparations.

Genital herpes is caused by the herpes simplex virus (HSV). Both types of the virus (1 and 2) are capable of causing genital disease, although type 2 is predominantly spread by sexual contact. Although a relatively common condition, infected pregnant women are at risk of serious systemic sequelae. Additionally, risks to the foetus include poor growth rates and death. Diagnosis has traditionally consisted of viral culture, although PCR techniques are now increasingly available. Site-specific viral swabs incorporating viral transport media to prevent bacterial overgrowth are the specimens of choice.

Chlamydia is the most common STI and may affect up to one in ten sexually active men and women, particularly those under the age of 25. There are over 100 000 cases per year in the UK. It is caused by the bacterium

Chlamydia trachomatis. The organism is present in the semen and vaginal fluid of infected men and women. Infection can often be asymptomatic, but if left untreated can result in painful complications and infertility. In women, spread to other reproductive organs can cause PID, which may result in long-term pain, blocked fallopian tubes, infertility and ectopic pregnancy. Men may experience urethritis, orchitis and reduced fertility.

Traditionally, culture in McCoy cell lines and then detection of either inclusion or elementary bodies was deemed to be the gold standard. This was often supplemented with antigen detection using enzyme immunoassay (EIA) techniques. Both have now been largely superseded by nucleic acid amplification testing (NAAT), for example strand displacement amplification (SDA) or ligase chain reaction (LCR). Although more expensive, it offers far greater sensitivity and specificity than antigen detection and is a lot less labour intensive than culture. Furthermore, diagnosis does not heavily rely on invasive genital sampling, as routine urine samples, especially in males, or self-taken swabs can often suffice.

Syphilis is caused by the spirochaete bacterium *Treponema pallidum* (Figure 7.8). Once regarded as an 'historical' infection', it has re-emerged as an STI and has been responsible for several recent outbreaks in the UK. The organism cannot be grown in routine culture.

Although sexual transmission predominates, IV drug use, blood transfusion and congenital transmission from mother to unborn child are also possible. Primary syphilis begins with one or more painless sores called chancres. These are highly infectious and, as they are commonly sited on the genitals, are the main route of sexual transmission. The sores disappear after 3 to 6 weeks and, if untreated, the disease will usually progress further. Secondary symptoms occur 2 to 10 weeks after the appearance of the chancre, and include skin rash, sore throat, tiredness and headaches. These symptoms typically disappear, even without treatment, although there may still be transient symptoms. A **latent** stage may follow where the patient suffers few symptoms and often feels 'cured'. Unfortunately, without treatment the infection may progress to tertiary syphilis. This can result in a variety of conditions including heart disease, dementia and death.

Since all pregnant women are screened for syphilis, laboratories should be capable of testing large numbers of specimens, the majority of which will be negative. Confirmation of positive results may be conducted locally or by regional reference centres. Discrepancies should always be referred. The initial serology screen consists of an EIA to detect both IgM and IgG. This is highly sensitive, but not quite as specific as TPPA/ TPHA (*Treponema pallidum* particle agglutination or *T. pallidum* haemagglutination) assays, which should be employed on patients' samples that are IgM/IgG positive or samples from contacts of positive patients. Final confirmation requires the VDRL/RPR test (Venereal Disease Reference Laboratory/rapid plasma reagin) to try to assess whether infection is likely to be recent or has been adequately treated.

SELF-CHECK 7.10

Why do you think sexually transmitted infections are on the increase, despite advertising campaigns, etc?

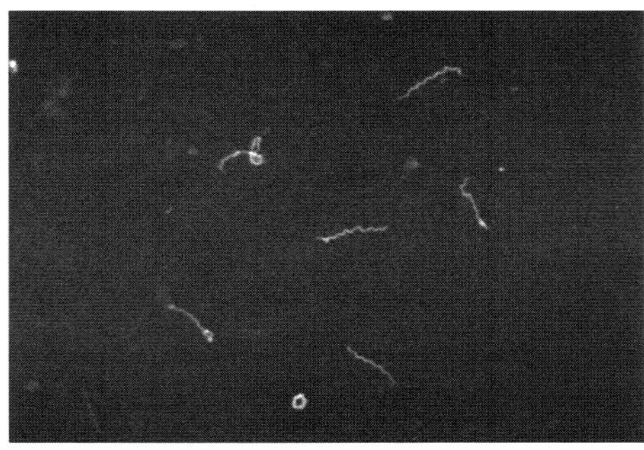

FIGURE 7.8

Photomicrograph showing the characteristic spiral shape of the causative organism of syphillis (*T. pallidum*). (Courtesy of CDC/W.F. Schwartz.)

SELF-CHECK 7.11

Why are tests to detect antibodies to *T. pallidum* used rather than trying to grow the organism in culture?

Chancroid is a sexually transmitted infection caused by *Haemophilus ducreyi*. Painful genital ulceration is typical. The infection is traditionally regarded as a 'tropical disease', although distribution now appears to be spreading. Diagnosis is often made on clinical grounds; culture is often difficult. The organism is quite fastidious and requires specific media, not normally stocked by routine laboratories, lending the disease to molecular diagnosis.

Non-sexually transmitted genital tract infections

Candida albicans is responsible for about 85% of cases of vaginal candidiasis. Normally present in small numbers, candidiasis occurs when yeasts are able to proliferate and predominate over normal vaginal commensals. This may occur after antibacterial chemotherapy, during immunosuppression or pregnancy or whilst using oral contraception, all of which are known to disrupt the normal vaginal flora. Non-*albicans* species known to cause candidiasis include *C. krusei*, *C. tropicalis* and *C. glabrata*.

Clinically, the patient may experience itching, dysuria and/or a creamy white discharge, although some women undergo soreness and swelling. Over-the-counter antifungal therapy creams are now widely available, but some only target *C. albicans*. Subsequently, less common yeast infections may often persist due to treatment failure. All genital swabs should be cultured to an appropriate medium for the isolation of yeasts (Sabouraud dextrose media). Use of the recently available chromogenic *Candida* media will help determine when a non-*albicans* species has proliferated.

Cross reference
Chapter 11.

Vaginitis may be caused by *Candida* spp, *T. vaginalis*, β-HS, *S. aureus* and coliforms amongst others. In young girls implicated organisms also include *H. influenzae*, although the cause may also be poor hygiene or simple irritation.

Ureaplasma urealyticum and *Mycoplasma hominis* are naturally stable bacteria that lack a cell wall. Whilst they may be present in small numbers within normal vaginal flora, they have been implicated in many infections of the genital tract of both males and females. Mycoplasmas have been associated with bacterial vaginosis (see below), postpartum endometritis, salpingitis and PID. Both have been implicated in non-specific urethritis (NSU) in men, a condition that may be sexually transmitted in nature but cannot be attributable to a particular cause. Both organisms are rarely investigated as they are difficult to isolate and require specific media and technical expertise. For *M. hominis* it is essential that a serum-enriched clear medium is used so that the colonies can be visualized under a plate microscope after growth and they present with a 'fried-egg' appearance due to the lack of a cell wall. *U. urealyticum* colonies (previously known as T-strain mycoplasmas where 'T' equals tiny) are too small to be visualized using this methodology and often a chemical reaction within the media is relied on for their detection.

Bacterial vaginosis (BV) occurs when changes to the vaginal environment lead to a reduction in lactobacilli and an increase in other organisms, especially anaerobic bacteria (e.g. *Mobiluncus* species). BV is generally accepted as a polymicrobial condition and other organisms have also been implicated, including *G. vaginalis*. For this reason BV was previously described as *Gardnerella vaginalis* syndrome. The fact that this organism may be recovered from many asymptomatic women ensures that BV is now the more commonly used term. It is not fully understood what causes this imbalance, but changing soap, bath foam and douching have all been implicated. A fishy smelling discharge is the most common symptom associated with BV and complications during pregnancy including **amnionitis,** premature labour and PID have been noted.

Clinical diagnosis relies on the following criteria (three of four):

- Thin grey-white discharge
- Vaginal secretions > pH 4.5
- Amine test positive (fishy amine smell present when vaginal secretions mixed with potassium hydroxide)
- Clue cells present on microscopic examination

Diagnosis is a contentious area - microscopic examination of vaginal smears for 'clue cells' is highly specific for BV, although not critical. Clue cells are epithelial cells that have large numbers of Gram-variable bacilli attached (*G. vaginalis* and *Mobiluncus* spp). Although a Gram film is recognized as the most appropriate microscopic technique, acridine orange staining offers the advantage of providing screening for both 'clue cells' and *Trichomonas vaginalis*. Gram staining, however, allows both Gram variability and reduction in lactobacilli to be assessed easily and is required if Nugent's or Hay's criteria are used for BV diagnosis.

Key Point

The diagnosis of bacterial vaginosis is still controversial and not standardized throughout microbiology departments. There are two criteria for diagnosing BV (Nugent's and Hay's) and the reader is referred to both publications to see how divisions exist in diagnosis (see Reference list).

Lancefield Group B streptococcus is a transient vaginal commensal. During pregnancy, however, the organism may infect the amnion which potentially can cause sepsis and meningitis in the neonate. Similarly, vertical transmission can occur at birth due to vaginal colonization. Testing for carriage tends to follow local policy and often involves screening introital and rectal swabs. Subsequent treatment during pregnancy and/or childbirth should follow both national and local policy. The presence of Lancefield Group A streptococci in HVS swabs during pregnancy is now quite uncommon but was the scourge of previous generations as many pregnancies resulted in **puerperal sepsis** and subsequent death in the mother either during or within days of delivery. Although uncommon, the significance of *Streptococcus* Group A in an HVS sample shouldn't be underestimated.

Normally, the isolation of *S. aureus* from a vaginal swab may be of doubtful significance, but the isolation of toxin-producing *S. aureus* from women of child-bearing age can be hugely significant, as it has been associated with the condition **toxic shock syndrome** (TSS). TSS has been linked to certain factors, particularly tampon usage, and is characterized by pyrexia, erythema

CASE STUDY 7.2 Toxic shock syndrome

The late 1970s saw the introduction of a new range of tampons in the US, driven mainly by demand from American women for increased absorbency. Procter and Gamble subsequently launched the 'Rely Superabsorbant' brand which appeared to contain the entire menstrual flow without leakage or constant replacement. However, in 1980 epidemiologists noted a rise in numbers of a newly described condition, TSS, mainly in menstruating women. Although other highly absorbent tampons were also later implicated, the Rely brand, consisting of carboxymethylcellulose and polyester, appeared to filter the toxin produced by *S. aureus* and significantly increase the risk of TSS. These products and materials are no longer used in tampons and, consequently, rates of TSS in women using tampons have fallen dramatically.

and hypotension. To produce toxin, *S. aureus* firstly requires a protein-rich environment. This is initially provided by the menstrual blood flow. In addition, both a neutral vaginal pH and increased oxygen levels are necessary; menstruation causes a drop in vaginal pH, whilst an inserted tampon alters the normally anaerobic vaginal environment.

Vaginal swabs from cases of suspected listeriosis, women who have experienced intrauterine death, miscarriage, or post mortem foetal tissue should be investigated for the presence of *Listeria monocytogenes*. It has been shown to be the cause of maternal septicaemia and serious morbidity and mortality to the foetus. Other samples warranting investigation include products of conception (POC), placental tissue/swabs and neonatal screens.

Intrauterine contraceptive devices (IUCDs) or coils have been associated with cases of PID, sometimes as a result of a polymicrobial infection. If signs and symptoms point to PID and/or an inflammatory response in an IUCD-wearing female, the organism *Actinomyces israelii* should also be considered. The removed IUCD should be inspected macroscopically for 'sulphur granules' - small, hard 'growths' on the surface of the coil that are essentially clumps of organisms. They do not actually contain any sulphur, but are yellow and sulphur-like in appearance. These can be crushed between two glass slides and Gram stained. *Actinomyces* species appear as branching Gram-positive bacilli. Extended anaerobic culture up to 10 or 14 days is essential for the recovery of this pathogen on primary isolation. Metronidazole can be added to specific culture media, as they are intrinsically resistant, unlike anaerobes. Unfortunately, differentiation from *Propionibacterium* spp. is still an issue, although the rapid indole test helps.

Other infections of the female genital tract include cervicitis, endometritis, salpingitis, PID and Bartholinitis (infection of the secretory Bartholin's glands on the vaginal wall). Generally, most of the organisms discussed in this section could cause any of these conditions, sometimes as complications and sequelae of vaginitis. In addition, it is important to remember that ascending infection during pregnancy can have adverse effects on the unborn child, as well as the mother.

Infections of the male genitourinary tract include prostitis, orchitis, epididymitis and **balanitis** (inflammation of the prepuce). In addition to *N. gonorrhoeae*, causative agents include *C. trachomatis*, coliforms and *Pseudomonas*, β-HS and *S. aureus*. *Candida* species are capable of causing penile candidiasis, although this is less common than vaginal thrush.

7.8 *Neisseria gonorrhoeae*: virulence factors

Although *N. gonorrhoeae* does not produce toxins, the organism has a wide range of virulence factors enabling it to cause disease, including three outer membrane proteins (PI, PII and PIII). Initially, the bacterium will adhere to microvilli of non-ciliated columnar epithelial cells (of the cervix). Adherence is accomplished using fimbrae and specific proteins (PII or Opa); penetration of the cells and multiplication of the organism then follows.

The porin protein, PI (Por), is believed to mediate penetration of host cells - the organisms enter cells by **endocytosis**, are delivered to the base of the cells via vacuoles and leave by **exocytosis**. Bacteria are then able to progress to subepithelial tissue. PI is also believed to help gonococci survive post-phagocytosis. The PI proteins are antigenic and each strain of *N. gonorrhoeae* exhibits only one type of PI; a trait used by laboratories to help confirm the identity of *N. gonorrheae*. Antibodies to a further membrane protein, PIII or Rmp, are believed to block the natural immune response.

Within the Gram-negative cell wall, *N. gonorrhoeae* exhibits large amounts of **lipooligosaccaharide** (LOS) which is believed to play a key role in pathogenesis. It is able to stimulate the production of TNF (tumour necrosis factor) which causes cell damage (e.g. to fallopian tubes) and can induce the production of proteases and phospholipases that are thought to help cause disease. Its presence is also known to cause a significant inflammatory response and is believed to be responsible for many of the symptoms of gonorrhoea. The organism is also capable of extracting iron from its host which is crucial for growth, multiplication and cellular invasion.

7.9 Sample processing

Samples specifically for Chlamydia and viral investigation incorporating appropriate transport media are submitted to the microbiology laboratory, but are not considered in this section. In most laboratories these are forwarded to reference laboratories for processing.

Processing of genital specimens for bacteriology will vary according to local policy. Variables such as clinical details and age will dictate how specimens are investigated and are therefore critical for accurate diagnosis. Generally, all genital specimens being screened for STIs are cultured for *Candida* species and *N. gonorrhoeae*. Selective media are advisable for the isolation of yeasts and gonococci to help suppress the growth of commensal organisms. Sabouraud dextrose agar and gonococci (GC) selective media containing an antifungal agent are usually employed, respectively. Gonococci are fastidious organisms and require an incubation temperature of 35–37°C supplemented with 5–10% CO_2 to grow and may take 48 h. Figure 7.9 shows the characteristic Gram-negative diplococci of *N. gonorrhoeae* in a urethral smear. Note how many of the organisms are intracellular.

Nucleic acid amplification testing (NAAT) is available that detects both *C. trachomatis* and *N. gonorrhoeae* from one sample. As mentioned previously, the sensitivity of *C. trachomatis* detection has increased with the introduction of molecular techniques. Likewise these techniques appear to be particularly sensitive for the detection of *N. gonorrhoeae* and can detect extremely low bacterial loads. It is accepted that occasional strains of *N. gonorrhoeae* may be inhibited by concentrations of vancomycin present in GC selective media. Infection with these strains may therefore produce discrepant results between culture and NAAT testing and need further investigating, including the use of enriched non-selective media. Unfortunately, at present NAAT testing has not been validated for either rectal or pharyngeal swabs, which makes detection of *N. gonorrhoeae* from these specimen types on non-selective culture media almost impossible, due to overgrowth of normal flora or presence of other commensal *Neisseria*, respectively. An important drawback in the application of NAAT testing is currently its inability to produce susceptibility test results.

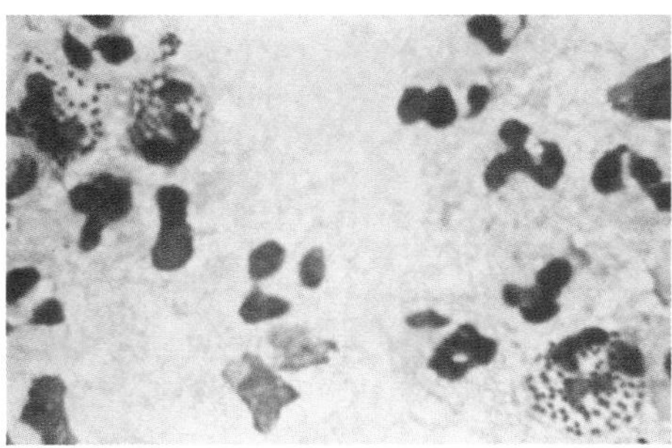

FIGURE 7.9
Photomicrograph of a urethral discharge showing intracellular Gram-negative diplococci characteristic of *N. gonorrhoeae*. (Courtesy of CDC/Brenda Novak.)

In addition, whilst some laboratories examine all HVS samples for TV (either by microscopy or by culture), others rely on specific clinical requesting, which could lead to either underreporting or at least delayed detection. All female specimens should be examined microscopically for BV. Specimens of urethral discharge should also be Gram stained for Gram-negative cocci.

Any postoperative samples should be examined for pathogens. All male genital specimens should be cultured to media that will also support the growth of staphylococci, β-HS, coliforms, *Pseudomonas* and anaerobes. Similarly, swabs from prepubescent, postmenopausal or pregnant women should be cultured in this way, as should those detailing PID or other complications discussed earlier. Swabs from cases of vaginitis in young girls should be investigated for *H. influenzae* using chocolate agar.

Enrichment, selective and/or differential media may be employed for the isolation of:

- Lancefield Group B streptococci in appropriate swabs from pregnant women
- *Actinomyces* species from IUCDs
- *Listeria* species from intrauterine death, miscarriage or suspected listeriosis
- *Haemophilus ducreyi*, although usually clinically diagnosed

Susceptibility testing of significant isolates, such as *N. gonorrhoeae*, assists with the treatment of patients, but also may help determine resistance surveillance. Each year the HPA Sexually Transmitted Diseases Reference Laboratory (STBRL) in Colindale, London, collects isolates of *N. gonorrhoeae* over a 3-month period from sentinel sites across the UK (Gonococcal Resistance to Antimicrobials Surveillance Programme - GRASP). Trends in antimicrobial resistance are published and distributed to appropriate UK health departments annually. At present, due to their rarity, any isolate of *N. gonorrhoeae* exhibiting resistance to antimicrobials such as azithromycin or cefixime should be forwarded to this reference facility for confirmation of the result.

7.10 Medicolegal issues

Specimens are sometimes collected that may have medicolegal implications. These include genital specimens from victims of alleged rape and sexual abuse. Local policies apply, but specimens of this type usually undergo a 'chain-of-evidence' procedure, where signatories are required at every stage of the process from collection of the specimen through to reporting of results, including transport. Final results are signed by registered biomedical scientists and then counter-signed by a more senior member of staff. Traceability of these samples is essential, and helps ensure tampering of specimens and results is impossible, as serious legal proceedings may follow. Lot numbers and expiry dates of culture media, kit reagents and other consumables are also recorded to help validate test results.

All possible pathogens are considered, but the following describes the investigation of *N. gonorrhoeae*:

Due to the delicate nature of the infection, precise identification of *N. gonorrhoeae* is critical in all patients, not just medicolegal cases, and so laboratories often follow the same protocols for identification of *N. gonorrhoeae* as is required by law.

N. gonorrhoeae should grow as colonially typical (grey, translucent) colonies on GC selective media. The bacteria should be oxidase-positive, Gram-negative cocci. Furthermore, they should be confirmed using two identification methods. Typically, laboratories firstly employ biochemical methods, including fermentation testing of glucose, maltose and lactose and/or the detection of preformed enzymes. A second test, usually antigenic, is employed to detect surface proteins (e.g. PI) as they are highly specific. Cultural methods of detection and subsequent identification have always been regarded as the 'gold standard', both for gonococci and Chlamydia. At present, molecular detection of *N. gonorrhoeae* is untested fully and hence is not accepted in the court of law.

SELF-CHECK 7.12

Can you describe the methods available for the laboratory diagnosis of gonorrhoea?

After reading this chapter you should know and understand:

- Most organisms associated with the skin are commensals.
- The skin and mucosal surfaces form an effective barrier, but breaching can lead to infection.
- Skin infections include cellulitis, folliculitis, boils, abscesses and ulceration.
- The most common causes of skin infections include *S. aureus*, β-haemolytic streptococci and anaerobes.
- The Gram stain can provide important information to clinicians, but can sometimes mislead.
- Selective media need to be used to allow the target pathogen to be isolated from a mixture of 'normal flora'.
- A variety of virulence factors and toxins are produced by pathogens to produce infection.
- Mucosal infections include conjunctivitis, otitis media and pharyngitis.
- These may be caused by *H. influenzae*, *S. pneumoniae* and Lancefield Group A streptococci.
- Selection of the correct media is critical in the isolation of pathogens. This will sometimes involve the use of selective media, e.g. for *N. gononhoeae*.
- Genital infections include gonorrhoea, Chlamydia and syphilis.
- Specimen quality and choice of site can be critical for accurate diagnosis of STIs.

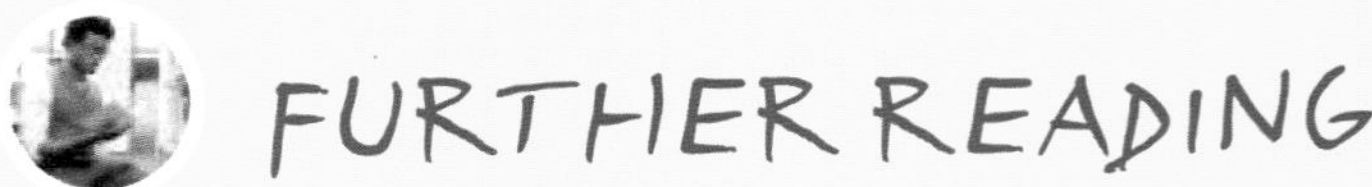

- **Bowler PG, Duerden BI, Armstrong DG. Wound microbiology and associated approaches to wound management. *Clinical Microbiology Reviews* 2001; 14: 244–269.**

  Excellent text for information on microbiological aspects of wound and soft tissue infections.

- **Hall GS. Molecular diagnostic methods for the detection of *Neisseria gonorrhoeae* and *Chlamydia trachomatis*. *Reviews in Medical Microbiology* 2005; 16: 69–78.**

  Review article describing the current methodologies for the diagnosis of important STDs.

- **Pellati D, Mylonakis I, Bertoloni G, Fiore C, Andrisani A, Ambrosini G, Armanini D. Genital tract infections and infertility. *European Journal of Obstetrics, Gynaecology and Reproductive Biology* 2008; 140: 3–11.**

  Article showing how infection of the genital tract can be associated with infertility and other complications.

- **Mims CA. *Medical Microbiology*, 3rd edn. Mosby, 2004.**

  Excellent text on infectious diseases and the process by which organisms mediate infection and insight into their pathogenic properties.

- **Centre for Disease Control and Prevention (CDC). http://www.cdc.gov/std/.**

  Excellent resource for reports and other publications related to human health.

- **The Path of Least Resistance. Department of Health, 1998.**

7.1 Should all laboratories not perform molecular testing for *Chlamydia* so that a positive result is diagnosed as soon as possible?

7.2 Do you think only molecular methods should now be used to detect the presence of *N. gonorrhoeae* in clinical samples? If not why not?

7.3 Why is *S. aureus* such a common occurrence in an infected wound?

7.4 Why are selective media so commonly used when isolating pathogens from wound and genital swabs?

7.5 Do you think there should only be one laboratory method for the diagnosis of bacterial vaginosis? Explain your reasoning.

8

Examination of cerebrospinal fluid and fluids from sterile sites

Derek Law

Although there are many body sites that are not sterile, such as the respiratory tract, genital tract and intestinal tract, there are also various internal body sites that are considered microbiologically sterile. These sites can become infected by bacteria or viruses via various routes causing overt clinical symptoms. However, other factors, such as trauma, inflammation and carcinoma, may also cause similar symptoms. These sites are frequently sampled for microbiological and other investigations. This chapter will describe the various fluids commonly received in the laboratory and describe the methods for processing these. The fluids to be considered in this chapter are cerebrospinal fluids, synovial, pleural, ascitic, pericardial and peritoneal dialysis fluids.

Learning objectives

After studying this chapter you should be able to:

- Understand the different fluids received in the laboratory
- Know how to safely handle the samples
- Understand the role of cell counts in processing and diagnosis
- Know the appropriate culture methods for fluids
- Know the common pathogens isolated from different fluids
- Recognize when further or other investigations are required

8.1 Investigation of cerebrospinal fluids

The meninges are a system of membranes that envelope and protect the central nervous system (brain and spinal column). They consist of three layers: the dura mater, the arachnoid mater and

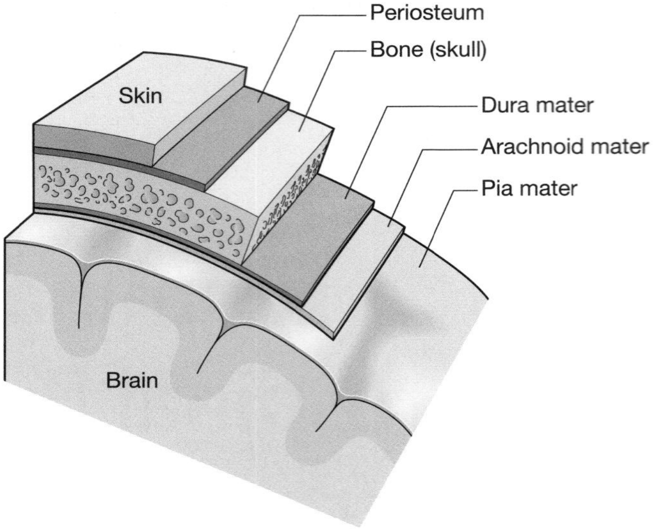

FIGURE 8.1
Diagrammatic cross section of the skull showing the arrangement of the protective coverings of the brain. The meninges consist of three layers, the dura mater, arachnoid mater and pia mater. Cerebrospinal fluid is found between the arachnoid mater and the pia mater - the subarachnoid space.

the pia mater. Cerebrospinal fluid (CSF) is a clear fluid that is found in the space between the pia mater and the arachnoid mater and is found in the ventricles of the brain and surrounding the spinal cord. Its main function is to protect the brain by acting as a cushion against shock.

Figure 8.1 shows the relationship between the various membrane layers and the brain and the skull. The pia mater is attached to the brain and follows the contours of the brain; the dura mater is attached to the skull. Cerebrospinal fluid is found in the gap between the pia mater and the arachnoid mater - the subarachnoid space.

Inflammation of the meninges is known as **meningitis**. Meningitis can be caused by bacteria, viruses or fungi and is an extremely serious condition owing to the proximity of the infection and inflammation to the brain and spinal cord. The potential for serious neurological damage or death requires prompt medical attention and the laboratory plays an important role in helping to establish or refute a diagnosis of meningitis. Various other conditions can mimic meningitis, particularly neurological conditions such as stroke and **subarachnoid haemorrhage**, and these must be differentiated from meningitis as the management of the patient is quite different in these cases.

We will first consider the signs and symptoms of the different types of meningitis as this can provide information on the possible cause of the infection.

Key Point

Inflammation of the meninges is called meningitis; meningitis can be caused by infection with bacteria, viruses or fungi. Meningitis is an extremely serious illness and requires prompt medical attention.

Bacterial meningitis

Although bacterial meningitis can be caused by many different organisms, the vast majority of cases are caused by a small number of bacterial species; Table 8.1 shows the bacterial species

TABLE 8.1 Common CSF pathogens according to patient type.

Age/type of patient	Common organisms
Neonates	Group B streptococci, *E. coli*, *Listeria monocytogenes*
Infants	*N. meningitidis*, *H. influenzae*, *Strep. pneumoniae*, *M. tuberculosis*
Children	*N. meningitidis*, *Strep. pneumoniae*
Adults	*N. meningitidis*, *Strep. pneumoniae*
Immunosuppressed patients	*Cryptococcus neoformans*, *Listeria monocytogenes* Potentially other organisms

most commonly found in cases of meningitis. The patient populations affected by the different species of bacteria are also shown as there are clear differences in infecting organism based on patient age.

Bacterial meningitis is usually much more severe than viral meningitis and without urgent antibacterial treatment carries a high mortality; however, even with antibiotic therapy the patient may survive with various neurological disorders such as deafness or **cerebral palsy**.

Cross reference

Septicaemia is described in Chapter 5, Blood Culture.

The symptoms of meningitis can be variable; initially the disease may start as a flu-like illness and further symptoms develop over the course of a few days, although they can have a very sudden onset. The main symptoms are severe headache, fever, **photophobia**, neck stiffness, pyrexia and in some cases a rash, and there is often drowsiness and a decreased level of consciousness. One of the most common causes of meningitis is *Neisseria meningitidis* (meningococci). Meningococcal septicaemia may accompany meningococcal meningitis and occurs when bacteria enter the blood stream and multiply uncontrollably.

photophobia

Photophobia is a dislike of bright lights.

endotoxin

Endotoxin is a component of Gram-negative bacterial cell walls. When released into the blood stream it causes fever.

Endotoxin is released which initiates various organ dysfunctions and disturbances of blood clotting.

Blood leaks from capillaries into the skin causing a **purpuric rash**. Meningitis and septicaemia often occur together but meningitis can occur without septicaemia and septicaemia can occur without meningitis.

purpura

Purpura is the appearance of red or purple discolorations on the skin, caused by bleeding underneath the skin.

Meningococci colonize the oropharynx of a small proportion of healthy people; in this situation the patient remains healthy but can act as a carrier and transmit the organism to others. In an individual, transition from carrier state to disease can occur; the factors responsible for the switch from carrier to invasive disease are not known. Spread of organisms from person to person is favoured by overcrowding and is more frequent in people living in close proximity, such as students and army recruits. Infections are also most common in these groups. Infections also occur more frequently in patients who have had their spleen removed **(splenectomy)**.

splenectomy

Splenectomy is removal of the spleen. Because the spleen is involved with the immune system, splenectomy patients are more susceptible to certain types of infection including meningitis.

Meningococcal septicaemia can occur with meningitis or it can occur on its own. It is a very serious illness; a purpuric rash is one of the most common symptoms. The release of endotoxin from the meningococcus results in disorders of blood clotting. This can lead to death of tissue on extremities, which in severe cases can eventually lead to **gangrene**.

complement

Complement is a series of proteins found in blood; when activated the proteins have an antibacterial action and help the body to fight infection.

Streptococcus pneumoniae (pneumococcal) meningitis can occur in any age group. In many cases the meninges are infected by spread from an ear infection. Pneumococcal meningitis is more common in patients who have had a splenectomy or who have a **complement** deficiency.

Group B streptococcal infection occurs principally in babies and two distinct forms exist: early onset and late onset. The early-onset form is acquired during birth (the organism is usually acquired from the mother's vagina during birth) and symptoms develop in the first few days of life. The late-onset form develops weeks to months later and infection is acquired from other sources. In both cases pneumonia and septicaemia are common.

Other causes of meningitis include *M. tuberculosis*; meningitis with this organism is rare in some countries but should be considered in patients from high-risk areas. It has a slow onset and signs and symptoms are not as clear cut as those of more typical, acute bacterial meningitis.

Key Point

Meningitis can affect all age groups; however, different bacterial pathogens are isolated from cases of meningitis in different age groups.

SELF-CHECK 8.1

What are the commonest causes of bacterial meningitis?

Viral meningitis

Viral meningitis is a relatively benign illness and is more common than bacterial meningitis. Many different viruses can cause meningitis; the most common are the enteroviruses and mumps virus. The illness is often mild, with the patient feeling unwell, developing flu-like symptoms and a headache. Many cases are not severe enough for medical attention and go unrecognized. In more severe cases where neck stiffness and other meningitis symptoms such as photophobia develop, the patient may seek medical attention and be referred to hospital. Most patients with viral meningitis make a full recovery, although some may be left with residual side effects.

Key Point

Compared with bacterial meningitis, viral meningitis is a milder illness and many cases do not require medical attention.

Cryptococcal meningitis

Cryptococcal meningitis caused by the fungus *Cryptococcus neoformans* is found mainly in patients with HIV infection. It also occurs in patients not infected with HIV but who have other cell-mediated immune defects.

Processing of CSF samples

Diagnosis of meningitis is a medical emergency and samples must be processed without delay and results reported back to the ward as soon as they are available. This section deals with how a CSF sample is processed. In a patient with suspected meningitis it is necessary to obtain a sample of CSF for analysis by **lumber puncture**. CSF samples are taken by inserting a needle between the third and fourth lower vertebrae and capturing some of the fluid. Figure 8.2 shows a diagram of a CSF sample being withdrawn. Notice that the needle is inserted between the vertebrae and pierces the lining of the spinal column to withdraw the CSF around the spinal cord. Samples for bacteriology are normally taken first and the sample is often split into three separate samples for reasons described later. Samples should be labeled 1, 2 and 3 if this is the case. Other samples are often taken for immunological investigation and biochemical investigations - glucose and protein measurements are important in helping differentiate different forms of meningitis. Samples should be transported to the laboratory without delay and once in the laboratory should be processed with urgency.

lumbar puncture
Lumbar puncture is the procedure used to obtain a CSF sample from a patient.

Key Point

CSF analysis by biochemistry, immunology, and cytology departments plays an important role in the diagnosis of meningitis and other neurological disorders, and the results of these investigations must be examined along with microbiological data.

There are various procedures involved in processing CSF samples all of which are important in arriving at a correct diagnosis and guiding the correct therapy for the patient:

- Appearance of sample
- Cell count
- Centrifugation
- Gram stain
- Culture
- Further tests based on microscopy results.

We will now discuss these stages of processing a CSF sample.

Appearance of CSF

The appearance of the sample is important as it gives an early indication of whether a patient has bacterial meningitis and helps in deciding further steps in processing. The

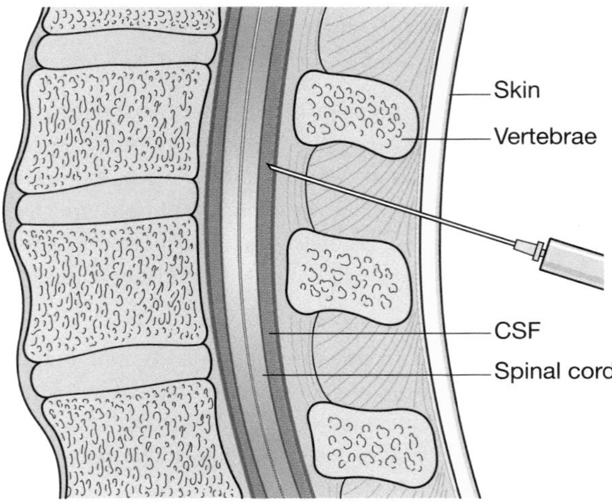

FIGURE 8.2
Diagram showing how a CSF sample is taken. The needle is inserted between the vertebrae and a sample of the fluid surrounding the spinal cord is aspirated.

sample should be in a clear, clean container and examined for turbidity, coloration and clots. Normal fluid is clear and colourless; in cases of bacterial meningitis large numbers of white blood cells are present in the CSF giving a turbid appearance. In cases of stroke or subarachnoid haemorrhage (SAH) the fluid may be stained with red cells; in some cases the degree of blood staining may be very heavy and a blood clot may form. Where an SAH occurred several days previously, the CSF may be yellow in appearance due to the presence of lysed red blood cells. This is known as **xanthochromia**; however, an accurate diagnosis of xanthochromia and hence an SAH can only be made by biochemical measurement of the blood pigments in a CSF sample.

xanthochromia
Xanthochromia is the term used to describe the yellow coloration of CSF samples caused by lysed blood. Confirmation of xanthochromia requires biochemical investigations.

traumatic tap
Traumatic tap is where a capillary or blood vessel is 'nicked' when a CSF sample is taken. Samples become blood stained. Serial samples can help identify whether the blood is due to a traumatic tap.

Xanthochromia may only be evident after centrifugation of the sample. The presence of a fine spider's web-type clot is often indicative of tubercular meningitis. If three samples are received the appearance of each should be recorded. Occasionally, when a CSF is taken the needle nicks a capillary when inserted; this is known as a **traumatic tap**. In such cases the appearance of the three samples differs; the amount of blood in each sample diminishes. If blood is present because of a haemorrhage then each sample will look similar. In some cases of meningitis the presence of blood from a traumatic tap may mask the turbidity due to meningitis. The appearance of the CSF should be recorded and reported.

Cross reference
See *Clinical Biochemistry* for measurement and characterization of xanthochromia.

Key Point

The appearance of a CSF sample can help in the diagnosis of meningitis and other illnesses. Cloudy fluids indicate a raised white cell count and possible infection. Xanthochromia may indicate a possible stroke. Many CSF samples received in the laboratory are clear and colourless but must still be processed.

Key Point

Cell counts on serial CSF samples allow differentiation of blood staining due to a traumatic tap from blood staining due to a stroke or recent subarachnoid haemorrhage. In the case of a traumatic tap the last sample provides the most information as to whether the patient has meningitis.

Cell count

The cell count is the next stage in processing the sample. In this stage we are looking for the total number and type of cells present in a CSF sample. We are interested in counting the numbers of red and white blood cells and in differentiating the white blood cells into polymorphonuclear cells (polymorphs) and lymphocytes. This gives important clues to whether infection is present and what the infecting organism is. Under normal conditions the number of white cells in a CSF sample is very small, typically $<5/mm^3$ in children and adults, and $<20/mm^3$ in neonates.

Cross reference
See Bain, *Blood cells – a practical guide*, in the References section, for morphology and function of the various types of white blood cell.

In cases of bacterial meningitis the white cell count increases dramatically and counts frequently exceed $1000/mm^3$ although in the very early stages of the illness numbers may be low. The cells are typically polymorphs. In viral and tuberculosis meningitis the white cell count is lower and cells are mainly lymphocytes, although in the early stages of viral meningitis a proportion of the cells may consist of polymorphs.

Key Point

The differential cell count is an important first stage in the diagnosis of meningitis. The number of cells and the type of cells (polymorphs or lymphocytes) are important indicators of whether meningitis is present and the possible cause of meningitis. It is often possible to make a provisional diagnosis on these results, and a normal cell count will often eliminate meningitis as a possible diagnosis.

For accuracy in counting cells, a **haemocytometer** counting chamber is used; a small drop of fluid is placed in the counting chamber which is then examined under a microscope. There are many types of counting chamber, but all have a grid with large and small squares marked on the underside. The grid is set up such that an area encompassing a fixed number of small or large squares is equivalent to a cubic millimetre of fluid. Counting the cells in that area gives the number of cells per cubic millimetre of fluid. It is essential that the user is familiar with the type of counting chamber used in their laboratory. Figure 8.3a shows a typical counting chamber used for counting cells. The fluid is loaded under the coverslip. Figure 8.3b shows a typical grid under

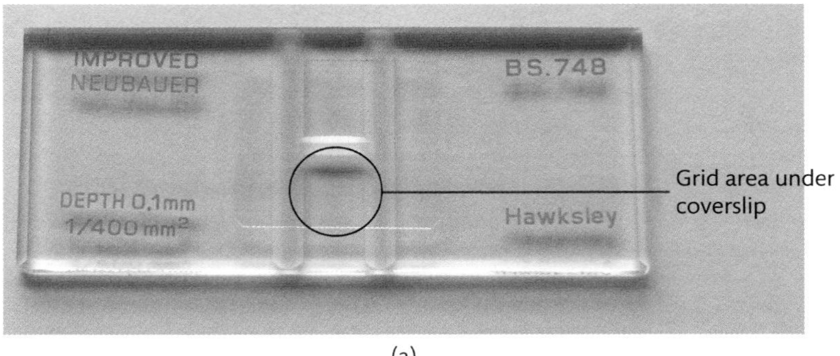

(a)

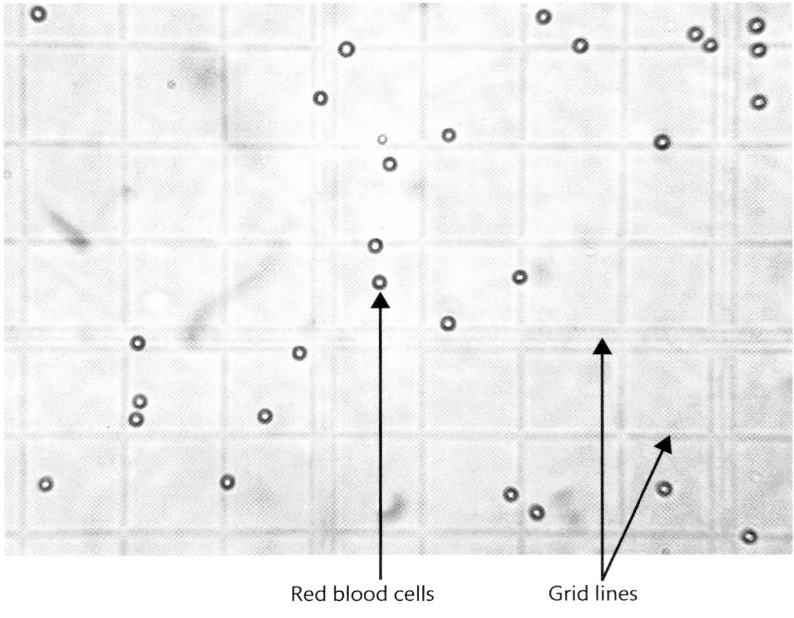

(b)

FIGURE 8.3

(a) Photograph of a counting chamber used for counting cells in CSF samples. (b) Photomicrograph of counting chamber with a blood stained CSF sample; the grid lines are clearly visible as are numerous red blood cells.

the microscope; notice the single and triple lines which mark out distinct areas of the grid. In the sample some red cells can be seen.

Clear or slightly turbid fluids can usually be added directly to the chamber and using a microscope with ×10 and ×40 objectives the cells can be enumerated by counting the number of cells in sufficient marked squares to give the number of cells per cubic millimetre. Red cells and white cells should be enumerated separately. It may also be possible to differentiate the different types of white cells based on the appearance of the nucleus. However, training and experience is necessary to help identify different cell types.

Where the cell count is high (number of cells >200/mm^3) or where the sample is turbid or blood stained, samples will often require some dilution in sterile saline before loading into the counting chamber, to get the cell numbers to countable levels. Cell counts then must be multiplied by the dilution factor to arrive at the correct count. In some cases it is possible that several dilutions of the CSF are required to produce a suspension that is countable. Typical dilutions for blood stained samples are 1 : 20, 1 : 50 or 1 : 100. This allows the red cell count to be obtained.

When a sample is blood stained it is also important to determine the ratio of red cells to white cells to determine if the white cell count is raised above normal due to a combination of white cells in the CSF and white cells in the blood . This is achieved by carrying out a separate white cell count, after first destroying the red blood cells. This is achieved by diluting the CSF sample in a white cell counting fluid, consisting of acetic acid and methylene blue. This fluid lyses red cells but does not affect white cells and stains their nuclei blue allowing them to be counted and allowing differentiation of polymorphs and lymphocytes. For white cell counts, typical dilutions of CSF are 1 : 2 or 1 : 10. Again, the relevant dilution factor must be taken into consideration when calculating the cell count per cubic millimetre. Typically, the RBC : WBC ratio in normal blood is between 300 and 500 and a lower ratio indicates a raised white cell count.

For CSF samples where serial samples have been received it is necessary to determine the red and white cell counts in each sample to assess any sample to sample variation. However, if the first sample has no or very few cells then there is little value in counting the remaining samples. However, if the red cell count is high in the first sample the successive samples should be counted: a decreasing count is indicative of a traumatic tap, whereas similar counts indicate that the red cells originated in the CSF.

It is not appropriate to carry out cell counts on heavily blood stained samples which have clotted, as such counts are inaccurate.

Key Point

Counting chambers are used to accurately determine the cell count in CSF samples. Dilution of samples is often required when cell counts are high or when samples are blood stained. The use of red cell lysing fluids is useful for accurate white cell counts in blood stained fluids. Accurate determination of cell counts in blood stained fluids can be time consuming if separate dilutions and counts are required for red and white blood cells.

SELF-CHECK 8.2

Why is it important to differentiate between polymorphonuclear cells and lymphocytes in CSF samples?

Centrifugation of CSF

Once the red and white cell counts have been determined, the sample must be centrifuged - this only applies to samples where the volume is >0.5 mL. Centrifugation serves an important function in concentrating organisms into a small volume of fluid, increasing the sensitivity of both culture and Gram stain. Centrifugation should be at 3000 g for at least 10 min. After centrifugation the supernatant should be removed with a sterile pipette to a clean, sterile, labelled container. The appearance of the supernatant should be recorded. The presence of xanthochromia in the supernatant is an indicator of a subarachnoid haemorrhage. The deposit should be resuspended in about 0.5 mL of the supernatant fluid and this is used to perform the next stages of sample processing. All supernatants should be retained and stored at 4°C in case further tests are required, such as viral culture.

HEALTH & SAFETY

Appropriate precautions should be taken when centrifuging samples. Sealed buckets should be used and after centrifugation samples should be opened inside a safety cabinet.

When working with cultures of *Neisseria meningitidis*, work should be carried out in a safety cabinet as there is a risk of contracting meningitis from aerosols formed when performing Gram stains or preparing suspensions for identification tests or agglutination tests.

Occasionally, samples may be sent from patients with a possible diagnosis of Creutzfeld-Jacob disease (CJD) or other related illnesses. Such samples should be handled in a safety cabinet, and gloves should be worn during all procedures. All materials should be incinerated after use.

Microscopy

In some laboratories all CSF samples are Gram stained, in others Gram stains are only carried out on samples containing >5 WBC/mm^3. Whatever the local protocol, a Gram stain is prepared from the resuspended deposit rather than the fluid itself. It is helpful to mark the slide where the drop is placed to help focus the microscope. If tuberculosis meningitis is suspected a smear for acid-alcohol fast bacilli (AAFB) staining should be prepared at the same time. To increase sensitivity, a CSF AAFB smear usually has three drops of liquid applied, each added when the others have dried.

Gram stains and AAFB stains are carried out as described earlier and examined as normal.

Cross reference

Gram staining is described in Chapter 2.

The Gram stain serves two functions - it stains bacteria present and also helps differentiate white blood cells into polymorphonuclear cells or lymphocytes. Although this can be done during the cell count it is sometimes easier to carry out on a Gram stain as the cellular structure is easier to see. A careful search for bacteria is required; organisms may be intracellular and may be difficult to see. *Haemophilus influenzae* stains poorly with neutral red counterstain and dilute carbol fuchsin is often favoured as a counterstain for CSF samples. Because CSF is a normally sterile liquid the presence of bacteria in a sample is indicative of meningitis. Medical staff should be immediately alerted if bacteria are seen on a Gram stain.

Cross reference

AAFB staining is described in Chapter 9.

Once microscopy has been carried out, the results of the appearance, cell count and Gram stain can be reported. Because of the urgency of CSF samples, processing time to this stage should be kept to a minimum; samples not requiring dilution, for cell count and Gram stain can often be performed within 30 min of sample receipt.

SELF-CHECK 8.3

Will the biomedical scientist always see bacteria in a Gram stain of an infected CSF?

Culture of CSF

Culture of CSF samples is relatively simple if there are no organisms seen on the Gram stain; a drop of the deposit from the centrifuged sample is inoculated onto a chocolate agar plate incubated in CO_2 for 48 h. Some laboratories inoculate aerobic and anaerobic blood agar plates but anaerobes are extremely rare in CSF samples and in most cases anaerobic plates are not required. Similarly, some laboratories inoculate a portion of the deposit into an enrichment broth. However, the benefits of enrichment culture are not clear. It is rare to isolate a pathogen from an enrichment broth when the culture on agar media is negative, and contamination can be problematic. If organisms are seen on the Gram stain then further media may be inoculated, such as aerobic and anaerobic blood agars. Direct susceptibility tests and key identification tests should also be set up dependent on the identity of the organism on the Gram stain, for example if Gram-positive cocci in pairs are seen then an optochin sensitivity test should be set up to confirm the identity of *S. pneumoniae*.

Cross reference
Chapter 4, Section 4.3.

Cross reference
Chapter 2.

Culture plates should be examined at 24 h and 48 h before being discarded. Enrichment broths should be incubated for at least 48 h and then plated onto chocolate agar incubated in CO_2 and aerobic and anaerobic blood agar.

The pathogens encountered in CSF are dependent on age and immunological status of the patients. Refer to Table 8.1 for a list of the pathogens commonly found in CSF samples.

Key Point

Meningococci and other meningitis pathogens are capsulate, that is they are surrounded by a polysaccharide capsule which makes them virulent. The capsules can be used to identify the organism and they also have a role in vaccination.

HEALTH & SAFETY

Because of the infection risk all work with potential *N. meningitidis* colonies should be carried out in a Class II safety cabinet. This includes preparing Gram stains from colonies, preparing suspensions for agglutination tests or biochemical identification tests.

Characteristics and identification of common meningitis pathogens

It is essential that the laboratory worker is familiar with the standard methods for identifying the common organisms encountered in CSF.

Tests routinely used in the laboratory are for *N. meningitidis*, Gram stain, oxidase reaction and fermentation tests of glucose and maltose or various enzymatic tests available in commercial kits. *N. meningitidis* occurs in several different serogroups based on the polysaccharide capsule which surrounds the bacterial cell. Antisera or latex agglutination kits are also available to identify the capsule type of *N. meningitidis* which can provide further confirmation of the identity. If antisera are not available isolates should be sent to a reference laboratory as serotype determination is important for epidemiological purposes. The most common serogroup is Group B, which causes approximately 60% of cases, followed by Group C, which causes about 40% of cases. Other serogroups, such as Group A, W135, are rare. This epidemiology can, however, vary considerably between geographical regions.

S. pneumoniae can be identified using Gram and colonial appearance and optochin susceptibility (if typical Gram-positive cocci are seen in the CSF then optochin susceptibility tests should be set up on the direct culture). Polyvalent antisera can be used to confirm the identity.

H. influenzae can be identified by its Gram-stain appearance, its colonial appearance and the characteristic poor growth on blood agar but good growth on chocolate agar. Dependence on X and V factors is used to establish the identity. Agglutination tests with type b antisera can be used to confirm the serotype. Vaccination of children against *H. influenzae* type b has resulted in a marked decrease in cases of infection with this organism.

Group B streptococci can be identified by colonial appearance and haemolytic reaction. Rapid confirmation can be carried out very quickly using a Lancefield grouping test.

E. coli can be provisionally identified by colonial appearance. Most strains from neonates possess the K1 capsule antigen and can be identified using appropriate antisera. Final confirmation of identity is through use of commercial identification kits.

L. monocytogenes can be identified by typical morphology on Gram stain and haemolysis reaction on blood agar. Further confirmation can be obtained by observing typical tumbling motility in liquid cultures grown at 30°C, a positive aesculin reaction and growth on *Listeria* spp. selective media.

It is possible to isolate almost any organism from a CSF sample and an organism must be considered potentially significant until confirmatory tests or clinical information suggest otherwise. In the absence of markers such as a raised CSF white cell count, growth of small numbers of organisms such as coagulase negative staphylococci or some Gram-positive bacilli, especially when isolated from enrichment broths, are usually regarded as contaminants. But each result must be decided on a case by case basis. Contamination of samples may occur when the sample is taken or during processing in the laboratory. CSF is a nutritious growth medium and organisms will readily grow in CSF samples.

In cases of meningitis, blood cultures should also be taken as they may become positive earlier than the CSF sample and help guide antibiotic therapy.

SELF-CHECK 8.4

Why is it important to fully identify the organism infecting the CSF?

Antigen testing

Microscopy and culture are the main means of processing CSF samples and a diagnosis of bacterial meningitis is typically made by a high white cell count with a predominance of polymorphonuclear cells and the presence of bacteria on a Gram stain. However, there are occasions when white cell counts may be raised but bacteria may not be seen; this can occur for several reasons. Firstly, if the patient has meningitis but antibiotics have been given bacteria may be dead and/or lysed and therefore not detected on microscopy; secondly, in cases of early meningitis organisms may not be present in sufficient numbers to detect on Gram stain.

In such cases antigen detection may be a useful test to carry out on samples using commercially available antigen detection kits due to the following:

- Many of the bacterial pathogens that cause meningitis produce capsules - polysaccharide coats around the cell that protect the bacteria from **phagocytosis** by white blood cells. These capsules are important for virulence of the organism and are highly antigenic.
- Antibodies raised against the organisms are coated onto latex particles. When these particles are mixed with antigen present in CSF or from organisms growing on an agar plate they agglutinate.
- Antigen tests are often positive when microscopy and culture are negative due to previous antibiotic therapy, and can therefore give a positive diagnosis and help guide appropriate therapy.

Key Point

Antigen tests are often positive when microscopy and culture are negative due to previous antibiotic therapy, and can therefore give a positive diagnosis and help guide appropriate therapy.

CLINICAL CORRELATION

Because prompt administration of antibiotics can be life saving in cases of meningitis, patients may arrive at hospital having received antibiotics. In such cases organisms may be scanty or absent on Gram stain. The use of antigen detection kits in such situations can help identify the causative organism, as sufficient antigen will be present in the sample to give a positive reaction.

Molecular testing

PCR
Polymerase chain reaction (PCR) is a technique involving amplification of specific DNA sequences which can be used to rapidly identify pathogens.

Some laboratories now offer **PCR** tests on CSF samples as a means of rapid diagnosis. These are useful when available to confirm the identity of an organism seen on Gram stain, and also in patients with a strong suspicion of having meningitis but who have received antibiotics and no bacteria can be seen on Gram stain. They are an alternative to antigen detection kits. Specific examples would include detection of meningococcal meningitis on patients receiving antimicrobial therapy or for a range of viral pathogens, such as enterovirus. Normally, such requests would be forwarded to a reference laboratory for analysis.

Viral meningitis

As described earlier, viral meningitis is a milder illness than bacterial meningitis. The typical CSF picture of viral meningitis is shown in Table 8.2. There is an increase in white cells, which are typically lymphocytes, glucose levels are normal and protein levels may be raised. No organisms are seen on Gram stain. If viral meningitis is suspected based on clinical findings and results of CSF analysis, the sample should be sent for virological analysis. This typically involves culture of the sample on a variety of cell lines. The CSF supernatant is suitable for virological analysis and should be transported to the virology laboratory on ice, along with full patient and clinical details. Common viruses isolated from CSF samples include mumps virus, echo virus and Coxsackie viruses.

TABLE 8.2 **CSF parameters in health and disease.**

	CSF appearance	CSF protein	CSF glucose	CSF cell count (per mm^3)	CSF Gram stain	Additional features
Normal	Clear and colourless	0.2–0.4 g/L (neonate <1.7 g/L)	60–80% of plasma glucose	<20 in neonates and <5 in older children and adults	No organisms	
Bacterial meningitis	Cloudy and turbid (if severe)	Raised >1.5 g/L	Glucose level is <50% of the plasma level	Cell count is high (100 to 1000+) and mostly neutrophils	Organism usually seen, morphology may help identification	Antigen tests are useful if organisms are not seen
Viral/aseptic meningitis	Clear	Raised or high end of normal	Glucose level is usually within normal limits	Cell count is high (100 to 1000+) and mostly lymphocytes	No organisms	Viral culture or PCR should be set up
Tuberculous meningitis	Clear or slightly cloudy, may have a cobweb appearance	Raised >1.5 g/L Protein is high (much higher than bacterial meningitis)	Glucose level is <50% of the plasma level	Cell count is high (100 to 1000+) with mainly lymphocytes	Negative	AFB smear should be carried out or PCR in reference laboratory
Cryptococcus meningitis	Slightly turbid	Raised	Normal to low	Moderate number of lymphocytes	Yeasts	India ink may show capsulate yeasts; cryptococcal antigen positive
Subarachnoid haemorrhage	Blood stained (although not always)	Raised or high end of normal	Glucose level is usually low	High number of RBCs	No organisms	Xanthochromia present
Guillain-Barré syndrome	Clear	Markedly raised	Glucose level is usually low	Normal levels	No organisms	
Multiple sclerosis	Clear	Raised	Glucose level is usually within normal limits	Mild pleocytosis with mononuclear cells	No organisms	

Tuberculosis meningitis

Tuberculosis meningitis is a slow chronic illness where the patient complains of headaches and malaise with, eventually, decreasing levels of consciousness. The course of the disease is slow and may take weeks to develop.

- CSF samples are typically slightly turbid and may contain a clot. The protein levels are markedly raised and glucose levels are low. The cell count is raised and is predominantly lymphocytic. It is sometimes difficult to distinguish tuberculosis meningitis from viral meningitis on the basis of cell count and biochemical data.
- A Ziehl-Neelsen (ZN) stain or auramine stain can be carried out to detect *M. tuberculosis* in the sample, but is not always positive.

- Culture is the most sensitive method of detection and if tuberculosis meningitis is suspected a portion of the deposit should be submitted for culture. Inoculation into liquid media such as Kirchner's medium is preferred as positive results are available earlier than using solid culture media. Further, contamination is not usually a problem.

Cross reference

See Chapter 9 for culture methods for TB.

- There is now growing use of PCR for diagnosis of TB, and CSF samples are highly amenable to PCR as the sample volume is small and inhibitors are generally absent, and because the sample is clean there is no interference with environmental organisms.
- When there is a strong suspicion of tuberculosis meningitis, serial samples may be taken over the course of several days to increase the chances of detection and isolation.

Key Point

Tuberculosis meningitis is a rare form of meningitis. It is difficult to diagnose as the illness is chronic and symptoms are not well defined. Microscopy (AAFB stain) and culture are standard diagnostic techniques, although the sensitivity of microscopy is low. Molecular-based techniques, such as PCR, are being used more often as they can provide rapid, specific and sensitive results.

Cryptococcal meningitis

This is caused by the yeast *Cryptococcus neoformans* and infection typically occurs in immunosuppressed patients or patients with other cell-mediated immunity defects. When meningitis is suspected in an immunosuppressed patient, infection with *C. neoformans* should be considered. Cells of *C. neoformans* are surrounded by a capsule which distinguishes this organism from common yeast pathogens such as *Candida* spp. Capsulate yeasts can be easily recognized using India ink stain. By mixing a sample (either a CSF sample or colony emulsified in saline) with India ink and observing under the microscope it is possible to visualize the yeast cell surrounded by a clear capsule. Look at Figure 8.4; it shows the centrifuged deposit of a sample of CSF from an HIV patient with headache and photophobia. The deposit has been mixed with India ink which stains the background; note the yeast-shaped cells surrounded by a clear halo - the capsule. This highly suggestive of *C. neoformans*.

Diagnosis of cryptococcal meningitis can be made by examining a Gram stain of the CSF deposit for the presence of typical yeast cells, with confirmation using an India ink preparation of the CSF deposit to detect capsulate yeasts.

Additional confirmation can be made by detecting cryptococcal antigen in the CSF. The capsular material is shed from the yeast and can be detected in an agglutination test in a similar way to the detection of bacterial antigens in CSF. The amount of antigen present can be measured in a semiquantitative fashion by diluting CSF in saline and carrying out antigen tests on diluted samples - the highest dilution to show a positive reaction is the titre. This has importance as it has been shown that the titre can be used to monitor the efficacy of therapy. Samples are often taken on a weekly basis during therapy. A falling titre indicates successful therapy, whereas a steady or rising titre suggests a poor response to therapy.

Cross reference

See Chapter 11 for methods of yeast identification and susceptibility testing.

CSF samples from immunosuppressed patients with raised white cell counts should be cultured on Sabouraud's agar incubated at 37°C and 30°C for up to 4 weeks, regardless of whether yeasts are seen. Any yeast isolated from the sample should be fully identified and have susceptibility tests carried out.

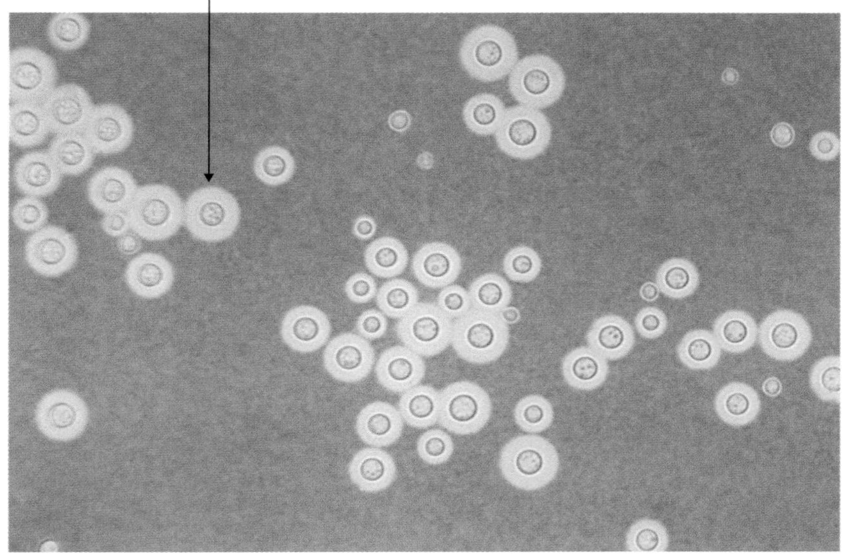

FIGURE 8.4
Photomicrograph of a CSF deposit from an HIV patient stained with India ink. Yeast cells surrounded by a clear capsule are clearly visible. (Photo kindly provided by CDC/ Dr Leanor Haley.)

CLINICAL CORRELATION

Cryptococcal antigen can be used as a tool for diagnosing infection with C. *neoformans*. It can also be used to monitor therapy - declining antigen titre in CSF indicates successful therapy. Many HIV patients with cryptococcal meningitis take antifungal drugs for life to prevent recurrences of infection.

Neurosurgical patients

Acute bacterial meningitis with the exception of neonates typically occurs in patients in the community and is caused by the bacteria described previously. However, meningitis can occur in hospitalized patients and frequently occurs in patients who have had some form of neurosurgical procedure carried out. Following surgery there are often shunts placed into the CSF to remove fluid; infection is frequently acquired through the shunts.

The microbiology of meningitis in neurological patients may be different to that seen in community patients. Common infecting organisms include *Staphylococcus aureus*, coagulase negative staphylococci, *Pseudomonas aeruginosa* and members of the Enterobacteriaceae. CSF samples from such patients are commonly blood stained and/or turbid and have raised white cell counts in the absence of infection.

Samples from neurosurgical patients are processed in the same way as other CSFs, in that they have a cell count, centrifugation, microscopy and culture, but cultural procedures that reflect the different pathogens should be employed. For example anaerobic cultures may be set up and chromogenic media can be inoculated to help give rapid identification of bacteria. If bacteria are seen on Gram staining then medical staff must be informed immediately. Treatment of infections may require the use of antibiotics injected directly into the CSF—**intrathecal** antibiotics. This produces high local levels of drug and can be more effective at eradicating infection. Often daily samples of fluid are received for analysis - declining white cell counts indicates successful treatment of infection.

intrathecal
Intrathecal administration of antibiotics is the administration directly into the CSF. This rapidly produces good levels of drug in the CSF, but toxicity may be a problem.

hydrocephalus
Hydrocephalus means water on the brain. It is a build up of CSF around the brain and it causes enlargement of the head in neonates.

Some patients may have a condition known as **hydrocephalus**, in which excess CSF is produced or it is not absorbed; this condition is most common in infants. Treatment is usually by insertion of a shunt which drains CSF away from the brain either into the abdomen or into the heart. Occasionally, the shunt has to be removed because it becomes infected. This can produce meningitis-like symptoms of infection. CSF may be obtained from such patients and processed as described as above. The actual shunt may be removed and a new shunt inserted. The old shunt should be cultured to determine if it is infected. Any obvious pus on the outside of the shunt should be removed with a swab and cultured. The interior of the shunt should be rinsed with sterile saline and the rinse material cultured. Typical pathogens are coagulase-negative staphylococci, *S. aureus*, *Ps. aeruginosa* and members of the Enterobacteriaceae.

There are various neurological conditions that may give similar symptoms to meningitis, for example **Guillian-Barré syndrome**, multiple sclerosis and CVA (stroke). CSF samples may be obtained from such patients; cell counts and biochemical data help distinguish these conditions from infective meningitis.

Key Point

The microbiology of meningitis in neurology/neurosurgical patients differs considerably from that of meningitis in community patients. Hospital pathogens such as staphylococci, Enterobacteriaceae and *Pseudomonas* spp. are common.

Treatment of meningitis

Meningitis is a serious infection and requires prompt and aggressive therapy. It is recognized that commencing therapy as soon as possible outweighs the need for a microbiological diagnosis. This means that where there is a strong suspicion of meningitis, treatment will commence before a CSF sample can be taken. Treatment is with very high doses of cefotaxime or ceftriaxone given intravenously; these antibiotics are active against the majority of bacteria causing meningitis including penicillin-resistant pneumococci and meningococci. They also penetrate well into the CSF. Treatment may be modified by results of susceptibility tests, and it is important that such tests are carried out on all isolated pathogens.

prophylaxis
Prophylaxis means the administration of antibiotics to prevent disease developing.

Close contacts of patients with meningococcal meningitis may be given **prophylactic** antibiotics to reduce the risk of them developing meningitis.

Cross reference
Antifungal drugs are described in Chapter 11.

Tuberculosis meningitis and cryptococcal meningitis require long-term treatment with specialized antitubercular or antifungal drugs, respectively.

SELF-CHECK 8.5

If a patient was suspected of having bacterial meningitis would the clinician wait until susceptibility test results are known before starting treatment?

Prevention of meningitis - the role of vaccination

Because meningitis is caused by a limited number of bacterial species it is possible to protect against some cases. In several countries, meningococcal meningitis is caused primarily by group B meningococci and the remaining cases are caused by group C meningococci.

- **Vaccination** is available against group C meningococci based on the polysaccharide group C capsule antigen. This is given to babies to prevent infection in the early years of life and a booster is often given in the teens as a second peak of infection occurs among students. There is currently no vaccine against group B meningococci, the predominant cause of meningitis in many countries; this is primarily due to the poor immunogenicity of the group B antigen. Intensive efforts are underway to identify antigens that are able to induce antibody to meningococci regardless of the serogroup.
- A vaccine is also available for pneumococcal meningitis which covers many of the serotypes that are associated with meningitis. This is given to babies and gives protection not only against meningitis but against other pneumococcal diseases.
- *H. influenzae* type b (Hib) used to be the most common cause of meningitis in young children; however, the implementation of a vaccination campaign against this organism has seen the number of cases reduced dramatically.

vaccination
Vaccination is the administration of material (an antigen) to produce immunity to a disease, usually by production of antibodies.

All cultures of *N. meningitidis*, *S. pneumoniae* and *H. influenzae* isolated from CSF or blood should be submitted to a reference laboratory for typing. This identifies the current serotypes in circulation and ensures that current vaccines provide coverage against the common types. The emergence of new types not covered by vaccines may require a change in the vaccine composition.

Education of the public about the signs and symptoms of meningitis is important in ensuring that cases receive urgent medical attention.

Key Point

The antigens used in meningitis vaccines are derived from the capsules of the bacteria. Vaccination only gives immunity to the same capsular type. Group A and C meningococcal vaccines are available but give immunity only against the same serotype. There is currently no vaccine for group B meningococci. Research is ongoing to develop a vaccine which provides immunity against all serogroups of meningococci.

CASE STUDY 8.1 Meningococcal meningitis

An 18-year-old student had a 1-day history of feeling unwell with a temperature; he began to develop a headache and went to bed. The next morning his friends tried to wake him but he was unresponsive so they called a doctor. When the doctor arrived he noted an unresponsive patient with pyrexia and purple blotches on his skin. The doctor immediately administered intravenous penicillin and called an ambulance. In the Emergency Department room a sample of CSF was taken which was found to be turbid. In the microbiology laboratory a cell count was performed; results were 1622 WBC/mm^3 with 95% polymorphs and 5% lymphocytes. No organisms were seen. Because of the high white cell count but absence of organisms, antigen tests against a panel of bacterial antigens were carried out. The antigen test was positive for meningococcal group C antigen. The patient was started on intravenous ceftriaxone. The CSF was cultured and the following day a light growth of *N. meningitidis* was isolated. This was found to be group C using appropriate antisera. Identification of the organism was confirmed using commercial kits. The student's flatmates were given prophylactic rifampicin to prevent them developing meningitis. The patient spent a week on the ICU but made an uneventful recovery.

- **Why did the doctor immediately administer penicillin?**
- **What other condition is this patient likely to have besides meningitis - what samples would be taken to confirm this and why might they be negative in this patient?**
- **What is the relevance of the positive antigen test when the microscopy is negative?**

8.2 Other sterile body fluids

There are other body fluids that are typically sterile but which may become infected. This section of the chapter will focus on the microbiology of these fluids. The fluids we are going to consider are synovial fluid, pleural fluid, ascitic (peritoneal) fluid and pericardial fluid. Processing of all of these samples is very similar and they can be considered together. Processing typically follows the same stages as those used to process CSF samples - a cell count is performed as this gives an indication of the likelihood of infection, the sample is centrifuged, then centrifugation microscopy and culture are performed on the deposit.

Synovial and bursa fluids

arthritis
Arthritis is the name used to describe inflammation in a joint. Septic arthritis is the name used to describe inflammation of a joint caused by infection.

The joints of the body are lubricated by synovial fluid, a viscous yellow fluid. Occasionally this can become infected, causing septic **arthritis**. The main symptoms of this are pain, decreased movement in the joint, redness in the skin over a joint and fever. Fluid can be aspirated from the joint via a needle and sent to the laboratory for analysis. Normal fluid is straw coloured and slightly viscous, but in cases of infection and other inflammatory disorders, such as rheumatoid arthritis, it becomes cloudy due to the presence of white blood cells. Bursae are small sacs of fluid that help lubricate ligaments as they pass over bones. They can become inflamed through trauma or by pressure, but they can also become infected. The main symptoms of infected bursae are pain and tenderness in a joint. Common infecting organisms are *S. aureus* and streptococci.

Pleural fluid

The lung is surrounded by two membranes - the pleural membranes. Pleural fluid is found between the two membranes, and an increase in the volume of pleural fluid is known as a pleural effusion. Effusions may arise because of infection such as pneumonia, TB or carcinoma, or due to conditions such as heart failure. A collection of pus in the pleural cavity, which occurs as a complication of bacterial infection, is known as an empyema. A wide variety of organisms may be involved in empyema, including anaerobes and other organisms causing lower respiratory tract infections, such as *S. pneumoniae*. Empyema is commonly a polymicrobial infection.

Fluid can be withdrawn from the effusion, and is sometimes used to assist the patient, as large volumes of effusion may compromise lung function.

The most common organisms found in pleural fluids are *S. pneumoniae* and *S. aureus*.

Pericardial fluid

The heart is surrounded by a membrane - the pericardium; inflammation of the pericardium is known as **pericarditis** and results in an increase in the volume of fluid in the pericardial sac. When a pericardial effusion occurs it can compromise cardiac function; aspiration is therefore important in improving cardiac function and allows analysis of fluid for infective causes. The most common infective agents in pericarditis are viruses such as enteroviruses and Coxsackie viruses; bacterial causes of infection include *S. aureus*, *S. pneumoniae* and *M. tuberculosis*.

Ascitic or peritoneal fluid

Ascites is an accumulation of fluid in the peritoneal cavity. The most common cause of ascites is cirrhosis or liver disease although cancer may be a cause in rarer cases. Bacterial infection is, however, extremely rare.

Fluid is aspirated from the abdomen and sent to the laboratory for analysis.

SELF-CHECK 8.6

What are the common organisms likely to infect sterile sites?

Processing of fluids

The appearance of the fluids can vary enormously. Normal fluids are yellowish in colour and clear; when infected they become turbid because of ingress of white blood cells. Trauma sometimes causes damage and these fluids are frequently blood stained and become clotted.

As well as microbiological investigations such as microscopy and culture, samples may also be sent for other investigations such as histology for malignant cells and biochemistry for a variety of investigations dependent on the sample and likely diagnosis; viral cultures may also be carried out.

Cell count

Although cell counts can be carried out in a quantitative manner using a counting chamber, because of the viscous nature of some of these fluids, semiquantitative counts are carried out in some laboratories.

Quantitative counts are carried out in the same way as for CSF samples; dilution and red cell lysis may be required, depending on the appearance of the sample.

Semiquantitative estimates of cells are carried out on well-mixed, uncentrifuged samples by taking a small (30–50 μl) sample and placing it on a microscope slide, placing a coverslip on top and examining the slide under the microscope using phase contrast. Red and white cells are easily distinguished and the numbers are recorded in a semiquantitative manner, that is +/– , +, ++, +++. In general, a report of + cells represents approximately 10 cells per high power field.

Bacteria are not always detected in such preparations. Following the cell count the fluid must then be centrifuged to concentrate the organisms into a smaller volume of fluid, increasing the sensitivity of the Gram stain and culture. It must be noted that clots can be present, which will make it impossible to provide a quantitation of cells present.

Key Point

Cell counts may be carried out in a semiquantitative or quantitative manner. Both red and white blood cells should be quantitated.

Microscopy and culture

After centrifugation the supernatant is removed to a clean sterile container, and the deposit is resuspended in a small volume of fluid. The supernatant can be used for biochemical investigations if necessary. The deposit is then used to inoculate culture plates and carry out a Gram stain if required. If any bacteria are seen, they should be immediately reported to a medical microbiologist. If organisms are seen on the Gram stain, additional media and direct sensitivity tests should be set up appropriate to the organisms seen. Typical culture plates for fluids with no organisms seen on Gram stain are a chocolate plate incubated in CO_2 and aerobic and anaerobic blood agar plates. These are incubated for 48 h and discarded if negative. Enrichment cultures using a nutrient rich broth, such as cooked meat broth or serum broth, can also be inoculated. This can be subcultured onto the chocolate plate in CO_2 and aerobic and anaerobic blood agar plates after 48 h incubation.

Additional tests

There are several additional investigations that may be appropriate for certain sample types, and cytology may be appropriate on pleural fluids as various types of carcinoma may manifest as a pleural effusion.

One of the most common causes of swollen, inflamed joints is gout; this is a metabolic disorder characterized by elevated levels of uric acid in the blood. Crystals of uric acid get deposited in the joints, where they are ingested by white blood cells which burst, releasing their contents which then attack the linings of the joint, bringing about pain and inflammation. In cases of gout, the synovial fluid contains white blood cells and uric acid crystals, which are often present inside the white cells. In some hospitals, rheumatology departments screen for uric acid crystals; in other hospitals it is the microbiology laboratory that carries out this investigation.

Identification of crystals requires specialized microscopic equipment and experience. Identification of uric acid crystals is carried out by placing a drop of a centrifuged deposit of a joint fluid on a microscope slide and examining the slide under dark field microscopy. Crystals appear as bright, regular-shaped objects which can be located extracellularly or they can be intracellular, inside white blood cells. Confirmation of the identity of the crystal is carried out using a polarizing filter - uric acid show as crystals with a particular colour under polarized light. Pyrophosphate crystals, which occur in synovial fluid as a cause of pseudogout, are a different colour under polarized light and can therefore be distinguished from uric acid crystals. Figure 8.5 shows a sample of knee fluid from a patient with gout viewed under polarized light; against the pink background (due to the polarizing filter) the long crystals of uric acid appear blue. Also present in this sample are some red blood cells.

The type and quantity of crystals should be reported. Occasionally, other types of crystal, such as oxalate crystals, may be detected in synovial fluids. These have a different appearance to uric acid or pyrophosphate crystals; their identification can be difficult and samples where the presence or identity of crystals is unclear are best referred to an expert for analysis.

Key Point

Gout is a metabolic illness characterized by deposits of uric acid crystals in various joints. Microscopy of aspirated fluid is one of the main methods of diagnosis. In cases of gout, uric acid crystals are seen in the fluid and inside white blood cells. They display a distinct coloration in polarized light, distinct from that observed with pyrophosphate crystals which cause pseudogout.

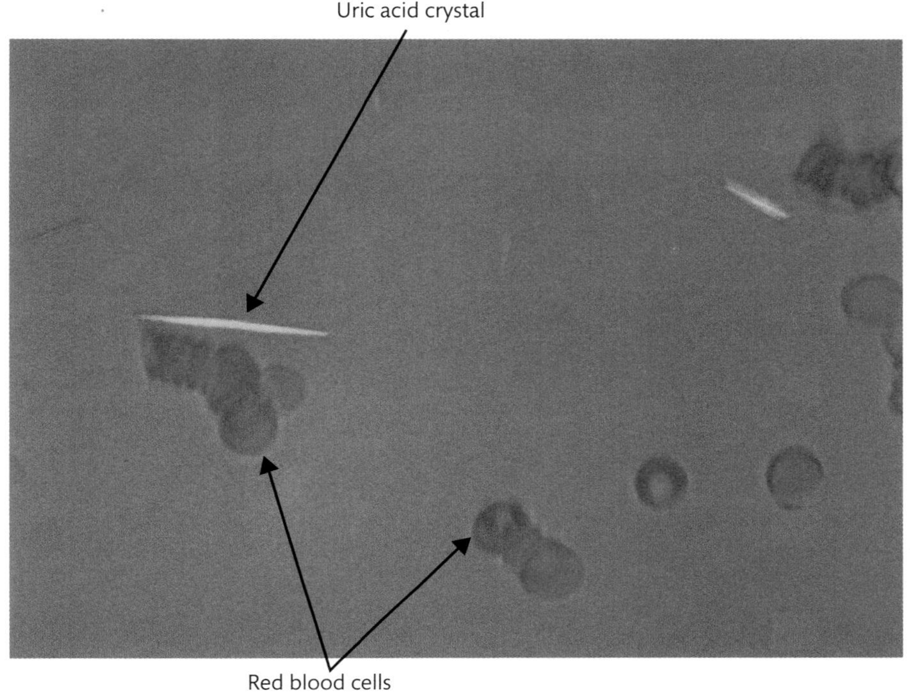

FIGURE 8.5
Photomicrograph of knee fluid from a patient with gout viewed under phase contrast with a polarizing filter. Long crystals are obvious which show blue coloration, indicative of uric acid. (Photo kindly provided by Ed Uthman.)

TB culture

TB culture may be requested on certain samples, particularly pleural fluids. The centrifuged deposit is used to prepare a smear for AAFB and the deposit is sent for TB culture. Microscopy with such samples is not very sensitive and culture should always be performed.

Occasionally, fluids may be received which contain large numbers of white blood cells. Such samples may be better treated as pus samples and additional culture plates set up, for example selective anaerobic plates.

Cross reference
Chapter 9, Section 9.4.

Continuous ambulatory peritoneal dialysis (CAPD) fluids

Patients with kidney failure are unable to rid their body of the waste products normally eliminated by the kidneys. Dialysis is frequently used as a means of eliminating the waste products. Haemodialysis is the most conventional form of dialysis but requires that the patient is attached to a haemodialysis machine for long periods of time, restricting many of their daily activities.

Continuous ambulatory peritoneal dialysis (CAPD) is an alternative treatment strategy for patients with renal failure, and has many advantages over haemodialysis; patients are mobile and active during dialysis and live a relatively normal life without the requirement for dialysis through a machine. Dialysis takes place by instilling 1–2 litres of fluid into the patient's abdominal cavity via a catheter. Waste products diffuse into the dialysis fluid, which is replaced every 6 h. In this way waste products are removed from the patient's blood.

Although CAPD has many advantages over peritoneal dialysis, it has one main disadvantage, peritonitis - infection of the peritoneal cavity. This is a relatively common problem with CAPD and although not life-threatening causes significant pain and discomfort for the patient, and may necessitate haemodialysis until peritonitis is cured. Most infections arise by contamination of the catheter when fluid is instilled into the peritoneal cavity or when it is drained out. The main symptoms of infection in CAPD patients are abdominal pain and pyrexia; the dialysis fluid, which is normally clear and colourless, becomes cloudy due to the presence of white blood cells. The most common organisms causing CAPD peritonitis are coagulase-negative staphylococci; these are usually derived from the patient's own skin. Other organisms frequently isolated from CAPD are *S. aureus*, members of the Enterobacteriaceae and *Pseudomonas* spp.; however, many different types of organism have been isolated as significant pathogens from CAPD fluids.

Key Point

Dialysis is a method used to get rid of waste products that build up in the blood of patients with kidney (renal) failure. There are two main forms - haemodialysis and continuous ambulatory peritoneal dialysis (CAPD). CAPD has many advantages for patients but peritonitis is the main disadvantage.

Processing CAPD fluid

When a patient on CAPD presents with symptoms of peritonitis, a sample of peritoneal fluid is drained into a sterile container for processing. Care must be taken not to contaminate the sample when it is being taken as common skin organisms such as coagulase-negative staphylococci are also frequent pathogens in CAPD fluids. Processing of CAPD fluids follows the same steps as other fluids: cell count, centrifugation followed by Gram stain and culture.

Cell count

Uninfected CAPD fluids are clear and colourless or with a very faint haziness; infected fluids become very cloudy due to the presence of white blood cells. The first part of the process is to enumerate the cells.

The cell count on a CAPD fluid is carried out in a counting chamber, in the same way as a CSF sample. Occasionally, with exceptionally high white cell counts the sample may require diluting in saline before counting. Occasional red blood cells may be seen and their presence should be recorded. Heavily blood stained samples are sometimes received and the red cells can be counted on a diluted sample and white cells counted on a sample diluted with red cell lysing fluid (see above for CSF processing). Counts of >100 WBCs per mm^3 are usually indicative of infection.

Despite the cloudiness of infected CAPD fluids the number of organisms present is often low; the fluid must therefore be concentrated by centrifugation as described previously. Following centrifugation the pellet should be retained and as much of the supernatant as possible removed and retained.

Gram stain

A small amount of the pellet should be used to prepare a Gram stain. The Gram stain should be examined carefully as the organisms may be scarce and in some cases may be inside the white blood cells. It is not uncommon to find no organisms even in very turbid samples and in more than 70% of samples no organisms are seen. If organisms are detected in a sample their presence should be reported to medical staff so that appropriate treatment can be started.

Key Point

The Gram stain is positive in only about 30% of cases of CAPD peritonitis.

Culture of CAPD fluid

There is no defined method for culturing CAPD fluids and different techniques are used in different laboratories. Because organisms are often found inside the white cells and either do not grow or grow slowly, techniques are used to lyse the white cells and release the intracellular organisms. This can be achieved by adding **Triton X** or **saponin** to the deposit; however, addition of a small volume of water to the deposit followed by mixing can bring about lysis of white cells, and is less toxic to fastidious or delicate organisms. Following lysis a drop of the deposit can then be inoculated onto aerobic and anaerobic blood agar plates and a chocolate agar plate incubated in CO_2. The remainder of the sample can be inoculated into an enrichment broth, although some laboratories favour inoculating blood culture bottles, especially if automated detection systems are used as this can speed up detection of infection.

The agar plates should be examined at 24 and 48 h and any enrichment broths subcultured after 48 h. Samples inoculated into blood culture bottles are treated as blood cultures and discarded after 5 days if negative and processed as a positive blood culture when they become positive.

If organisms are seen on Gram stain then appropriate direct sensitivity tests may be inoculated.

Key Point

Improved positivity rates can be obtained by using procedures which lyse cells or enrichment techniques such as inoculation of the deposit into blood culture bottles.

Although culture has a higher positivity rate than microscopy, up to 30% of cultures may be negative in patients with cloudy fluids and signs of infection. If the patient continues to be unwell and fluid is persistently cloudy with an elevated white cell count, additional work can be carried out on samples. Larger volume samples should be obtained and these should be centrifuged and pooled to give more material for further investigations. These could include inoculation of larger volumes into blood culture bottles, culture for yeasts at 37°C and 30°C on Sabourauds agar, prolonged incubation of standard culture plates for fastidious organisms and mycobacterial cultures.

Key Point

Although CAPD fluids may be cloudy and have high white blood cell counts, microscopy is frequently negative and many samples do not yield a pathogen on culture. Where samples have a high white cell count but no organisms are seen, antibiotics will often be administered to the patients. Many culture-negative infections will respond to antibiotics, suggesting that infection is caused by bacteria.

SELF-CHECK 8.7

Why are lysis techniques useful for processing CAPD samples?

Treatment of CAPD peritonitis

In the majority of cases infection remains localized to the peritoneal cavity and does not spread to other parts of the body. Treatment is often by administering antibiotics into the peritoneal fluid. Because the majority of infections are caused by Gram-positive organisms, such as coagulase-negative staphylococci or *S. aureus*, vancomycin, an antibiotic that is active against Gram-positive but not Gram-negative organisms, is commonly used as a treatment. This will often be used even if bacteria are not seen on the Gram stain if the patient has signs of infection and the CAPD fluid is cloudy and contains white cells. If Gram-negative bacteria are seen, gentamicin or tobramycin are used as alternatives. Susceptibility testing of any isolated bacteria should be carried out to provide alternative treatments to vancomycin, gentamicin or tobramycin.

Prevention of CAPD infection

Infection is the most common problem in CAPD patients; the risk of infection can be minimized by strict adherence to correct procedures and the use of aseptic technique. CAPD is best suited to well motivated patients who understand the procedures and the practices required to minimize infection.

After reading this chapter you should know and understand:

- Meningitis and the bacterial pathogens involved
- Pathogens infecting the CSF differ depending upon patient type
- The importance of the CSF cell count and the role of other departments in the diagnosis of meningitis
- How to process CSF samples in the laboratory and the requirement for urgency when processing these samples
- Why certain antibiotics are used to treat meningitis and why only a limited number are useful
- The importance of vaccination to protect from meningitis
- The other sterile body sites and why fluids are taken from these sites
- Fluid samples require different media to be inoculated depending on the sample site
- The importance of non-cultural methods such as antigen detection to provide a diagnosis

FURTHER READING

- **Meningitis Research Foundation. http://www.meningitis.org/**

  Educational site which provides information and patient experiences of meningitis.

- **Merck Online Medical Manual. http://www.merck.com/mmhe/index.html**

  Excellent general source of medical information, containing much relevant information on meningitis and other neurological conditions.

- **CAPD dialysis. http://www.homedialysis.org/learn/types/capd/**

  American site providing information and graphics showing the process of CAPD plus other related material.

DISCUSSION QUESTIONS

8.1 Why are certain CSF pathogens associated with certain patient groups and what can be done to prevent this?

8.2 What is the role of other departments in the diagnosis of meningitis?

8.3 How useful is the Gram stain in providing an initial diagnosis of an infection in a sterile fluid?

9

Investigation of respiratory samples

Louise Hill-King

Respiratory infections are among the most common causes of morbidity. Clinicians send respiratory samples, such as sputum, to the microbiology laboratory in order to identify the causative organisms and to obtain guidance for administering appropriate treatment.

This chapter describes the laboratory investigations that are used for lower respiratory tract samples. Infections of the nose and throat are described in Chapter 7.

Cross reference
Chapter 7.

Learning objectives

After studying this chapter you should be able to:

- Describe the anatomy of the respiratory tract
- List the organisms that constitute normal respiratory flora
- List the common respiratory pathogens
- Describe the routine tests used for respiratory samples
- Discuss infections associated with cystic fibrosis
- Describe the tests used to diagnose tuberculosis (TB)
- Discuss the health and safety issues associated with respiratory samples

9.1 The respiratory tract

Overview of the respiratory tract

In order to appreciate the impact that respiratory infections have on the body we must first gain an understanding of how the healthy respiratory tract works. Its prime function is to supply oxygen to the body and remove carbon dioxide from it.

Look at Figure 9.1 to appreciate the anatomy of the lower respiratory tract.

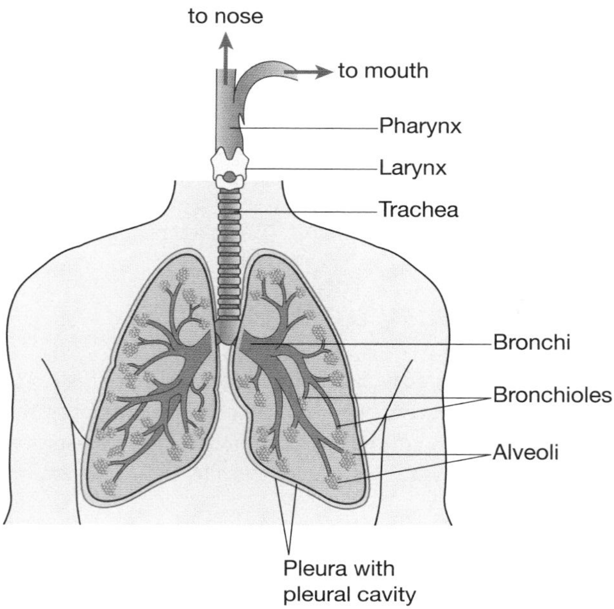

FIGURE 9.1

Schematic illustration of the respiratory tract.

After passing through the nose or throat, air travels through the larynx (voicebox) and trachea (windpipe) into the lower respiratory tract. The rest of the tract is referred to as the bronchial tree because of the way the airways branch into increasingly smaller airways. The smallest branches end in tiny balloon-type sacs called alveoli. This is where the gas exchange of oxygen (into the blood) and carbon dioxide (out of the blood) takes place. A network of capillaries surrounds each alveolus where the membranes are so thin that molecules of oxygen and carbon dioxide can diffuse across them.

Respiratory tract membranes

Each lung is surrounded by a double-layered serous membrane, called the pleura. The two layers are lubricated by the pleural fluid.

There are two features of the respiratory tract lining which are crucial in preventing infection. Firstly, mucous glands produce mucus, a sticky substance which traps dust particles and organisms. Secondly, the motion of **cilia** moves the mucus, along with its trapped particles, upwards to the pharynx where it is swallowed.

Biological methods of removing organisms include the action of antibodies (in mucus), enzymes (e.g. lysozyme) and **macrophages** (in the alveoli).

cilia

Cilia are hair-like beating structures which protrude from respiratory epithelial cells.

macrophage

Macrophages are white blood cells (WBCs) which engulf then digest foreign bodies including organisms.

Key Point

The lower respiratory tract is a branching series of tubes which deliver oxygen to the blood and remove carbon dioxide from it.

BOX 9.1 Smoking damages cilia

The cleansing action of the respiratory cilia is diminished in smokers. Smoke has a paralysing effect on cilia so their beating becomes slower. The mucus, containing trapped bacteria, is not removed from the respiratory tract swiftly enough. This increases the risk of infection.

colonize
Organisms are said to colonize a body site when they live there without causing harm.

normal flora
Normal flora or human microbiotica is the collective term for the organisms colonizing a given body site.

α-haemolytic, β-haemolytic
Bacterial colonies can be described as α-haemolytic, β-haemolytic or non-haemolytic depending on their action on the red blood cells in the culture medium.

Normal flora of the respiratory tract

Many of the body's mucous membranes are covered with colonizing bacteria. Since the respiratory tract is open to the external environment, many organisms are continually entering the tract. We have already discussed various ways in which the body removes these organisms. Some, however, manage to **colonize** the inner surface of the tract. We refer to such organisms as **normal flora** or human microbiotica.

The composition of each person's normal flora varies and is influenced by many factors including age, sex, nutrition, genetics and environmental exposure. Typically, normal respiratory flora is a mixture of non-haemolytic and **α-haemolytic** streptococci, *Neisseria* spp., *Haemophilus* spp. and *Moraxella* spp. Importantly, organisms that are pathogenic for the lower respiratory tract are often part of the normal flora of the upper respiratory tract.

Disruption of the normal flora

The composition of the normal flora can gradually change with time. Some situations, however, cause an abrupt change which may or may not result in disease. Examples of such disruption include:

- Viral infections
- Antibiotics
- Injury
- Invasive instrumentation

Hospitalization, especially when patients are on antimicrobials and mechanically ventilated, tends to result in the replacement of the pre-existing normal flora by Gram-negative bacilli such as *Escherichia coli*, *Pseudomonas aeruginosa*, *Acinetobacter* spp. and *Enterobacter cloacae*. Since antibiotics kill much of the normal flora, resistant organisms are able to flourish.

Key Point

The respiratory tract is colonized by many organisms which do not usually cause harm. Anything that affects the body's ability to maintain a healthy equilibrium between these organisms can lead to infection.

Infections of the lower respiratory tract

Infections result when organisms cease to merely colonize our cells and manage to invade them, causing damage. Alternatively, infections occur when a new organism is encountered.

Infections of the lower respiratory tract are generally described according to the region that they affect rather than by the infecting organism. For example:

- Bronchitis: inflammation of the bronchi
- Bronchiolitis: inflammation of the bronchioles
- Pneumonia : inflammation of the alveoli.

Common bacterial pathogens

It would be difficult to list every organism that is capable of causing a lower respiratory tract infection (LRTI). Many cases are caused by viruses. In this chapter, we shall concentrate on bacterial causes of LRTI, although we shall also consider the main fungal pathogens.

Let us consider the bacteria that can cause pneumonia. The range of common pathogens varies depending on factors such as:

- Where the pneumonia was acquired: in hospital or in the community
- Whether the patient has an underlying disease
- The patient's age.

Cross reference
Chapter 13.

Community-acquired pneumonia

The most common bacterium to cause community-acquired pneumonia is *Streptococcus pneumoniae*. Other frequent culprits are *Haemophilus influenzae*, *Mycoplasma pneumoniae*, *Staphylococcus aureus* and *Moraxella catarrhalis*. With the exception of *M. pneumoniae*, these organisms are part of the normal upper respiratory flora.

S. aureus is associated with secondary infections following influenza or measles. The viruses responsible for these primary infections damage the respiratory tract lining. This makes the patient more susceptible to infection by bacteria which were previously colonizing the tract.

Typical and atypical pneumonia

M. pneumoniae mainly affects children and young adults. It is the second most common non-viral cause of community-acquired pneumonia. Unlike the typical pneumonia caused by *S. pneumoniae*, *H. influenzae*, *S. aureus* and *M. catarrhalis*, with acute onset and production of sputum, the pneumonia caused by *M. pneumoniae* is described as **atypical pneumonia** with minimal, if any, sputum production and chronic onset.

Typical and atypical pneumonia are caused by different groups of organisms; atypical pneumonia is associated with **intracellular organisms**, that is organisms which can only multiply within the cells of their host. In addition to *M. pneumoniae*, causes of atypical pneumonia include:

- *Chlamydophila psittaci* (acquired from parrots, causes psittacosis)
- *Coxiella burnetii* (acquired from farm animals, causes Q fever)
- *Legionella pneumophila* (acquired from water storage systems, often associated with foreign travel, causes Legionnaire's disease).

Typical and atypical pneumonia can be differentiated by non-microbiological techniques such as chest X-rays. As we shall learn later, microbiological testing does not always identify a pathogen so other diagnostic tools are important.

Hospital-acquired pneumonia

Although pneumonia in hospitalized patients may be caused by the same organisms as community-acquired pneumonia, it is often caused by different organisms.

As we saw earlier, the composition of the normal flora is altered by such interventions as antibiotics and invasive instrumentation. Hospitalized patients are more likely to harbour resistant organisms such as meticillin resistant *S. aureus* (MRSA). Intensive care patients frequently require mechanical ventilation. As a consequence of this, they tend to be colonized with Gram-negative bacilli more usually associated with the intestines. These organisms can then cause pneumonia. If an infection develops more than 48 h after admission to hospital, it is described as **nosocomial infection.**

Nosocomial pneumonia is increasingly caused by bacteria that are resistant to multiple antibiotics. Many of these organisms were previously thought to have low pathogenicity but now cause particular problems in intensive care wards. *Stenotrophomonas maltophilia* and *Acinetobacter baumannii* are examples of such organisms.

Neonates (babies less than 1 month old) with pneumonia are usually infected by organisms from their mother's normal vaginal flora. Examples include *E. coli* and *Streptococcus agalactiae* (Group B *Streptococcus*).

Predisposing factors

Underlying diseases may increase the risk of developing pneumonia. Situations that affect consciousness, such as strokes, seizures or alcoholism, increase the likelihood of **aspiration pneumonia**. This occurs when the epiglottis fails to cover the larynx during swallowing. Food and saliva, which would normally be directed down the oesophagus, enter the trachea and can lead to pneumonia. This can give rise to infections with organisms normally found in the mouth. Examples are:

- Actinomycosis (caused by *Actinomyces* spp.)
- Mixed anaerobic/streptococcal infections.

A major predisposing factor for respiratory tract infection is the immunological status of the individual. **Immunocompromised** patients are at the greatest risk of infection as they do not possess the necessary mechanisms to resist infection.

immunocompromised, immunodeficient, immunosuppressed
Immunocompromised, immunodeficient and immunosuppressed all refer to cases where the patient's body is unable, or less effectively able, to resist infection. This can result from drug action or disease.

In addition to the organisms that can infect immunocompetent individuals, immunocompromised patients can develop pneumonia with organisms that would not ordinarily cause disease. Examples of this include:

- Nocardiosis (caused by *Nocardia* spp.)
- *Mycobacterium avium* complex (MAC) infections
- *Pneumocystis* pneumonia (PCP), caused by *Pneumocystis* (formerly *Pneumocystis carinii*).

MAC organisms will be discussed in Section 9.4. Sometimes the isolation of unusual pathogens from a patient can alert the clinician to immunodeficiencies that were previously undetected.

Lung abscesses

Abscesses are pus-filled cavities surrounded by inflamed tissue. They can form in the lungs as a complication of aspiration pneumonia or they can arise from blood-borne spread of organisms from other sites. Lung abscesses can be caused by a mixture of organisms, commonly anaerobes and members of the *Streptococcus anginosis* group (formerly the 'milleri' group). If there are multiple small abscesses the condition is called **necrotizing pneumonia**.

Whooping cough

Whooping cough, or pertussis, is a two-stage disease. The first stage, which lasts around 10 days, is fairly non-specific and symptoms resemble those of a common cold. During this stage, the bacterium *Bordetella pertussis* colonizes the cilia in the upper respiratory tract. It multiplies there and can be spread to other people by coughing or sneezing. The second stage, which is mediated by toxins, is more specific. Patients develop fits of prolonged coughing ending with a characteristic 'whoop'.

A mild form of whooping cough is caused by another bacterium, *Bordetella parapertussis*.

Common fungal pathogens

Fungi are sometimes isolated from lower respiratory tract samples. Fungi are discussed in detail in Chapter 11 but it is worth considering the most common respiratory fungi here. Fungi are divided into two groups: yeasts and moulds.

Cross reference
Chapter 11: Mycology.

Yeasts

Yeasts are common components of normal respiratory flora. Since the majority of lower respiratory samples are collected in a way which contaminates them with upper respiratory flora, it is not surprising that many of them are found to contain yeasts, primarily *Candida* spp. It is thought, however, that these merely reflect overgrowth of yeasts in the upper respiratory tract and that genuine candidal pneumonia is rare.

Moulds

The most common types of mould found in the respiratory tract are *Aspergillus* spp. *Aspergillus* spores are inhaled from the air and do not generally cause disease but can be pathogenic for immunosuppressed patients or those with pre-existing lung disease. When *Aspergillus* grows in the lungs it forms a ball of fungus called an **aspergilloma**.

A particular problem in hospitals dealing with immunosuppressed patients is that disturbance of the environment, as occurs with building work, increases the circulation of fungal spores in the air. This increases the likelihood of fungal lung infections.

BOX 9.2 Overgrowth by Candida spp.

Although yeasts commonly colonize the upper respiratory tract they are usually present in relatively low numbers. Some medical interventions disturb the balance of the normal flora and allow the number of yeast cells to markedly increase. The resultant white coating of areas such as the tongue is called oral thrush. Drugs that are commonly associated with this problem include antibiotics and cancer drugs. Apart from being unpleasant, this situation makes interpretation of sputum cultures difficult.

There is a group of fungal respiratory infections that are relatively rare in the UK but need to be considered if patients have a relevant travel history. These include:

- Histoplasmosis
- Blastomycosis
- Coccidiomycosis
- Paracoccidiomycosis.

P. jiroveci is an unusual organism. It is now believed to be a fungus but has previously been classified as a protozoon.

Parasites

Parasitic infections of the respiratory tract are rare in the UK but may be seen in immigrants or returning travellers. Parasites that may pass through the lungs include *Ascaris lumbricoides*, *Strongyloides stercoralis* and *Paragonimus westermani*. One of the body's responses to parasitic infection is the proliferation of **eosinophils**. Eosinophils are a type of WBC associated with allergy. They may be detected in sputum microscopically.

Key Point

Lower respiratory tract infections may be caused by viruses, bacteria, fungi or parasites. Bacterial infections are usually caused by organisms from the upper respiratory tract. Hospital patients tend to be infected with different organisms to those infecting patients from the community.

9.2 Processing of lower respiratory samples

Sample types

The most common respiratory sample received by microbiology laboratories is sputum. The advantage of this sample is the ease with which it can be obtained, especially if the patient has a **productive cough.**

productive cough
A productive cough results in the expectoration of sputum.

Ideally, patients should rinse their mouth with water prior to providing a sputum sample and should avoid adding saliva to their sputum. Unfortunately, such advice is often lacking and the poor quality of many samples significantly diminishes the validity of the results. Small children are unable to produce sputum so cough swabs are sometimes taken as an alternative.

We have already considered the problems associated with the contamination of lower respiratory samples by upper respiratory flora. One approach to solving this problem is to physically bypass the upper tract in order to take a sample directly from the lower tract. Bronchoscopes are used for this purpose.

BOX 9.3 Bronchoscopy samples

A bronchoscope is a thin, flexible, fibre-optic telescope which is passed into the bronchi via the nose, mouth or an endotracheal tube. Sterile saline is introduced through the bronchoscope and then aspirated into a sterile container to send to the laboratory.

The two most common sample types collected in this way are:

- **Bronchial washings–from large airways**
- **Bronchoalveolar lavages (BALs)–from small airways.**

Pleural fluids are also sent for microbiological investigation. As we learned earlier, the two layers of the pleura are lubricated by the pleural fluid. The volume of this fluid is normally around 15 mL. If it accumulates, either by overproduction or by failure to drain away, we call this a **pleural effusion**. Samples are collected by inserting a needle between the ribs through the back. There are many causes of pleural effusion, including infection. If pus is present in the fluid, the condition is called **empyema;** this is likely to be due to bacterial infection.

empyema
Empyema is a collection of pus and necrotic tissue in a body cavity, e.g. the pleural cavity.

In cases of suspected whooping cough, testing can be performed on nasopharyngeal aspirates, nasopharyngeal swabs or **pernasal** swabs. The organism is only present in high numbers during the first 3 weeks of the illness. Later stages of the illness require blood tests for specific antibodies.

pernasal
A pernasal swab is taken by passing a thin, flexible swab through the nose to the back of the throat.

Key Point

The majority of lower respiratory tract samples received by microbiology laboratories are sputum. Cough swabs, bronchial washings, bronchoalveolar lavages, pleural fluids and lung biopsies are also received.

Sample processing

When sputum is received in the laboratory, its macroscopic appearance is noted. This is to assess how **purulent** the sample is and to note any bloodstaining.

purulent
A purulent sample is a sample containing a lot of pus.

Terms used to describe sputa include:

- Purulent (pus-like)
- Mucoid (mucus-like)
- Salivary (saliva-like)
- Mucopurulent (part mucoid, part purulent)
- Mucosalivary (part mucoid, part salivary)
- Bloodstained (in addition to any of the above).

Bloodstained sputum samples are indicative of **haemoptysis,** which may be present in cases of pulmonary TB.

haemoptysis
Haemoptysis literally means 'the spitting of blood'.

Laboratory policies regarding poor sputum samples vary. Some laboratories do not process salivary samples. Other laboratories are more stringent and do not process mucoid samples either. These decisions reflect the likelihood that such samples originate from the upper respiratory tract. Any organisms that are present merely reflect the patient's normal upper respiratory flora and it would be misleading to imply that they were responsible for causing the lower respiratory symptoms.

Additional information can be obtained by examining sputum microscopically. Gram staining may be used for this. A good sputum sample will contain many WBCs and few epithelial cells. In contrast, a poor sputum sample will contain few WBCs and many epithelial cells. In general >25 epithelial cells per low-power field on microscopy is considered to be an unacceptable sample for culture.

The decision to discard a sputum sample isn't always appropriate. If a patient is immunocompromised, shows signs of atypical pneumonia or is suspected of having TB, then their samples should be processed, regardless of their macroscopic or microscopic appearance.

SELF-CHECK 9.1

Why is it inappropriate to consider a mucoid sample from an immunocompromised patient to be unsuitable for culture?

Homogenization of sputum samples

Sputum is viscous due to the presence of pus and mucus. Also, the organisms in sputum are distributed unevenly throughout the sample. In order to overcome these limitations, it is common practice to use digestion agents to liquefy the samples, and agitation to homogenize them. Examples of digestion agents are dithiothreitol (DTT) and *N*-acetyl-L-cysteine (NAC). These compounds do not affect the viability of respiratory pathogens.

Since the number of organisms causing an infection should be significantly higher than the number of contaminating organisms, dilution of the sample is performed. This removes much of the normal flora but retains the significant organisms. The dilution factor used to achieve this is 1 : 1000.

Concentration of lower respiratory samples

In contrast to the dilution of sputum described above, pleural fluids and samples that have been obtained by bronchoscopy, such as BALs, need to be concentrated. Contaminating organisms should not be present in these samples and any pathogens present will have been diluted by the saline used in bronchoscopy.

Concentration is achieved by centrifugation. The supernatant is removed and the deposit is used for microscopy and culture.

Key Point

Respiratory samples, particularly sputum, need to be pretreated before they are suitable for culture.

SELF-CHECK 9.2

Why do some lower respiratory samples need to be diluted before testing whilst others need to be concentrated?

Microscopy techniques

A variety of staining techniques are used on lower respiratory samples to visualize organisms and/or blood cells. The most common procedures are:

1. Gram staining
2. Immunofluorescence for *Legionella* spp.
3. Immunofluorescence or histochemical staining for *P. jiroveci*
4. Differential staining of WBC for eosinophilia
5. Acid-alcohol fast bacilli (AAFB or AFB) staining.

We shall consider AAFB staining in Section 9.4. We shall look at the rest in turn here.

Gram staining

We have already seen how Gram staining can be used to assess the quality of sputum samples. It can also be used diagnostically to predict which organisms are present. This is only presumptive identification and its value depends on the organisms seen. Table 9.1 summarizes the common organisms that cause pneumonia and describes their appearance on a Gram stain.

TABLE 9.1 The Gram stain appearance of common lower respiratory pathogens.

Organism	Gram appearance	Further information
S. pneumoniae	Gram-positive cocci	Diplococci*, lancet-shaped
S. aureus	Gram-positive cocci	Clusters ('bunches of grapes')
M. catarrhalis	Gram-negative cocci	
N. meningitidis	Gram-negative cocci	
H. influenzae	Gram-negative bacilli	Slender bacilli or coccobacilli
Enterobacteriaceae	Gram-negative bacilli	Large
P. aeruginosa	Gram-negative bacilli	Large
S. maltophilia	Gram-negative bacilli	Large
L. pneumophila	Gram-negative bacilli	Stain poorly, pleomorphic†
B. pertussis	Gram-negative bacilli	Coccobacilli‡
Candida spp.	Gram-positive	Large, oval-shaped
M. pneumoniae	No Gram reaction	No cell wall to stain

* pairs of cocci.
† short when young but long when old.
‡ short with rounded ends.

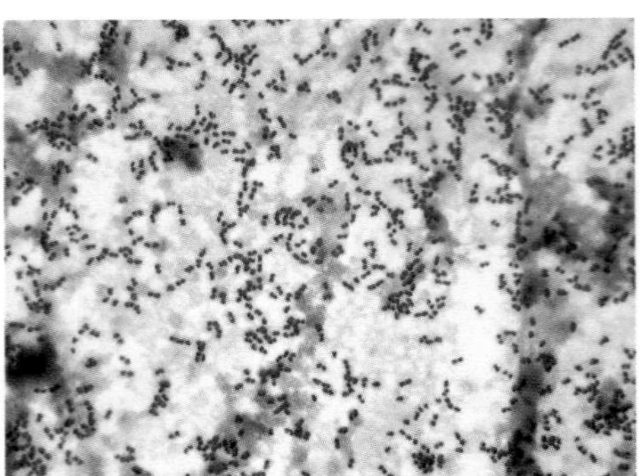

FIGURE 9.2
Gram stain of *S. pneumoniae* showing characteristic Gram-positive cocci in pairs. (Courtesy of Mike Grover.)

A useful feature of *S. pneumoniae* is its distinctive appearance on a Gram stain. Remember that this is the most common bacterial cause of community-acquired pneumonia. Its cells are seen as Gram-positive diplococci. If you look at the dark purple cells in Figure 9.2 you can see that the cocci, arranged in pairs, are not spherical but slightly elongated with pointed ends. This appearance is referred to as lanceolate (lancet-shaped). In addition to this characteristic shape, a Gram stain of *S. pneumoniae* can reveal a zone of clearing around each pair of cells. This is the polysaccharide capsule surrounding the cells.

The majority of other bacteria associated with pneumonia are insufficiently distinctive for a Gram stain to have much predictive value; they may look identical or very similar to other organisms which are not pathogenic. Yeast cells are significantly larger than bacterial cells, have an oval shape and appear Gram-positive.

We learned earlier that *M. pneumoniae* is the second most common cause of community-acquired bacterial pneumonia. An unusual feature of mycoplasmas is their lack of cell walls. Since Gram staining relies on the dyes binding to the cell wall, mycoplasmas are not detectable by this method.

Microscopy for detection of Legionella spp.

L. pneumophila, which causes Legionnaire's disease, can be detected in lower respiratory samples using immunofluorescence. This technique utilizes specific antibodies which bind to specific *Legionella* antigens in the sample. Bound antibodies are detected by the fluorescence of an attached dye. The two methods used for this are referred to as DFA (direct fluorescent antibody) and IFAT (indirect fluorescent antibody test). However, these tests are associated with a poor predictive value for *Legionella* infection.

Microscopy for PCP

induced sputum
A sputum sample can be induced by inhaling nebulized saline.

Microscopy is the principal means of diagnosing PCP, as *P. jiroveci* does not grow on routine culture media. The optimal sample types are BAL, **induced sputum** or lung tissue.

Immunofluorescent microscopy, similar to that used for *Legionella* spp., may be used for *P. jirovecii*. Alternatively, a variety of histochemical stains can be used including:

- Methenamine silver nitrate
- Calcofluor white
- Giemsa

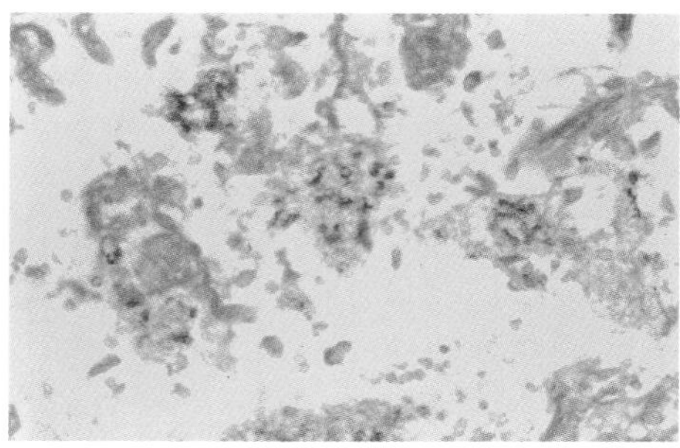

FIGURE 9.3
P. jirovecii cysts staining black with methanamine silver stain. (Courtesy of Dr Mona Elmahallawy.)

Departments other than microbiology, such as histology, often perform PCP testing. Figure 9.3 shows cells of *P. jiroveci* stained black by methenamine silver stain.

SELF-CHECK 9.3

If PCP testing is requested on a sample, what does this imply about the patient?

Differential staining of WBC (eosinophils)

As their name suggests, eosinophils take up the acidic stain eosin. More specifically, the stain is concentrated within the granules of their cytoplasm, giving them a characteristic granular appearance which differentiates them from other WBCs. Their presence in a respiratory sample indicates an allergic response.

Key Point

Microscopy of lower respiratory samples can be used either to indicate the quality of a sample, to give an indication of likely pathogens or to specifically detect pathogens.

Common isolation media

Most of the bacteria and fungi associated with LRTIs can be isolated using routine culture techniques. The main exception is *M. pneumoniae*. Although isolation media for mycoplasmas exist, they are not commonly used in laboratories; diagnosis of *M. pneumoniae* infection usually relies on serological tests. Reference laboratories offer molecular tests for respiratory samples, where necessary.

Cross reference
Chapter 3: Culture media.

Cross reference
Chapter 13: Molecular testing.

The exact selection of culture media varies between laboratories but will always reflect the spectrum of organisms associated with LRTIs. A combination of **selective** and **non-selective** agars is generally necessary.

non-selective, selective
Non-selective agars support the growth of a wide range of organisms; they do not contain inhibitory substances. Selective agars contain substances which inhibit the growth of certain organisms in order to enhance the isolation of others.

Commonly, three basic media are used for lower respiratory samples:

- Chocolate agar
- Blood agar
- Cysteine lactose electrolyte deficient (CLED) agar or McConkey agar.

FIGURE 9.4
Blood agar plate showing characteristic draughtsman-like colonies of *S. pneumoniae*. (Courtesy of Deane Rhone.)

Between them, these support the growth of:

- *S. pneumoniae*
- *H. influenzae*
- *S. aureus*
- *M. catarrhalis*
- Enterobacteriaceae
- *Pseudomonas* spp.
- *S. maltophilia*
- *Candida* spp.
- *Aspergillus* spp.

In order to recover *S. pneumoniae* and *H. influenzae* directly from clinical samples, the incubation atmosphere needs to be enriched with carbon dioxide. In this atmosphere *S. pneumoniae* produce characteristic 'draughtsman' like colonies on blood agar; look at Figure 9.4 to see these.

Chocolate agar can be made more selective for *H. influenzae* by adding an antibiotic, bacitracin.

bronchiectasis
Bronchiectasis is the irreversible dilation of the bronchi.

Although *S. aureus*, *Candida* spp. and *Aspergillus* spp. can grow on the media listed above, selective agar is indicated for immunocompromised patients and those with **bronchiectasis.** Several media that are selective for *S. aureus* are available. Sabouraud's agar is the most common fungal agar for respiratory samples. It contains nutrients which enhance the growth of fungi, and antibiotics to suppress the growth of bacteria. Selective **chromogenic agars** are also available for both *S. aureus* and *Candida* spp. These utilize colour to highlight the organisms of interest.

SELF-CHECK 9.4

If the pathogens grow on the basic isolation media, why are selective agars necessary?

Samples from cystic fibrosis patients need further selective plates; these are discussed later.

If Legionnaire's disease is suspected, or if a patient has signs of atypical pneumonia, then selective agar for *L. pneumophila* is necessary. Buffered charcoal yeast extract (BCYE) agar is available in various formulations, each containing a different cocktail of antimicrobial substances to inhibit the growth of other organisms. Pretreatment of samples by heat or dilution can also remove other organisms. The agars need to contain cysteine and iron salts since these are growth requirements for *L. pneumophila*.

Colonies of *L. pneumophila* have a ground glass, iridescent appearance. As *L. pneumophila* is a slow-growing organism, the plates need to be incubated for up to 10 days.

Sometimes, *Legionella* infections are diagnosed by an alternative test: the urinary antigen test (*L. pneumophila* serogroup 1 only). This is as sensitive as culture of respiratory samples, is highly specific and is simpler to perform. A disadvantage of the antigen test is the lack of epidemiological data which culture provides. These data can be valuable in outbreak situations.

Culture of *Bordetella* spp., for cases of suspected whooping cough, requires selective agar containing charcoal and blood. An antibiotic, cephalexin, is incorporated into the agar to inhibit normal respiratory flora. These plates can be made available to nursing staff so they can be inoculated directly as the organism often dies rapidly on swabs.

A high level of humidity in the incubation atmosphere is important for these organisms and the plates need to be incubated for up to 7 days. Colonies of *B. pertussis* are domed with a pearly appearance.

We have seen the importance of receiving a good quality sample, we have looked at ways of pretreating the sample and we have considered the range of media used to grow the organisms associated with LRTIs. Despite all of these requirements being satisfied, a significant proportion of samples do not yield a pathogen. Many of these cases are due to viral infection but others may be caused by organisms which we are currently unable to culture or are not yet recognized as pathogens.

Key Point

A combination of selective and non-selective agars is used to culture respiratory samples. Certain clinical details prompt the inclusion of further selective agars.

Confirmatory tests

Experienced biomedical scientists are able to presumptively identify the pathogens isolated from lower respiratory samples by examining the **primary culture plates.**

We have already seen how the distinctive microscopic appearance of *S. pneumoniae* is useful. Similarly, its macroscopic appearance can also be distinctive. Colonies of *S. pneumoniae* can undergo **autolysis** which results in the centre of the colony subsiding. These colonies are described as 'draughtsman' colonies.

Most presumptive identifications need to be confirmed, usually by biochemical tests. Some of these can be incorporated into the primary culture step. For example, addition of an **optochin** disc to a primary plate serves to differentiate *S. pneumoniae* from other α-haemolytic colonies.

If you study Table 9.2, you will see some of the other confirmatory tests which may be used for the identification of respiratory pathogens. There is a growing trend for laboratories to use automated instruments which perform identification tests as well as antimicrobial susceptibility tests but these have not yet superseded the need for manual identification techniques entirely.

primary culture plates
Primary culture plates are those onto which the clinical samples are directly inoculated. Organisms from primary plates can be inoculated onto further plates; these secondary plates are called subcultures.

autolysis
Autolysis is where an organism produces enzymes which lead to the breakdown of its own cells.

optochin
Optochin is another name for ethyl hydrocuprein hydrochloride. Most strains of *S. pneumoniae* are sensitive to optochin whilst most other α-haemolytic streptococci are resistant.

Cross reference
Chapter 2 describes identification tests in more detail.

TABLE 9.2 **Confirmatory tests for common lower respiratory pathogens.**

Organism	Confirmatory tests
S. pneumoniae	Optochin sensitivity, bile solubility, latex agglutination
S. aureus	Latex agglutination, DNase, coagulase, selective agar
M. catarrhalis	Butyl esterase, tributyrin
N. meningitidis	Commercial kit
H. influenzae	X/V dependency
Enterobacteriaceae	Commercial kit
P. aeruginosa	Oxidase, growth at 42°C, commercial kit
S. maltophilia	Oxidase, resistance to imipenem, aesculin, commercial kit
L. pneumophila	Subculture to BCYE +/– cysteine, autofluorescence, latex agglutination
B. pertussis	Agglutination with antisera
Candida spp.	Pseudopods, germ tube, selective agar

coliform
Coliform is a loose term used for bacteria which are '*Escherichia coli*-like'. They are members of the Enterobacteriaceae family which ferment lactose to produce acid and gas at 35–37°C.

In some instances, identification of the organism to species level may not be necessary. Many laboratories would report a yeast as '*Candida* sp.' or a member of the Enterobacteriaceae as a **'coliform'**.

Routine microbiology laboratories can identify common respiratory pathogens to the required level for most situations. Occasionally, further testing by a reference laboratory is necessary. Reasons for this include:

BOX 9.4 Confirmatory tests used for Haemophilus spp.

Colonies of presumptive *Haemophilus* spp. can be identified as *H. influenzae* by demonstrating their requirement for two growth factors: X factor and V factor. This is shown in Figure 9.5.

Colonies demonstrating requirement for only one of the factors may be another species of *Haemophilus*, e.g. *H. parainfluenzae* requires only V factor.

Both X and V factor are available from chocolate agar but only X factor is available from blood agar. The phenomenon of satellitism occurs on blood agar, where *H. influenzae* is able to grow around colonies of other organisms producing V factor, e.g. *S. aureus*.

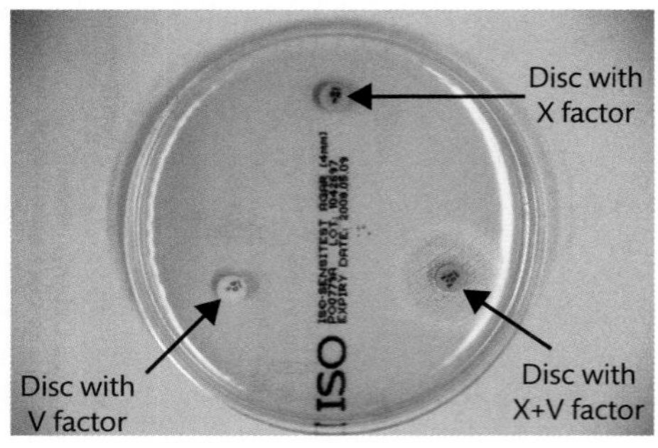

FIGURE 9.5
Iso-Sensitest agar showing *H. influenzae* only growing around a disc containing both X and V factor. (Courtesy of Louise Hill-King.)

- Outbreaks of infection requiring typing of organisms
- Unusual antimicrobial susceptibility patterns requiring confirmation.

Key Point

Bacteria isolated on culture plates need to be identified to genus or species level, usually by means of biochemical tests.

Antimicrobial susceptibility testing

Often the most important aspect of a microbiology report is the information regarding antimicrobials. The range of antimicrobials against which an organism is tested is determined by the type of organism. The choice of antimicrobials to include in the report is determined by factors such as:

- The severity of the disease
- Whether or not the patient is in hospital
- The patient's age
- Pregnancy
- Drug allergies, e.g. penicillin
- Local factors, e.g. known circulation of resistant strains.

Cross reference
Antimicrobial susceptibility testing methods are discussed in Chapter 4.

As with other infections, patients with an LRTI will frequently be prescribed a 'best guess' antibiotic, based on likely pathogens. Once the results are available, a different antibiotic can be prescribed if appropriate.

The disadvantage of partial identification is that correct interpretation of antimicrobial susceptibility test results often requires species-level identification. Several types of coliform are known to be naturally resistant to aminopenicillins. Ampicillin and amoxycillin, which are aminopenicillins, are commonly prescribed for LRTIs.

Antifungal susceptibility testing is not routinely carried out for respiratory infections so treatment is often empirical. *Candida albicans* is generally sensitive to the antifungal agents that are likely to be prescribed, whereas some other species of *Candida* are resistant. In this situation, the differentiation of '*albicans*' and 'non-*albicans*' yeasts is helpful.

Key Point

Bacteria isolated from lower respiratory samples are challenged with a range of antimicrobial substances in the laboratory. The results of these tests form the basis of recommendations for appropriate treatment.

9.3 Cystic fibrosis

Cystic fibrosis (CF) is an inherited disease. CF has such a profound effect on the patient's lungs that they experience respiratory infections which merit special attention. First, let us examine the symptoms that predispose them to these infections.

Symptoms and outcome of cystic fibrosis

CF occurs due to mutation of the CF transmembrane conductance regulator (*CFTR*) gene. This leads to an imbalance of sodium and chloride ions throughout the body, resulting in the dehydration of secretions from affected organs, including the lungs. These thick, dehydrated secretions block passages in the body and lead to the classical CF symptoms.

Earlier, we learned that mucus is continually cleared from the respiratory tract by cilia. In CF patients, the mucus is so thick that the cilia are ineffective, so the mucus collects in the airways. Trapped organisms multiply in this mucus and form biofilms. These biofilms provide a reservoir of organisms which can cause infections. The body mounts an inflammatory response to clear the organisms; this causes further damage. The airways respond to the damage caused by repeated episodes of inflammation by dilating irreversibly. This condition, which is not unique to CF patients, is called bronchiectasis.

With good health care, CF patients can now survive beyond 40 years but eventually the cumulative damage to the lungs culminates in respiratory failure and death. The life of some patients can be prolonged by lung transplantation.

> *Key Point*
>
> **Cystic fibrosis is an inherited disease which involves progressive lung damage associated with recurrent infections.**

Organisms associated with CF

Initially, pneumonia in CF patients is caused by the same organisms as in the rest of the population. Infections with typical respiratory pathogens, such as *H. influenzae*, tend to be followed by infection with *S. aureus*. Another, more complex, stage then develops which is characterized by *Pseudomonas* spp. including *P. aeruginosa*.

Once *P. aeruginosa* has colonized the CF respiratory tract it adapts well to its environment and starts to produce an **exopolysaccharide** (EPS, also called alginate). This slimy substance provides a matrix for biofilms and helps to anchor the organisms to the respiratory cells. It also serves to protect the organism from internal attack by the body's immune system, and from external attack by antimicrobial substances. When these organisms are cultured in the laboratory, they produce large, mucoid colonies which contrast with typical non-mucoid colonies of *P. aeruginosa*. Multiple strains of mucoid and non-mucoid *P. aeruginosa* can coexist in the CF lung.

Many other bacteria are found to be colonizing the lungs at this stage, including several species that are usually associated with environmental contamination of other samples. Their clinical significance is the subject of debate as they seem to flourish after the patient has been given antipseudomonal drugs. Are they simply filling a niche created by the eradication of other organisms or are they part of the disease process? Examples of these organisms are:

- *S. maltophilia*
- *Alcaligenes (Achromobacter) xylosoxidans*
- *Ralstonia* spp.
- *Burkholderia gladioli*.

There is a group of bacteria that are collectively known as the ***Burkholderia cepacia* complex** (Bcc). They are not associated with disease in the general population but cause respiratory infections in CF patients, particularly in the advanced stages. Some Bcc organisms seem to be more transmissible and cause more severe, even fatal, infections. Therefore, the importance of detecting and correctly identifying the Bcc organisms has social, as well as physical, implications for the patient.

The presence of Bcc organisms in the CF lung reflects the extensive use of antimicrobial substances throughout the patient's life. Bcc organisms are resistant to a wide range of antimicrobials and are therefore able to survive and multiply when other organisms are eradicated or suppressed. This situation also applies to other multiresistant organisms including atypical *Mycobacterium* spp. and fungi, particularly *Aspergillus* spp. *Mycobacterium* spp. are covered in more detail in Section 9.4.

Cross reference
Chapter 11: Aspergillus infection.

Key Point

Recurrent respiratory infections are a prominent feature of cystic fibrosis. Specific organisms, such as mucoid strains of *P. aeruginosa* and members of the *B. cepacia* complex (Bcc), are associated with the condition.

Culture of CF samples

Lower respiratory samples from CF patients need to be tested for the same organisms as routine samples but also need additional investigations to detect the organisms which are specifically associated with CF:

- *S. aureus*
- Mucoid and non-mucoid *P. aeruginosa*
- *S. maltophilia*
- *B. cepacia* complex
- Fungi

Cross reference
Chapter 3: Culture media.

These organisms are generally sought by adding additional selective agars to the routine culture protocol, for example mannitol salt agar (MSA) or chromogenic agar for *S. aureus*, *Pseudomonas*-selective agar for *P. aeruginosa*, *cepacia*-selective agar for Bcc organisms and Sabouraud's agar for fungi.

Key Point

A wider range of agars needs to be used for cystic fibrosis samples compared with other samples because of the wider range of organisms that can cause infection.

Identification of pathogens from CF samples

Presumptive identifications of colonies growing from CF samples need to be confirmed with biochemical tests, as for other samples. The most common confirmatory tests are shown in Table 9.2 but we need to consider some further tests for CF isolates here.

Cross reference

Microscopy techniques for fungi are described in Chapter 11.

For colonies of presumptive *Aspergillus* spp., microscopy using lactophenol cotton blue (LPCB) can demonstrate typical microscopic structures.

Resistance to imipenem and the ability to hydrolyse aesculin are useful markers for *S. maltophilia. S. maltophilia* is intrinsically resistant to imipenem whilst most *Pseudomonas* spp. are sensitive.

Identification of colonies growing on *cepacia*-selective agar can be problematic. Most laboratories use commercial identification kits but they have limited ability to differentiate Bcc organisms from other biochemically similar organisms. It is therefore advisable to refer possible Bcc isolates to a reference laboratory for confirmation.

SELF-CHECK 9.5

Why is it so important to correctly identify organisms of the *B. cepacia* complex?

Further identification of Bcc organisms

The *B. cepacia* complex comprises at least 10 types of bacteria with distinct genomic profiles. We refer to these as genomovars to reflect the way in which they are differentiated. Previously, these were biochemically indistinguishable and thought to be a single species which was called *B. cepacia* (formerly *Pseudomonas cepacia*). Each genomovar is designated by a Roman numeral and a name. These are shown in Table 9.3.

All Bcc genomovars have been isolated from the lungs of CF patients but genomovars II (*B. multivorans*) and III (*B. cenocepacia*) are more closely associated with disease. CF clinics are especially interested in genomovar III as it is associated with greater person-to-person transmission and higher virulence than other genomovars.

Genomovar typing is beyond the scope of routine microbiology laboratories so Bcc isolates are referred to a reference laboratory for typing.

TABLE 9.3 ***B. cepacia*** **complex genomovars.**

Bcc genomovar	Name
I	*B. cepacia*
II	*B. multivorans*
III	*B. cenocepacia*
IV	*B. stabilis*
V	*B. vietnamiensis*
VI	*B. dolosa*
VII	*B. ambifaria*
VIII	*B. athina*
IX	*B. pyrrocinia*
X	*B. ubonensis*

Key Point

B. cepacia complex organisms need to be referred to a reference laboratory for genomovar typing.

Antimicrobial susceptibility testing of CF isolates

If the isolates from a CF sample are typical respiratory pathogens, such as *H. influenzae* or *S. aureus*, then antimicrobial susceptibility testing is the same as for any other sample. If, however, the isolates are mucoid *P. aeruginosa*, Bcc organisms or other CF-associated Gram-negative bacilli then susceptibility testing is more challenging. Difficulties include:

- Mixtures of similar organisms can be difficult to separate
- Organisms can be slow growing
- Mucoid organisms can be difficult to manipulate
- Results can be difficult to interpret due to the lack of relevant data
- False susceptibility is common with disc testing
- A wide range of antimicrobials needs to be tested

Key Point

Antimicrobial susceptibility testing of CF isolates is problematic.

Antimicrobial strategies for CF

The treatment of CF patients requires a multidisciplinary approach, one aspect of which is antimicrobial therapy. The risk of infection is lowered by physiotherapy to remove excess mucus from the lungs.

Unfortunately, it is apparent that clinical response to antimicrobials does not correlate well with laboratory results for CF isolates. This is partly related to the complex environment of the CF lung and partly to the combination therapy that patients receive.

There are two reasons for giving antimicrobials to a CF patient. One is for the specific treatment of an active infection. The other is to manage the colonization of the lungs between active infections. The first of these reasons is generally guided by laboratory results but the second tends to be more empirical and based on local policies.

In CF, it is usual practice to prescribe more than one antimicrobial at a time. Ideally, the infecting organism should be tested against these substances in combination but, in practice, such testing is labour intensive and not often performed.

If a patient needs intravenous antibiotics then they require hospitalization. This is inconvenient for the patient and costly to the health service. For long-term administration of antimicrobials between active infections, **nebulized antimicrobials** are used to suppress the growth of *P. aeruginosa* (e.g. tobramycin or colistin).

nebulized
Nebulized drugs are inhaled using a device which creates a cloud of small particles. High drug concentration in the lung is achieved and minimal systemic absorption reduces toxicity.

Key Point

Long-term strategies for antimicrobial therapy are necessary for CF patients. Combination therapy and nebulized antimicrobials are widely used.

9.4 Mycobacteria

The family of bacteria that are commonly called mycobacteria are sufficiently different from the other bacteria we have described to merit separate discussion. They have distinctive cell walls containing a high concentration of **lipids.**

lipids
Lipids are fat soluble, naturally occurring compounds such as fats, oils and waxes.

These lipid-rich cell walls offer a high degree of protection to the cells and account for many of the properties that set mycobacteria apart. For example:

- Resistance to drying and osmotic lysis
- Resistance to many antibiotics and disinfectants
- Resistance to acids and alkalis
- Impermeability to stains
- Survival within macrophages.

fatty acids
Fatty acids are acids found in fats and oils. They consist of long hydrocarbon chains with a carboxyl acid group at one end.

The most notable lipids in the mycobacterial cell wall are **mycolic acids**. These are long **fatty acid** molecules with one long chain and one short chain. If you look at Figure 9.6 you can see how they are arranged in the cell wall, forming a **hydrophobic** (literally 'water-hating') barrier.

One of the mycolic acids in mycobacteria, **cord factor**, promotes the formation of strings of cells which give a cord-like appearance microscopically. This led early workers to liken them to the structures of fungal hyphae, hence the name mycobacteria (literally 'fungus-like bacteria').

Key Point

The lipid-rich cell walls of mycobacteria are different from those of other bacteria and give them distinctive properties.

Common mycobacterial types

Tuberculosis (TB) is the most significant mycobacterial disease worldwide and so the mycobacteria are often divided into those that cause TB and those that do not.

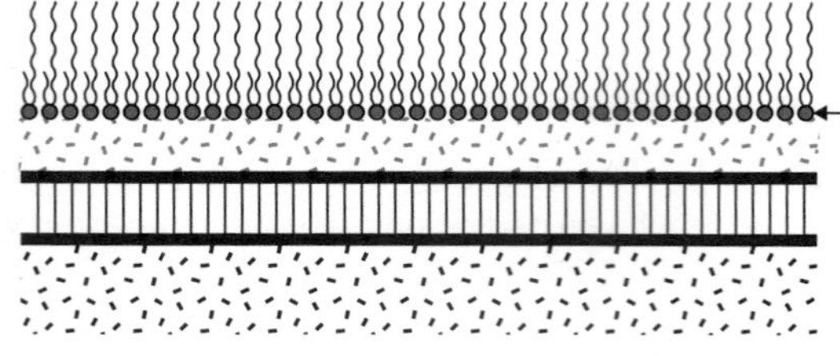

FIGURE 9.6
Structure of the mycobacterial cell wall showing the parallel arrangement of mycolic acid molecules.

The group of organisms that cause TB is known as the *Mycobacterium tuberculosis* (MTB) complex which comprises the following species:

- *M. tuberculosis*
- *M. bovis*
- *M. africanum*
- *M. pinnipedii*
- *M. microti*
- *M. caprae*
- *M. canettii.*

Most human cases of TB are caused by the first three of these.

The mycobacteria that are not part of the MTB complex are grouped together and variously called *non-tuberculous mycobacteria*, *atypical mycobacteria* and *mycobacteria other than tubercle bacilli*.

Mycobacteria are described as **rapid-growing mycobacteria** if they produce visible colonies on solid culture media within 7 days. Those taking longer than 7 days to produce visible colonies are described as **slow-growing mycobacteria**. All of the mycobacteria that cause respiratory infections are slow growers.

Of the non-tuberculous mycobacteria, members of the *M. avium* complex (MAC) cause the majority of respiratory infections. The MAC includes several species, the two most important of which are:

- *M. avium*
- *M. intracellulare.*

MAC infection is closely associated with immunocompromised patients, particularly those infected by human immunodeficiency virus (HIV).

Key Point

Most respiratory mycobacterial infections in humans are caused by members of the *Mycobacterium tuberculosis* (MTB) complex or the *Mycobacterium avium* complex (MAC).

BOX 9.5 The BCG strain of M. bovis

The Bacille Calmette Guerin (BCG) vaccine contains a live, attenuated strain of *M. bovis*. An isolate of *M. bovis* was subcultured every 3 weeks for 13 years by Calmette and Guerin in Paris. They noted that it had become less virulent in animals over that time and it was first used as a human vaccine in 1921.

Normally, people receiving the vaccine would not develop disease from it. Immunocompromised people, however, can develop active infection with the BCG strain of *M. bovis*.

Microscopy for *Mycobacterium* spp.

Since mycobacteria grow so slowly, microscopy results can be available several weeks before culture results. Positive microscopy results can therefore have a significant impact on patient management. Negative results are less influential; if the clinical presentation of a patient suggests TB then treatment will be given, regardless of negative microscopy results. As we shall see later, microscopy lacks the sensitivity of other tests.

Conventional staining techniques, as used for other bacteria, are of no use with mycobacteria because of their impermeability to dyes.

Fortunately, there are ways of overcoming this impermeability. The two most common staining techniques for mycobacteria both use **phenol**. Phenol (formerly carbolic acid) acts like a detergent; it reduces the hydrophobic effect of the lipids, enabling the dye to penetrate the cell wall.

HEALTH & SAFETY

Phenol

Phenol is a toxic substance. Although toxic by inhalation, its distinctive smell alerts workers to its presence before dangerous quantities are inhaled. Toxicity to skin is more likely to be a problem; there is a risk of burns during the preparation of phenolic stains.

Historically, phenolic disinfectants were widely used for decontamination procedures in TB laboratories. They have now been superseded by less hazardous alternatives.

In the **Ziehl–Neelsen** (ZN) technique, the dye fuchsin is combined with phenol. This mixture is called carbol fuchsin. The use of phenol in the other common technique is more obvious as it is referred to as the **auramine–phenol** stain. Figure 9.7 shows the typical fluorescence of mycobacteria upon staining with a fluorescent dye.

The use of heat can also facilitate dye penetration and this is employed in the ZN technique.

To distinguish mycobacteria from other bacteria we use a further characteristic of the mycobacterial cell wall: resistance to acids. Addition of acid/alcohol mixtures to most stained bacteria would result in the removal of any bound dyes, that is decolorization. However, dyes that are bound to mycobacteria are resistant to acid/alcohol attack so decolorization does not occur. Therefore, we describe these organisms as being **acid-fast** (or acid-alcohol-fast) and mycobacteria are referred to as acid-fast bacilli (AFB) or acid-alcohol-fast bacilli (AAFB).

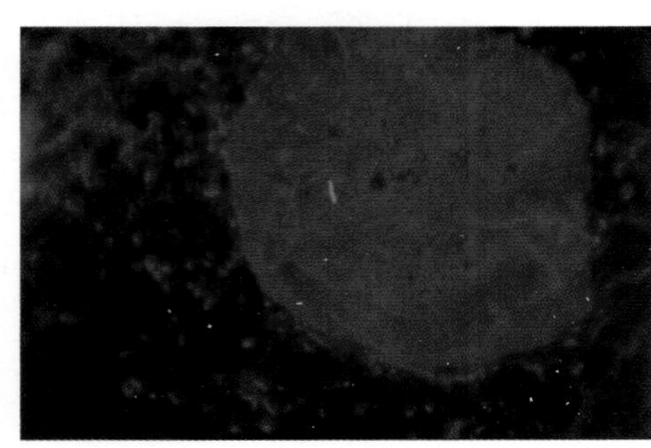

FIGURE 9.7
Acridine orange stained *M. tuberculosis* in sputum showing yellow fluorescent bacillus. (Courtesy of Mike Grover.)

BOX 9.6 Non-mycobacterial acid-fast organisms

Some non-mycobacterial organisms are also acid-fast due to the presence of mycolic acids in their cell walls. In these organisms, the mycolic acids are shorter than those in mycobacteria and they constitute a lower proportion of the cell wall. Consequently such organisms, which include *Nocardia* spp. and *Rhodococcus* spp., tend to be weakly acid-fast.

Since respiratory secretions containing mycobacteria accumulate in the lungs overnight, early morning sputum samples will contain more mycobacteria than those taken later in the day. It is usual practice to collect sputum samples on three consecutive days for mycobacterial testing. Sputum samples can be smeared directly onto a glass slide for staining. Other samples, such as pleural fluids or BALs, require concentration by centrifugation.

Key Point

Microscopy techniques for mycobacteria exploit their acid-fast nature. Microscopy results may be available several weeks before culture results so they can influence early patient management.

Conventional culture for mycobacteria

Despite the advantages of simplicity and rapid results, microscopy for AFB requires 5000–10000 bacilli/mL of sample. Culture, though slow, is more sensitive, requiring only 10–100 bacilli/mL. In addition, isolation of the causative organism enables full identification, antimicrobial susceptibility testing and typing of the organism to be performed.

The aim of any culture method is to encourage the growth of the target organism(s) and discourage the growth of others. This is especially important for mycobacteria as they grow so slowly relative to other organisms. Although inhibitory substances are incorporated into mycobacterial culture media, they cannot sufficiently inhibit all contaminants; it is therefore necessary to **decontaminate** many samples before adding them to the culture medium.

decontamination
Decontamination of samples involves the killing of organisms other than the target organisms.

There are several methods for decontaminating samples. Unfortunately, all of the methods affect mycobacteria also. Optimizing the process is therefore aimed at killing *most* of the contaminants whilst preserving *most* of the mycobacteria. Some methods include the use of a digestion agent.

SELF-CHECK 9.6

Why are decontamination agents used for sputum samples submitted for TB?

Pleural fluids should not contain any contaminating organisms so do not require a decontamination step.

Solid and liquid culture media for mycobacteria

Three types of media are used for conventional mycobacterial culture:

- Egg-based solid media (e.g. Löwenstein–Jensen medium)
- Agar-based solid media (e.g. Middlebrook agar)
- Liquid media (e.g. Kirchner or Middlebrook broth).

Ideally, a combination of solid and liquid media should be used.

Solid media

In the UK, Löwenstein–Jensen (LJ) medium is the most commonly used solid medium. It includes:

- Glycerol (± pyruvate)
- Malachite green
- Eggs.

LJ medium is prepared in bottles, which are heated whilst tilted. The heat dehydrates the egg proteins so the medium solidifies. This process is called **inspissation**. Since a sloping medium is obtained, a large visible surface area is available for inoculation with the sample.

The dye, malachite green, is inhibitory to most bacteria but not to mycobacteria. It is included in the culture medium to prevent the growth of organisms that survive the decontamination step.

Within a couple of weeks of incubation at 35–37°C, typical *M. tuberculosis* appears on LJ medium as irregular, dry colonies which are buff-coloured (beige). They are 'rough, tough and buff'. Other mycobacteria can display different colonial appearances. You can see the typical colonial appearance of mycobacteria in Figure 9.8.

Solid media need to be incubated for 10–12 weeks before they can be deemed negative. Although the MTB complex organisms will grow within a few weeks, some other mycobacteria require this extended incubation period (e.g. *M. xenopi).*

Liquid media

Mycobacteria grow more quickly in liquid culture broths than on solid media. A disadvantage of conventional liquid media is that contaminated or mixed cultures cause difficulties.

> Key Point
>
> **Conventional culture methods for mycobacteria include solid and liquid media. They have the drawback of being particularly slow as visible colonies can take several weeks to grow.**

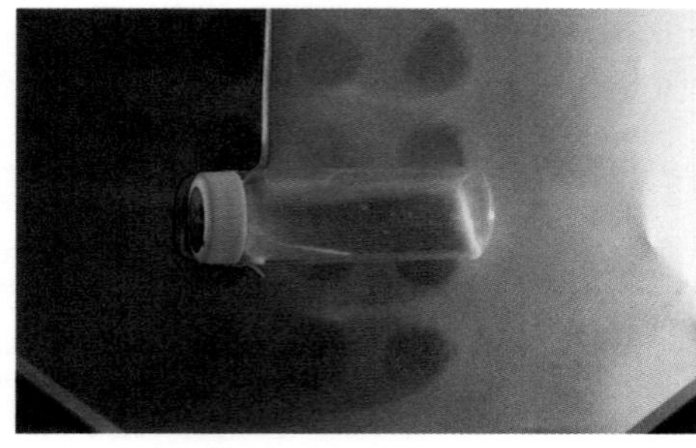

FIGURE 9.8
LJ slope showing colonial growth of *M. tuberculosis*. (Courtesy of Louise Hill-King.)

Automated mycobacterial culture methods

The use of automated systems for mycobacterial culture is now widespread in the UK. Various automated systems are commercially available. They are all based on liquid culture and detect the growth of mycobacteria more rapidly than conventional culture.

Growth of organisms consumes oxygen and produces carbon dioxide. This may be detected by changes in:

- Radioactivity
- Fluorescence
- Reflectance
- Pressure.

The use of radiolabelling is now discouraged due to waste disposal problems. The fully automated systems continuously monitor the culture bottles and flag new positives.

Despite the advantages of automated systems for mycobacterial culture, some disadvantages remain. These include the use of a single incubation temperature and the necessity for supplementary solid media.

Key Point

Automated systems for mycobacterial culture reduce the time to detection but still have limitations.

Identification of *Mycobacterium* spp.

Full identification of mycobacteria is not performed in routine laboratories. Testing is more efficient and more accurate when performed by reference laboratories. Following isolation of presumptive mycobacteria, laboratories issue a report to indicate that a *Mycobacterium* species has been isolated; some may add a comment relating to the likelihood of the isolate being typical *M. tuberculosis* or otherwise. A further report is issued once the organism has been fully identified.

Culture-based identification methods for mycobacteria are slow. They have largely been superseded by molecular techniques.

Some molecular techniques require an amplification step to generate multiple copies of the target sequence. If a technique involves amplification, it is referred to as a nucleic acid amplification test (NAAT).

BOX 9.7 Infection control measures for active MTB infection

Patients with MTB complex infections are infectious whereas other mycobacterial species are not associated with person-to-person transmission. Infection control measures such as patient isolation and contact tracing need to be considered for cases of active MTB infection.

genetic probe
A genetic probe is a nucleic acid sequence which is complementary to the target sequence.

An alternative approach for detecting known sequences is the use of **genetic probes.** If a target sequence is present, a labelled probe will hybridize with it. The label on the probe can then be detected.

Current UK guidelines recommend that a NAAT or a hybridization gene probe test should be performed within one working day of isolating MTB.

Key Point

Rapid molecular identification methods for mycobacteria have replaced conventional tests.

Antimicrobial susceptibility testing

TB treatment includes the long-term administration of multiple antimicrobial substances. This is arduous for the patient and expensive for the health service. It is therefore important to obtain accurate and timely information regarding the antimicrobial susceptibility of mycobacteria.

Key Point

Treatment for TB disease involves multiple antibiotics which need to be taken for 6 months.

Antimicrobial susceptibility testing of mycobacteria is generally performed by reference laboratories.

Automated liquid culture systems can be used to detect antimicrobial resistance by adding antimicrobial substances to the culture bottles.

Conventional techniques for detecting resistance to antimycobacterial drugs rely on growth, or inhibition of growth, of the organism. Since resistance arises from genetic mutations, another approach is to detect the mutations themselves. Many mutations associated with resistance have been identified and molecular tests to detect them have been developed.

BOX 9.8 TB treatment in the UK

The antimicrobial treatment of active TB comprises two stages:

1. **Initial stage: isoniazid, rifampicin and pyrazinamide for 2 months.**
2. **Continuation stage: isoniazid and rifampicin are continued for a further 4 months.**

Strains which are resistant to both isoniazid and rifampicin are said to be multidrug resistant (MDR). Treatment of patients with MDRTB involves the use of five drugs.

Rifampicin resistance is used as a marker for possible MDRTB. Recent statistics show that approximately 1% of UK isolates of *M. tuberculosis* are MDR but some other countries have a greater MDRTB problem, particularly in populations with a high HIV prevalence.

BOX 9.9 Antimicrobial susceptibility testing of mycobacteria

The methods for determining antimicrobial susceptibility of mycobacteria include:

1. **Absolute concentration method–determines minimum inhibitory concentration (MIC).**
2. **Resistance ratio method–compares MIC with that of a control strain.**
3. **Proportion method–determines the proportion of bacterial cells that are resistant.**
4. **Mycobacteriophage method–uses viruses that infect live mycobacteria but not antibiotic-killed mycobacteria.**

Key Point

Antimycobacterial susceptibility testing can be based on inhibition of growth or detection of genetic mutations. Tests are generally carried out at reference laboratories.

Typing of mycobacteria

All new MTB complex isolates need to be typed. This means using further tests that can discriminate between multiple isolates of the same species. Many typing methods are now based on genomic differences between the isolates (genotyping) rather than on differences in their behaviour (phenotyping).

The typing method that is currently recommended enables comparisons to be made nationally or internationally. It is known as mycobacterial interspersed repetitive units–variable number tandem repeats (MIRU-VNTR) typing and is performed in reference laboratories. A 15-digit profile is obtained and entered onto a national database.

Key Point

MIRU-VNTR typing is used to discriminate between different strains of *M. tuberculosis*.

Immunodiagnostic tests

Tests that are used for diagnosing latent TB infection do not involve detection of mycobacteria; they assess the host's cell-mediated immune response by detecting a **cytokine,** interferon-gamma (IFN-γ). They are used to detect latent disease in patients who are at risk of developing active disease, that is those whose immune function is diminishing due to disease, immunosuppressive drugs or age.

cytokine
A cytokine is a regulatory protein, released by cells of the immune system.

Detection of latent TB previously relied solely on the **tuberculin skin test** (TST). The TST involves the subcutaneous inoculation of a purified protein derivative of MTB.

Key Point

Latent TB can be detected using immunodiagnostic tests that are based on cytokine production.

9.5 Hazardous organisms

Many organisms are capable of causing LRTIs, most of which do not pose great risks to laboratory workers. Some, however, present a greater risk and need to be handled accordingly. *M. tuberculosis* falls into this category.

Classification of hazardous organisms

The Advisory Committee on Dangerous Pathogens (ACDP) is a UK advisory body whose responsibilities include:

- The classification of biological agents into **Hazard Groups**
- The definition of **Containment Levels** for laboratory work.

Biological agents are assigned to Hazard Groups 1, 2, 3 or 4. Group 4 is the most hazardous. Table 9.4 shows the relevant considerations for each of the groups.

M. tuberculosis, *M. bovis* and *M. africanum* are in Hazard Group 3, whilst many other mycobacteria are in Hazard Group 2, including *M. avium*, *M. intracellulare* and the BCG strain of *M. bovis*.

SELF-CHECK 9.7

Which species of *Mycobacterium* most commonly cause respiratory infections? To which Hazard Groups do they belong?

TABLE 9.4 **Assignment of biological agents to ACDP Hazard Groups.**

Hazard Group	Causes human disease?	Hazardous to workers?	Spread to community?	Effective treatment available?
1	Unlikely	No	No	N/A
2	Yes	Possibly	Unlikely	Usually
3	Yes (can be severe)	Yes (can be serious)	Possible	Usually
4	Yes (severe)	Yes (serious)	Likely	Usually not

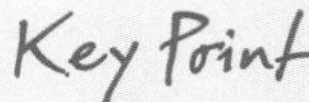

Organisms are classified by the Advisory Committee on Dangerous Pathogens (ACDP) as belonging to Hazard Group 1, 2, 3 or 4.

Containment Levels

The handling of clinical samples requires the use of facilities and working practices that are appropriate for the highest Hazard Group likely to be encountered. The ACDP defines four Containment Levels (1, 2, 3 and 4) for laboratory work; these directly correspond to the four Hazard Groups, for example Hazard Group 3 organisms require Containment Level 3 facilities and working practices. The general requirements for Containment Level 3 are shown below.

HEALTH & SAFETY

General requirements of Containment Level 3 are as follows.

Facilities:

1. **Spacious, self-contained room**
2. **Dedicated equipment**
3. **Easily cleanable surfaces**
4. **Negative air pressure**
5. **Microbiological safety cabinet**
6. **High efficiency particulate absorption (HEPA) filtration of extracted air**
7. **Sealable room to permit fumigation**
8. **Non-opening window**
9. **Wash basin near the exit.**

Working Practices:

1. **High standard of training and close supervision of staff**
2. **Dedicated laboratory coats (autoclaved)**
3. **Disposable gloves must be worn**
4. **The use of 'sharps' should be avoided**
5. **Manipulations of organisms/samples to be carried out in a microbiological safety cabinet**
6. **Documentation of procedures, maintenance records and names of staff using the facility**
7. **Clear evacuation procedures in the event of an accident.**

Key Point

Containment Levels 1, 2, 3 and 4 prescribe the facilities and working practices that are required for Hazard Group 1, 2, 3 and 4 organisms, respectively.

Microbiological safety cabinets

Containment Level 3 requirements include the use of a microbiological safety cabinet. There are three types of cabinet available: Class I, II and III.

All three types of cabinet protect the worker from hazardous aerosols by extracting air through a HEPA filter. The greatest protection is offered by Class III cabinets as they are totally enclosed; manipulations are achieved using gloves or gauntlets. In addition to protecting the worker, Class II cabinets protect the sample by recirculating some of the extracted air, creating a downward laminar airflow. Generally, Containment Level 3 work in clinical microbiology laboratories is performed in Class I cabinets.

The airflow patterns for the three classes of cabinet can be understood by studying Figure 9.9. The airflow speed in a safety cabinet must be regularly monitored and recorded. A typical class I safety cabinet is shown in Figure 9.9.

Key Point

Containment Level 3 work is generally performed in Class I microbiological safety cabinets.

Personal protective equipment

Another element of laboratory safety is the use of **personal protective equipment** (PPE). PPE creates a physical barrier between a worker and potential hazards. The most common forms of PPE that are used in microbiology laboratories are:

- Laboratory coats
- Disposable gloves
- Eye goggles.

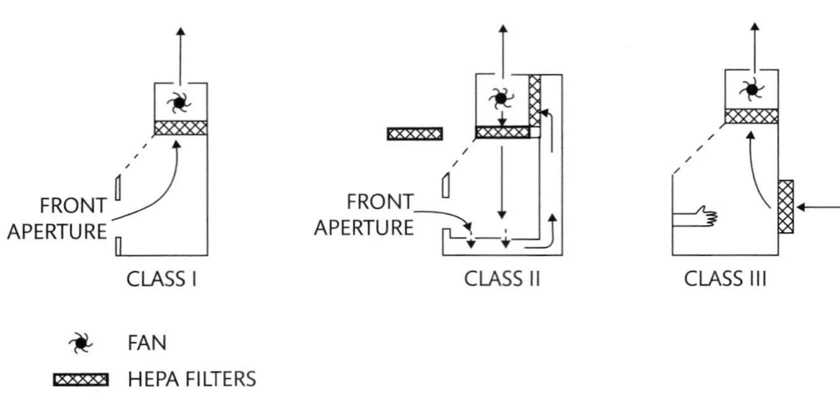

FIGURE 9.9
Airfow through microbiological safety cabinets.

The coats that are required for microbiological work need to fasten to the neck and have elasticated cuffs. Those that are used for Containment Level 3 work must be decontaminated by autoclaving before being washed. Concurrent use of a plastic apron overcomes the problem of laboratory coats not being waterproof.

The use of gloves is advisable, and often desirable, for Containment Level 2 work but is mandatory for Containment Level 3.

For procedures that are performed outside of the cabinet in a Containment Level 3 room, goggles provide eye protection from potential splashes.

Key Point

Personal protective equipment (PPE) creates a barrier between the worker and potential hazards.

Spillages

All staff who are authorized to work in a Containment Level 3 laboratory must be familiar with the protocol for spillages.

Small spillages, which are contained within a microbiological safety cabinet, only require decontamination of the surface.

Large spillages require a different course of action. All respiratory samples are treated as potentially harbouring *M. tuberculosis* so large spillages necessitate *evacuation of the room*. Microbiological safety cabinets are decontaminated by **fumigation** with formaldehyde vapour.

Key Point

Large spillages in Containment Level 3 laboratories require evacuation of the room, decontamination of surfaces and fumigation of safety cabinets.

CHAPTER SUMMARY

After reading this chapter you should know and understand:

- The respiratory tract is normally colonized by organisms which do not cause harm.
- The respiratory tract cleanses itself by trapping organisms in mucus and then removing the mucus by the action of cilia.
- Bacterial infections of the lower respiratory tract are usually caused by organisms descending from the upper respiratory tract and are the result of some impairment of the usual defence mechanisms.
- Laboratory investigations of lower respiratory samples (e.g. sputum) include microscopy and culture techniques.
- Interpretation of sputum cultures can be difficult due to the presence of upper respiratory organisms in badly taken samples.

- Cystic fibrosis patients experience recurrent respiratory infections due to the thick secretions resulting from their genetic defect.
- Cystic fibrosis patients are susceptible to infections with organisms that rarely affect other people, e.g. *B. cepacia* and mucoid *P. aeruginosa*.
- Mycobacteria have distinctive cell walls with a high lipid content.
- Microscopy for mycobacteria exploits their acid-fast nature.
- Traditional culture techniques for mycobacteria take several weeks.
- Automated systems for mycobacterial culture are more rapid.
- Identification, susceptibility testing and typing of mycobacteria mostly rely on molecular techniques at reference laboratories.
- Organisms are categorized into Hazard Groups 1, 2, 3 and 4.
- If samples are likely to contain organisms from Hazard Group 3 then work must be carried out under Containment Level 3 conditions.

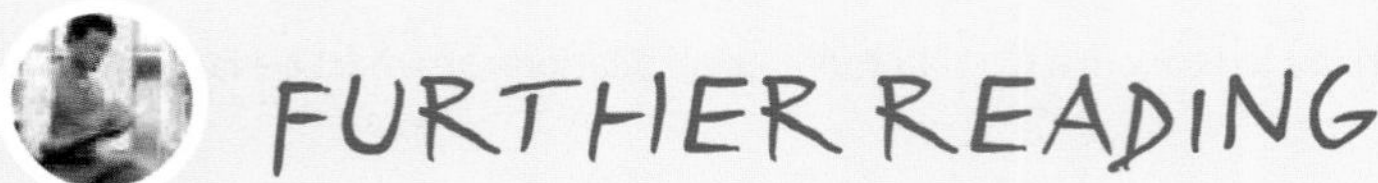

- **Jain A, Mondal R. Extensively drug-resistant tuberculosis: current challenges and threats. *FEMS Immunology and Medical Microbiology* 2008; 53:145–150.**

  Excellent review article on muti-drug resistant tuberculosis strains and impact on global health.

- **Meachery G, De Soyza A, Nicholson A *et al*. Outcomes of lung transplantation for cystic fibrosis in a large UK cohort. *Thorax* 2008; 63: 725–731.**

  Article discussing the survival outcomes for patients with CF who have undergone lung transplantation.

- **Patel RR, Ryu JH, Vassallo R. Cigarette smoking and diffuse lung disease. *Drugs* 2008; 68: 1511–1527.**

  Review article which investigates the complex relationship between cigarette smoking and lung disease.

DISCUSSION QUESTIONS

9.1 What happens to the respiratory tract when a person smokes? How does this help bacteria to multiply and cause a chest infection?

9.2 Some of the organisms that cause respiratory tract infections are part of the normal flora. What are the processes involved in changing from normal flora to infection-producing organisms?

9.3 Why do only a very limited number of mycobacteria commonly cause human respiratory tract infections?

9.4 Organisms of the *Burkholderia cepacia* complex only cause infection in a limited range of patients. Why is this so?

10

Investigation of gastrointestinal specimens

Kathy Nye

Gastrointestinal disturbance in the form of diarrhoea and/or vomiting is very common, even in western, industrialized societies with good sanitation. In fact, it is so common that the majority of people do not seek medical attention and get better on their own.

Although some cases may be caused by bowel disease or the physiological effects of dietary indiscretions, it is estimated that around 1 in 5 people (approximately 9.5 million people) in England will suffer from infectious intestinal disease every year. Of these, only 1 in 7 will present to their doctor and still fewer will have samples sent to a laboratory for examination. Even when the most rigorous techniques are used, no causative microorganism will be detected in half to two-thirds of cases. This may be because there are other infectious causes of gastrointestinal disease which are, as yet, unknown to us, or, perhaps, that known organisms are present in numbers too small to detect.

In this chapter, you will discover how microbiologists detect and identify the major, known, microbiological causes of gastrointestinal disease as well as something of their epidemiology, pathogenicity and clinical management. Whilst these are some of the most unpleasant samples received in the laboratory most biomedical scientists would agree that enteric microbiology is one of the most interesting areas to work in.

Learning objectives

After reading this chapter you should be able to:

- Recognize the major microbiological pathogens implicated in gastrointestinal disease
- Understand the principal techniques available for the detection of these pathogens
- Describe the main mechanisms involved in the production of disease
- Outline the methods available for their identification and differentiation
- List the options for patient management and control of infection
- Understand the technical limitations of current methods and how to make best use of reference facilities

10.1 The general principles of the investigation of gastrointestinal disease

The gastrointestinal tract begins at the mouth and runs for around 10 metres via the oesophagus, stomach, small and large intestines to the anus.

Conditions along the length of the gut change according to the digestive function of each particular region, as do the innate mechanisms designed to resist disease (Figure 10.1).

Almost all microbes that cause gastrointestinal disease are taken into the mouth, either directly via food or water, or indirectly via contaminated hands or inanimate objectives. Once ingested, they are quickly recognized as alien and various defence systems are triggered. These fall into five different categories, that is pH, the normal microflora, gut motility, immunity and enzymes.

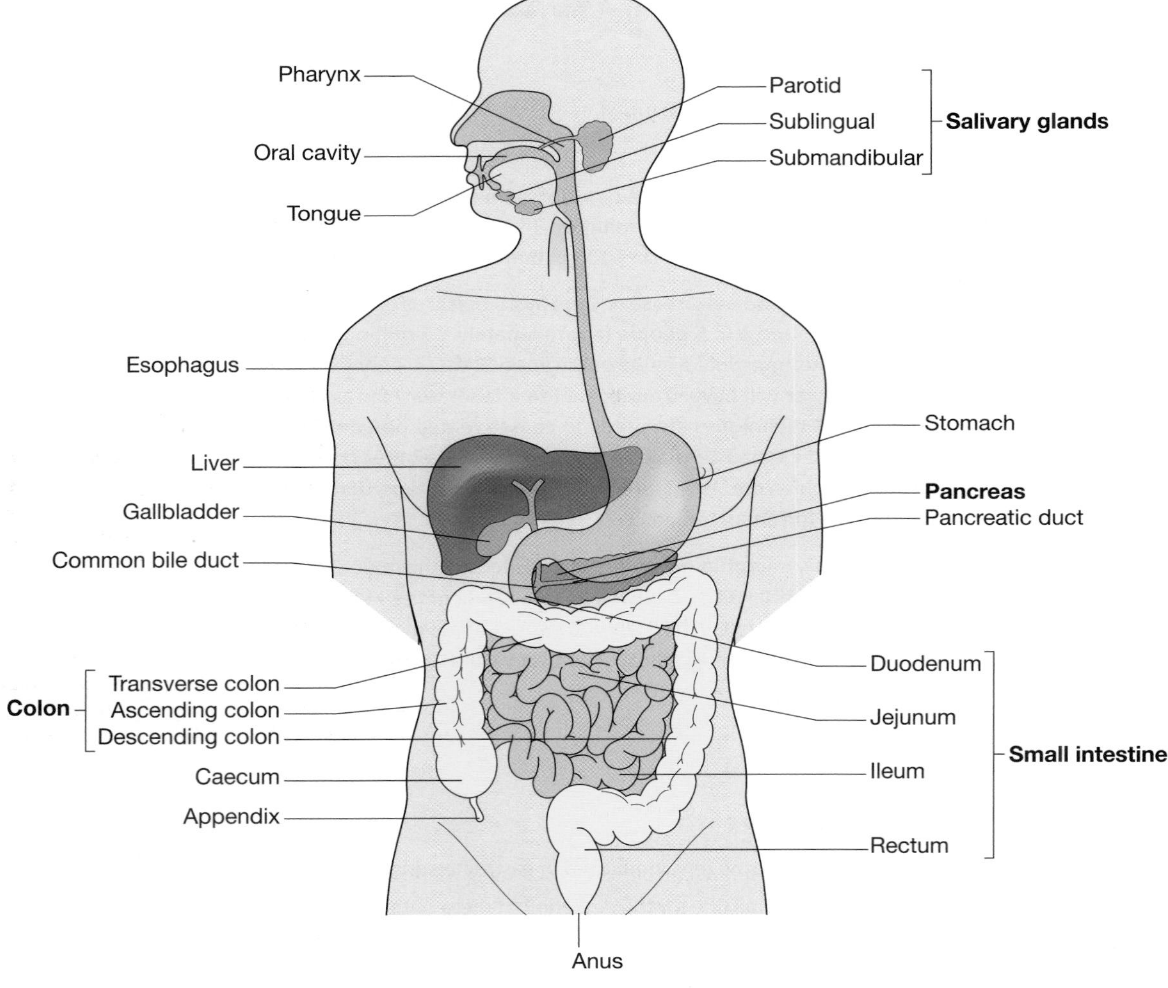

FIGURE 10.1

Diagram of the human gastrointestinal tract.

The single most important defence mechanism is the high level of acidity in the normal stomach (pH <4) which ensures that the vast majority of microbes pass no further into the system. Phagocytes (white cells) and immunoglobulins (e.g. IgA) are abundant in the lining of the gut and form a further line of defence. In addition the presence of digestive enzymes, which help create a hostile environment, and gut motility (**peristalsis**) keep the intestinal contents moving along.

Key Point

The body has numerous mechanisms to ensure that it is very difficult for a micro-organism to initiate gastrointestinal disease.

The normal microbial inhabitants of the bowel are also vitally important. These organisms, usually bacteria and protozoa, form a tightly knit and finely balanced community, surviving on the internal surface (**epithelium**) of the gut, along its entire length. By means of complex interactions, which are still poorly understood, they work to prevent harmful microbes from becoming established and boost immunity to disease.

SELF-CHECK 10.1

What host mechanisms are involved to ensure that harmful microbes don't cause gastrointestinal infections?

Species and numbers vary according to situation, that is there are small numbers of **lactic acid bacilli** in the stomach, around a million (mainly Gram-positive bacteria) per millilitre of fluid in the small intestine and more than 10^{11} mixed organisms per gram of faeces in the large bowel. This is a vast number!

Impairment of any of these defence mechanisms through age, illness, drugs, etc. renders that person more susceptible to all infectious intestinal diseases. However, the likelihood of any individuals developing such disease depends not only upon their own susceptibility, but also on the capacity of the organism to cause illness (its **virulence** and the dose ingested), and upon the opportunity for exposure, that is what the person ate or drank, where they have been and at what time of year. Many gastrointestinal infections have a definite seasonal variation and may be particularly prevalent in specific geographical areas.

virulence
Virulence is the ability of a microorganism to cause disease. Various factors are involved in this such as the ability to adhere to the gastrointestinal tract for example.

Once infected, it will take some time before a susceptible person develops signs of illness, which may be hours or days. This time, during which the infection takes hold, is called the **incubation period** and varies within generally accepted ranges for each pathogen.

incubation period
This is the length of time between exposure to the organism and when symptoms begin.

The most common symptoms are diarrhoea, nausea and/or vomiting and abdominal pain, although, in some circumstances, systemic upset such as fever, headache and general malaise may be a prominent feature. The nature of the diarrhoea (e.g. presence of blood, watery consistency, etc.) may give clues as to the cause.

Although in the developed world most people recover spontaneously from gastrointestinal infections, severe and sometimes life-threatening infections are still seen, particularly in those who are already vulnerable, for example, the very young, the elderly and those debilitated through illness. Even in the otherwise healthy, this type of infection accounts for many working days lost and consequently major economic losses.

In large parts of the world where nutrition and healthcare are poor, simple intestinal infections are still important causes of death, especially in young children.

SELF-CHECK 10.2

How severe are gastrointestinal diseases in the West compared to the developing world? What are the reasons for these differences?

When samples, usually of faeces, are submitted to the laboratory for investigation, it will not be possible or practical to attempt to look for all potential causes of infection. For this reason, the microbiologist must decide which causes are most likely, based on any information available. In practice, investigations are selected to cover the most common and severe infections with additional testing depending upon what is known about the patient and where they have travelled recently.

SELF-CHECK 10.3

Why do you think it is not possible to examine for all causes of gastrointestinal disease on every sample sent?

10.2 Bacterial pathogens associated with gastrointestinal disease

An overview of the main features of infection by the principal bacterial pathogens is provided in Table 10.1.

Campylobacter species

Campylobacters are one of the most frequently isolated and characteristic bacteria causing diarrhoea. They are found in the gut of a wide variety of domestic and wild animals and birds, but are particularly common in poultry. *Campylobacter* spp. are also found in surface waters where they may adopt a dormant state and survive for long periods. Although many species exist, most cases of human disease are caused by *C. jejuni* and, to a lesser extent, *C. coli*. The majority appear to occur sporadically, although outbreaks have been described which are often associated with eating under-cooked poultry meat or food which has been cross-contaminated from raw meat. Campylobacters are not normally transmitted from person to person. Figure 10.2 shows the characteristic 'S' shape of the organism when seen on electron microscopy.

Salmonella species

After campylobacters, salmonellae are the next most commonly isolated bacterial cause of infectious intestinal disease. Once again, they are found in a wide range of animals and birds which often carry them without any symptoms. Several thousand different types have been identified, but the two most commonly found in the UK are *S. enteritidis* and *S. typhimurium*.

Cases are often attributed to the consumption of under-cooked or contaminated food, but person-to-person spread is also important. Salmonellae usually cause gastrointestinal infection but some species are said to be 'host-adapted', that is they are able to cause severe, systemic disease in specific animal species. In humans, *S. typhi* and *S. paratyphi* are the agents of typhoid and paratyphoid respectively, known as the enteric fevers. After an incubation

TABLE 10.1 Clinical features and other factors associated with common bacterial enteric pathogens.

Bacterium		Usual incubation period	Main clinical features	Usual duration of symptoms	Source	Usual infectious dose
Campylobacter spp.		2–5 days	D, AP, F, B	2–7 days	F, W, A	10^2–10^6
Salmonella spp. (excluding enteric fever)		12 h–3 days	D, V, F	2–7 days	F, PP, A	10^5
Shigella spp.		1–7 days	D, B	Up to 2 weeks	F, W, PP	10^1–10^2
Clostridium perfringens		12–18 h	D, AP	24 h	F	10^5
Clostridium difficile		1–7 days	D,B	Highly variable	PP, via contaminated environment	N/A
Escherichia coli (VTEC)		1–6 days	D, B, kidney failure	4–6 days if uncomplicated	F, W, A, PP	<10^2
Bacillus cereus	Emetic illness	1–5 h	V, N (D, AP)	24 h	F	N/A
	Diarrhoeal illness	8–16 h	D, N (V & AP)	24 h	F	N/A
Other *Bacillus* spp.		1–15 h	D+/– N, V, AP	24 h	F	N/A
Staphylococcus aureus		2–4 h	V+/– AP/F	12–48 h	F	N/A
Salmonella spp. (enteric fevers)		1–3 weeks	F and general malaise	10–14 days	F, PP	10^5
Vibrio cholerae		2–3 days	D (watery)	1–7 days	F, W	10^8
Other *Vibrio* spp.		12–18 h	D	1–7 days	F	10^8
Yersinia spp.		3–7 days	D, AP, F	1–3 weeks	F, A	10^9

Clinical features key: D, diarrhoea; B, blood in faeces; V, vomiting; AP, abdominal pain; N, nausea; F, fever.
Source key: F, contaminated food; W, contaminated water; PP, person to person (faeco-oral); A, contact with animals.

period of around 2 weeks they produce a septicaemia with high fever and 'flu-like' symptoms. Diarrhoea may develop but is not the principal feature of the illness. Infection is acquired from other humans, either by direct person-to-person spread or indirectly via food and water contaminated by human sewage. Naturally, it is a greater problem in areas of the world where sanitation is poor and most cases identified in the UK are acquired abroad.

In vulnerable people, the food-poisoning salmonellae can sometimes cause a typhoid-like illness.

After infection with any salmonellae, the bacteria can be excreted in faeces for weeks or even months after clinical recovery. This is described as a **carrier state** and people carrying salmonellae remain a potential source of infection to others unless they pay scrupulous attention to personal hygiene.

carrier state

A carrier state is where an individual, who is not ill, can harbour an infectious organism which may cause disease if it is passed on.

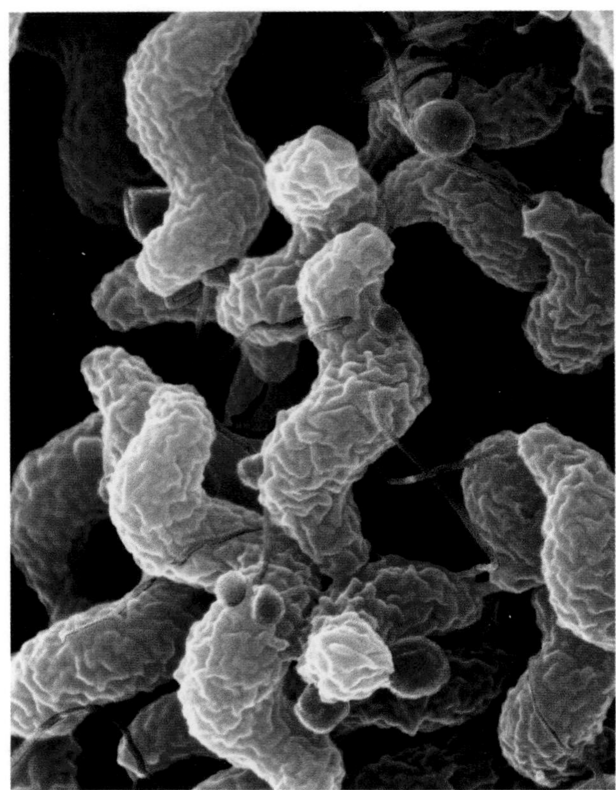

FIGURE 10.2
Electron micrograph of *Campylobacter jejuni*. Note its characteristic 'S' shape. (Courtesy of De Wood, Pooley, USDA, ARS, EMU.)

Shigella species

These bacteria are human pathogens and are not found in animals. A very low infective dose is required so they can be transmitted easily between people and via contaminated food and water. Four species exist, of which only *S. sonnei* is a significant cause of diarrhoeal illness in the UK. *S. sonnei* is the mildest form of shigellosis and tends to cause outbreaks involving young children in nurseries or schools where it is difficult to ensure good personal hygiene.

Shigellae cause classical bacillary dysentery, that is diarrhoea characterized by the presence of blood and mucus, accompanied by fever and abdominal cramps. *S. dysenteriae* causes the most severe disease and, in common with *S. flexneri* and *S. boydii*, is usually acquired in parts of the world where sanitation is poor.

Bacillary dysentery can be life threatening, particularly amongst populations already debilitated by malnutrition and illness. It has been said that in several military campaigns of previous centuries, bacillary dysentery killed more people than war-related injuries.

Clostridium perfringens

This anaerobic organism makes up part of the normal gut flora of man and many animals. It is able to form highly resistant spores in order to survive under adverse conditions. When large numbers of these bacteria are eaten, they form spores in the adverse environment of the small intestine and, in doing so, release a potent toxin which causes diarrhoea and abdominal pain. Illness is unpleasant, but short-lived and self-limiting.

Thorough cooking at high temperature will kill both vegetative bacteria and spores, but problems arise when meat-based foods are subjected to inadequate cooking temperatures, then kept warm for several hours. Warm holding temperatures (e.g. in a bain-marie) allow any surviving spores to develop and multiply quickly to levels sufficient to cause illness.

Escherichia coli

E. coli forms part of the normal gut flora of man as well as many animals and birds, but several strains can act as human pathogens, causing diarrhoeal illness by a variety of mechanisms. Strains recognized as pathogenic include enteropathogenic *E. coli* (EPEC), enterotoxigenic *E. coli* (ETEC), enteroinvasive *E. coli* (EIEC), enteroaggregative *E. coli* (EAEC) and diffusely adherent *E. coli* (DAEC).

Their isolation and characterization is difficult due to their similarity to their relatively harmless cousins, and, as most infections are mild and self-limiting, the majority of laboratories are not equipped to look for them. However, one group of *E. coli* strains is capable of producing a severe illness with bloody diarrhoea and complications including kidney failure and derangement of blood clotting. These are the verocytotoxin-producing *E. coli* or VTEC.

The most common VTEC is identified by its possession of the polysaccharide antigen O157. The infecting dose is very low, less than 100 organisms. VTEC are found in cattle and a number of other food animals. Infection is usually acquired by eating under-cooked or cross-contaminated meat or by direct contact with infected animals or their faeces. Contamination of water and milk may also occur and person-to-person spread is a frequent feature. A number of large, foodborne outbreaks have been described, with a particularly high rate of serious complications in those affected.

Bacillus species

These bacteria are extremely widespread in the environment and are able to form resistant spores to survive adverse conditions. Spores are found on many cereals (e.g. rice and wheat).

When the cooking process is insufficient to destroy all spores, for example, when boiling rice, and the food is left unrefrigerated post-cooking, the spores develop into vegetative bacteria, growing quickly to large numbers. During the period of exponential growth, toxins are formed which cannot be denatured by brief re-heating of the food. It is these toxins which cause disease. A classic example of this type of food poisoning occurs when boiled rice, left unrefrigerated overnight, is stir-fried briefly before consumption. The resulting illness is known as 'Chinese restaurant syndrome'. *B. cereus* is the most commonly implicated species and, in this case, two different toxins may be formed. The first, an **emetic toxin,** produces vomiting 1 to 5 h after ingestion, and the second, a diarrhoeal toxin, produces diarrhoea after 8 to 16 h.

emetic toxin
This is a toxin which causes severe often projectile vomiting and such a symptom can be characteristic of *B. cereus* food poisoning.

Both syndromes are unpleasant, but short-lived and rarely cause serious illness, and come to public attention mainly when outbreaks occur rather than sporadic cases.

Staphylococcus aureus

S. aureus is carried on the skin and in the nose of humans and many animals. Food may contain *S. aureus* for three main reasons: (a) because it was present in the food animal itself; (b) because of contamination from a food handler carrying *S. aureus*; or (c) because of cross-contamination from another source, for example raw to cooked meat. If conditions allow the organism to multiply to high levels in the food, heat-stable enterotoxins are formed which cause the abrupt onset of vomiting a few hours after ingestion. Foods most often implicated in this type of food poisoning are confectionery products, which require a lot of handling, and processed meats.

Vibrio species

Vibrios of many species are found in surface and sea waters around the world. Several species are capable of producing human disease, the most important being *V. cholerae*, which, as its name suggests, is the cause of cholera.

Cholera is acquired via contaminated water or food and produces a range of illness from a mild gastrointestinal upset to profuse, watery diarrhoea, which, if untreated, may kill the patient in a few hours due to fluid and electrolyte loss (Figure 10.3). Some patients, however, may have no symptoms at all, while others may recover and become chronic carriers. In times past, cholera has been responsible for a number of major **pandemics.**

pandemic
A pandemic is an epidemic that occurs over a wide geographical area such as a region or even worldwide.

Several variants of *V. cholerae* have been identified, which have subtly different clinical and epidemiological characteristics.

V. parahaemolyticus causes a milder, diarrhoeal illness as a result of eating raw or under-cooked seafood, or food rinsed by contaminated seawater. A number of other *Vibrio* species have also been implicated in the causation of diarrhoea but their precise role is less well documented.

Yersinia species

Y. enterocolitica is the main species associated with human gastrointestinal disease and often gives rise to symptoms similar to those of appendicitis. Only a small number of strains are capable of causing illness and must possess specific virulence factors in order to act as pathogens.

FIGURE 10.3
Image of a contaminated water supply with *V. cholerae*. Note the environment and how easy it is for transmission of the organism to occur under such conditions.

Y. enterocolitica is found in the normal gut flora of many mammals, including pigs, sheep and cattle, and it is thought that most cases result from eating undercooked meat. A particular feature of infection is its prolonged duration and potential to give rise to chronic, immune-mediated complications such as arthritis.

Clostridium difficile

C. difficile is another anaerobe, which is a normal component of the gut microflora in a proportion of the normal, healthy population, particularly in young children and the elderly. Some strains are capable of producing one or both of two potent toxins, known as toxins A and B, although having the capability to produce toxin does not necessarily mean that the toxins will actually be produced. The toxins can, in predisposed individuals, cause diarrhoea of varying severity, which may be prolonged and difficult to treat.

Predisposed individuals are those with some disturbance of the normal bowel flora, which allows the infection to become established. **Broad-spectrum antibiotics** often cause this disturbance, so it tends to be found most frequently in patients in health-care settings. Patients with *C. difficile* diarrhoea contaminate their environment widely with resistant spores capable of infecting other susceptible patients via the inanimate environment. These properties make *C. difficile* the foremost cause of hospital-acquired diarrhoea.

Cross reference
Chapter 12, Section 12.5.

SELF-CHECK 10.4

Which gastrointestinal infections are seen most in (a) developing countries and (b) developed nations?

10.3 Mechanisms involved in the production of disease by bacterial causes of intestinal infection

There are three principal mechanisms whereby bacteria cause intestinal disease: attachment to the gut epithelium, invasion of the intestinal epithelial cells and production of toxins. Organisms may utilize one or more of these mechanisms. Many are associated with the possession of specific **transmissible plasmids** coding for the production of virulence factors. For most organisms, mechanisms of pathogenicity are still not completely understood.

transmissible plasmids
A plasmid is an extra chromosomal DNA particle which can carry genes for antibiotic resistance and virulence factors, e.g. toxins. If they are transmissible then these genes can be acquired by other organisms.

- Epithelial adherence: this mechanism is best described in strains of pathogenic *E. coli* where the possession of certain surface antigens, or adhesins, allows the bacterium to attach itself to the surface of intestinal epithelial cells. Obviously, this gives the bacterium an advantage in terms of ability to colonize the gut, but also brings it into close proximity with the site of action of its toxins and creates a base from which to invade the deeper layers of the epithelium. Some strains of *E. coli* appear to be able to cause damage to the epithelium at the point of attachment, causing diarrhoea as a result.
- Epithelial invasion: although often accompanied by toxin production most of the major bacterial causes of gastrointestinal disease, i.e. *Campylobacter, Salmonella, Shigella, E. coli*

and *Yersinia*, are capable of epithelial invasion. This is, again, a plasmid-mediated property and usually depends upon modifications of the 'O' side chain antigens of cell wall lipopolysaccharides.

- Food poisoning salmonellae, like campylobacters, usually invade only the superficial layers of the epithelium lining the small intestine and, to a lesser extent, the colon, where they cause inflammation and an increase in gut motility. It is possible that they also produce toxins, although this has not been firmly established. Occasionally, invading bacteria may reach the bloodstream, causing a bacteraemia carrying infection to other parts of the body.
- In the enteric fevers, bacteria invade deeply via the lymphoid tissue (**Peyer's patches**) of the small intestine, rapidly leading to bacteraemia rather than diarrhoea.
- Shigellae and some strains of *E. coli* have a particular capacity to burrow into and destroy the superficial layers of the colonic epithelium leading to diarrhoea with blood and mucus, but they rarely penetrate beyond the mucosa, even in severe cases of dysentery.
- Toxin production: toxins produced by bacteria causing gastrointestinal disease fall into three groups: neurotoxins, enterotoxins and cytotoxins. Production is usually plasmid-mediated and most are polypeptide molecules.

The neurotoxins act on the nervous system of the host to produce symptoms, and, as they are often eaten as pre-formed toxins, have a very short incubation period. Examples of neurotoxins are found in *Staphylococcus aureus* (toxin B) and *Bacillus cereus* (emetic toxin).

adenylate cyclase

This is an intracellular enzyme which catalyses the conversion of ATP to cyclic AMP (cAMP). Increased cAMP results in an outpouring of fluids and diarrhoea and dehydration.

Enterotoxins have a direct effect upon the intestinal epithelium, mainly by causing an irreversible activation of **adenylate cyclase** in the epithelial cells. High levels of cyclic AMP accumulate, causing cell membrane proteins to permit loss of water from the blood into the gut. This fluid loss may be catastrophic, for example in cholera, where death is usually the result of fluid and electrolyte loss.

Apart from *V. cholerae*, other organisms producing varying amounts of enterotoxin are *Shigella dysenteriae*, *C. perfringens* and *B. cereus*, as well as some strains of pathogenic *E. coli*. *C. difficile* produces two toxins, A and B, of which A is a potent enterotoxin.

Key Point

Not all *C. difficile* strains produce toxins A and B, even though they may have the capacity to do so, but the majority of strains implicated in disease produce both.

Cytotoxins, as their name suggests, actually destroy intestinal epithelial cells causing inflammatory diarrhoea. Perhaps the best known of these is the highly potent Shiga toxin produced by *S. dysenteriae* type 1. *C. perfringens* and verocytotoxin-producing *E. coli* also produce a Shiga-like cytotoxin. *V. parahaemolyticus* and *C. difficile* produce different cytotoxins, although their effects are similar. In the case of *C. difficile*, the cytotoxin is known as toxin B.

A number of other virulence factors are known to exist, including the production of **mucinase**, which presumably aids access to the epithelial cell membrane by breaking down the slimy mucus protecting it, and bacterial motility (e.g. in *V. cholerae*), although there is still a great deal that is not understood about the precise contributions of various mechanisms to the production of disease.

10.4 Isolation media and how they work

The isolation of pathogenic bacteria from faecal material is extremely difficult. There are around 10^{11} bacteria in every gram of faeces, just over 95% of which are **obligate anaerobes.** The rest represent a wide variety of organisms, the majority of which are Gram-negative bacilli known collectively as the Enterobacteriaceae. Looking for pathogens, which are often present in relatively small numbers, is, therefore, like looking for needles in haystacks. This is made even more difficult by the fact that some pathogens, for example, *Shigella* spp., are closely related to components of the normal flora, that is *E. coli*. Still others, such as *Clostridium perfringens*, *Clostridium difficile* and *E. coli*, are recognized as part of the normal bacterial flora and only become pathogenic in certain circumstances. For this reason, it is rarely possible to be able to identify a pathogen after simply plating onto any single medium.

obligate anaerobes
These are bacteria which only grow when no oxygen is present.

The basic principle of isolation is to encourage the growth of the target organism (enrichment) by exploiting its metabolic characteristics while inhibiting as many other organisms as possible (selection). Some organisms may be present in such low numbers that it would be unlikely, on a purely statistical basis, to have much chance of isolating them from small amounts of faecal material.

Key Point

Sometimes it is like trying to find a needle in a haystack when you think about how many other bacteria are present in the GI tract.

As a result of this, for *Salmonella* and *Shigella* spp. it is usual to spend some time attempting to increase their numbers, relative to the normal flora in order to have a better chance of isolating them on selective culture media. In the case of *Salmonella* spp. this is achieved by inoculating an amount of faeces into a 'selective enrichment broth'.

Broth is chosen because it provides ideal conditions for bacterial multiplication. After a few hours of incubation, the broth can be subcultured onto a selective, solid medium. Obviously, simple measures such as aerobic incubation will suppress the growth of all obligate anaerobes, but a thorough knowledge of the action of various antibacterial chemicals and antibiotics is necessary to design a medium which suppresses all but the target organism.

There is no single medium that can select for all pathogens or even all members of the same genus of pathogens, for example some of the many species of salmonellae have sufficiently different characteristics that optimal isolation would only be achieved by using specific media for each.

After enrichment and/or plating onto selective media, there will usually still be a number of 'normal flora' organisms growing alongside any potential pathogens. Isolation media are designed to make it easier to differentiate these two groups by, once again, exploiting their different metabolic characteristics using one or more indicator systems.

Table 10.2 provides a summary of the media most frequently used in the UK for the isolation of enteric pathogens and their methods of identification.

Cross reference
Chapter 3: Culture media.

For *Bacillus cereus* and *Clostridium perfringens*, it is important to know roughly how many organisms are present in the faecal material, as both may be found in small numbers in normal,

TABLE 10.2 Guide to the isolation and identification of common bacterial enteric pathogens.

Bacterium	Enrichment (time)	Selective medium	Selective agents	Incubation (atmosphere/ temperature)	Colonial appearance	Identification
Campylobacter	N/A	CCDA (charcoal, cefoperazone desoxycholate) agar	Amphotericin, cefoperazone, sodium desoxycholate	Microaerophilic/ 42°C	Grey, moist, pink or metallic sheen often noted	Typical 'gull wing' appearance on Gram stain
Salmonella	Selenite cysteine or Rappaport-Vassiliadis broth (16–24 h)	XLD (xylose lysine desoxycholate) agar	Sodium desoxycholate	Aerobic/ 37°C	Red with black centre	Biochemical and serological profiles
Shigella	N/A	XLD (xylose lysine desoxycholate) agar	Sodium desoxycholate	Aerobic/ 37°C	Red	Biochemical and serological profiles
VTEC	N/A	CT-SMAC (cefixime, tellurite sorbitol McConkey) agar	Cefixime, potassium tellurite, bile salts	Aerobic/ 37°C	Colourless	Biochemical profile and agglutination with specific antiserum
V. cholerae	Alkaline peptone water (5–8 h)	TCBS (thiosulphate, citrate, bile, sucrose) agar	Ox bile, sodium chloride	Aerobic/ 37°C	Yellow	Biochemical profile
V. parahaemolyticus					Blue-green	
Yersinia enterocolitica	N/A for routine culture	CIN (cefsulodin, irgasan, novobiocin) agar	Cefsulodin, irgasan, novobiocin, sodium desoxycholate	Aerobic/ 30°C	Pale periphery, red centre-'bulls eye' colonies	Biochemical profile
Clostridium perfringens	N/A	Neomycin blood agar	Neomycin sulphate	Anaerobic/ 37°C	Large, convex, shiny +/– haemolytic	Motility test Nagler reaction Lactose/sucrose utilization
Clostridium difficile	N/A	CCEY (cycloserine, cefoxitin, egg yolk) agar	Cefoxitin, D-cycloserine	Anaerobic/ 37°C	Yellow, 'ground glass' appearance with serrated edge	Agglutination of colonies with specific antisera
Bacillus cereus	N/A	*B. cereus* selective agar	Polymyxin B	Aerobic/ 37°C	Peacock blue with zone of blue precipitation	Typical appearance on spare/ lipid staining

healthy people. Selective media can be used but it is best practice to dilute the faeces and then plate out the dilution in order to estimate the number of bacteria present in the sample. As some organisms may be present as spores, a more reliable estimate of both spores and vegetative bacteria can be obtained by additionally incubating the sample in 95% ethanol before dilutions are made so that vegetative bacteria are destroyed.

Key Point

Spores are resting cells associated with *Bacillus* and *Clostridium* spp. As conditions improve the spores 'generate' into vegetative cells and on media appear as the typical organism.

SELF-CHECK 10.5

Why are enrichment broths used for the detection of *Salmonella* but not *Shigella* spp. from faeces samples?

10.5 Identification of bacterial pathogens isolated from faeces samples

Cross reference
Chapter 2.

For a full description of identification tests the reader is referred to Chapter 2.

Presumptive identification of pathogens can be made by examining the characteristics of colonies growing on the selective isolation medium and by performing a few simple tests.

Colonies of bacteria appear very different based on the culture medium used. For example, on xylose lysine desoxycholate agar (XLD), *Salmonella* and *Shigella* spp. both appear as red colonies. Most *Salmonella* spp. also have a black centre due to the production of hydrogen sulphide (Figure 10.4). A small number of non-pathogenic bacteria produce identical colonies (e.g. *Proteus* and *Citrobacter freundii*). In order to narrow the identification down further, suspect colonies are tested for urease production. Salmonellae and shigellae do not produce urease, so only negative colonies need to be investigated further. Even at this stage (black colonies which are urease negative), it is still possible that the colonies are not pathogens, although the probability that they are will prove to be low.

Sorbitol McConkey's agar is used for the selective isolation of *E. coli* O157 strains. Commonly the medium is supplemented with cefixime and potassium tellurite to improve selectivity. *E. coli* O157 strains generally, in contrast to other *E. coli* strains, do not ferment sorbitol and thus appear as 'pale colonies'. Confirmation is through the use of serological tests and/or biochemical identification. Generally, if the isolate is suspected of being an *E. coli* O157 it would be forwarded to a reference laboratory to confirm the presence of toxin genes.

The medium thiosulphate, citrate, bile, sucrose (TCBS) agar is used for the selective isolation of pathogenic vibrios. The medium has a high pH and this helps to improve selectivity. Typically *V. cholera* strains are isolated as large yellow colonies with other vibrios such as *V. vulnificus* and *V. parahaemolyticus* as blue-green colonies.

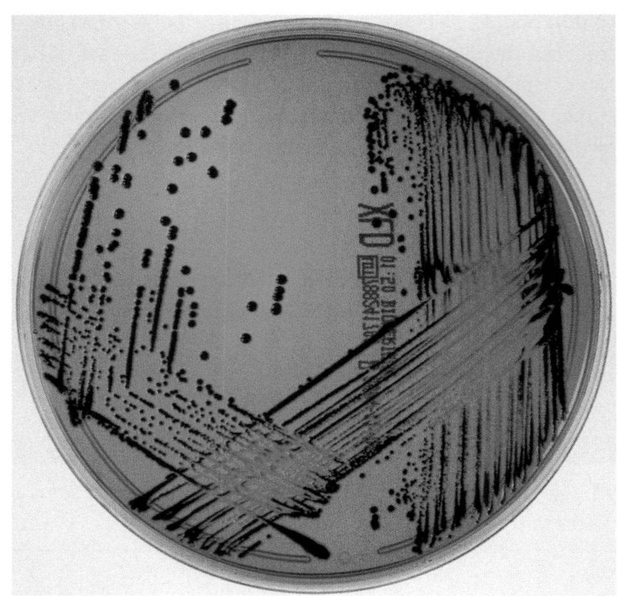

FIGURE 10.4
XLD plate showing the typical colonial appearance of *Salmonella typhimurium*. Note the typical black colonies due to the organism producing hydrogen sulphide. (Courtesy of bioMérieux.)

There are several media available for the isolation of *Campylobacter* spp. from stool samples. The variation is in the choice of selective supplement. Generally the supplement consists of an antifungal agent such as amphotericin B, and both Gram-negative and Gram-positive selective agents. A selective broth is available for enrichment of small numbers of organisms.

microaerophillic
Microaerophillic is a term applied to organisms such as *Campylobacter* spp. whereby the organisms require an environment containing a reduced oxygen level for optimal growth. Many microphiles also require an elevated concentration of carbon dioxide (>5%).

Since these organisms are **microaerophilles** it is essential that the plates are incubated in this environment as soon as possible. Following 24–48 h incubation, characteristic effuse wet colonies are seen often with a pink or metallic sheen. Selectivity can be increased by incubation at 42°C.

Confirmation is by characteristic curved or 'S' shaped appearance (Figure 10.2) on Gram stain. Human pathogenic *Campylobacter* spp. such as *C. jejuni* are typically oxidase and hippurate positive.

Table 10.2 provides a guide to the isolation and identification of common bacterial enteric pathogens together with typical colonial characteristics.

Cross reference
Chapter 2, Section 2.8.

The biomedical scientist can now proceed to a more complex set of tests, based on the metabolic and biochemical properties of the suspected pathogen. Each organism has a well established metabolic and biochemical profile which will enable confident identification in the majority of cases. These are often performed using kit systems (e.g. API).

SELF-CHECK 10.6

If a colony is red on XLD with a black centre does it mean it is a *Salmonella*? If not why not?

For food poisoning salmonellae and shigellae, these are confirmed by serological testing. Suspensions of the organism are mixed with a series of specific antisera raised against characteristic antigens. According to the agglutination patterns detected, species or, in the case of *Salmonella enterica*, serovars can be identified.

Key Point

It must be remembered that Gram-negative organisms can have similar antigens to those of *Salmonella* and *Shigella*, so cross reactions and false-positive reactions can occur.

Non-cultural methods for the detection of bacterial causes of gastrointestinal disease

In most cases of bacterial gastrointestinal disease, it is appropriate to use culture techniques in order fully to characterize the causative organism. However, this is a time-consuming method and may not always be sensitive enough to detect small numbers of pathogens. Two newer methods of improving isolation rates and detection times are outlined below.

Immunomagnetic separation (IMS)

Effectively a selective enrichment technique, this process has been used in the detection of several enteric pathogens, principally salmonellae. Specific antibodies are bound to latex beads which are then mixed with a suspension of faecal material. The beads are washed to remove the faecal debris, leaving any salmonellae bound to the antibody-coated beads which are then cultured in the usual way. Salmonellae present in the sample are thus concentrated and contaminating organisms are removed, allowing a significantly higher isolation rate to be achieved.

While IMS improves isolation rates for the selected pathogen, in most clinical situations, a range of organisms needs to be sought, as the cause is unknown. The additional cost of this technique and the fact that only limited, selected organisms can be detected render it of doubtful cost-effectiveness in routine clinical bacteriology.

Molecular bacteriology

DNA probes are available for the detection of some species (pathogenic *E. coli*). 'Signature' DNA sequences characteristic of the target organism are sought by incubating a preparation of the sample with a complementary oligonucleotide segment which binds to the target site. This section of hybridized DNA is then repeatedly replicated or amplified until it reaches levels able to be detected.

This technique is very sensitive but is expensive and demands considerable technical expertise. Once again, its specificity renders it unsuitable for screening clinical samples where a number of different pathogens may be involved. Similarly, it does not result in the isolation of an organism which can easily be fully characterized (e.g. susceptibility tests) and so has little place in the routine clinical laboratory. As techniques and cost-effectiveness improve, this technology may become more generally useful.

Molecular techniques are used by reference laboratories in the detection of genes coding for toxin production, for example VTEC and *S. aureus*.

Bacterial toxin detection

The detection of toxins in faecal material is not straightforward. This is because the faeces sample itself contains many millions of bacteria, food debris and toxic chemicals resulting from food breakdown or direct ingestion. Just think of how many chemicals are contained in the GI tract. Any one of these can interfere with the detection of a toxin.

Suspected infection with *C. difficile* is the only circumstance in which toxins are routinely sought directly from faeces. In this case, culture of the organism is used only when it is important to know whether infecting strains are related, for example, in outbreaks.

Some diseases, as we have discovered earlier in the chapter, are toxin mediated. This means the living bacterium doesn't need to be present as it could have been destroyed by cooking but the toxin remains to initiate illness. For other toxin-mediated diseases, the microbiologist will attempt to detect toxins either present in food implicated in the illness or produced by bacteria isolated from faeces or food. In certain circumstances, for example infection with VTEC and *S. aureus*, it is possible, using DNA probes, to demonstrate that the organism possesses the genes coding for toxin production even if the toxin itself is not detected.

The following methods are used for toxin detection.

Tissue culture

Cytotoxins, such as *C. difficile* toxin B and Shiga toxin, destroy epithelial cells and some of the earliest tests for these molecules were performed on living animals. A popular model was to use an isolated loop of rabbit ileum, into which the suspect sample was instilled. Cytotoxins would rapidly cause severe damage to the epithelium associated with bleeding and fluid secretion. As you can imagine this was not the most pleasant of tests to perform routinely.

This technique was developed into a tissue culture method; it became possible to grow a thin layer of cells on the inside surface of a clear bottle, bathed in a supporting fluid. When an extract containing suspected cytotoxin was added to the cell layer, any effects could be observed directly by low-power magnification. Toxin-affected cells would shrivel up into balls and die.

For many years, this detection method was accepted as the 'gold standard' for detecting *C. difficile* toxin B. However, faeces may contain chemicals that cause a cytotoxic effect as well as other toxin-producing bacteria. The test can be made more specific by retesting after the addition of an antitoxin, which neutralizes the effect of toxin B. Toxin B is unstable and any delay in transport to the laboratory or failure to keep the sample cool may result in false-negative results. Although some laboratories still use tissue culture, it is labour-intensive, time consuming and does not detect toxin A, which is more important in the causation of diarrhoea. Since *C. difficle* is an extremely important cause of morbidity and mortality the number of requests for examination for *C. difficile* is growing rapidly. As a result, less labour-intensive methods have superseded cell culture.

Cross reference
Chapter 12, Section 12.5.

Enzyme immunoassay

In this technique, specific antibodies to *C. difficile* toxins are conjugated with an enzyme capable of breaking down an indicator molecule (substrate). In a simple example, purified toxin is used to coat the wells of a microtitre tray. A suspension of the sample is added to the well, followed by a dilute preparation containing the antibody–enzyme conjugate. In a control well, the sample is replaced by distilled water or saline.

The conjugate will bind with any toxin present, that is both in the liquid in the well and on the tray surface. In the control well, it will only bind at the surface. The wells are then washed and a detector substrate added - usually a colour indicator or fluorescent compound. In the control well, a very strong signal will be detected, as all the conjugate would have bound to the surface. In the test well, the signal will be less as much of the conjugate would have bound to the toxin present in the sample.

Cross reference
Chapter 13, Section 13.5: Serology.

Various modifications of this test are used in the detection of *C. difficile* toxins A and B, *S. aureus* enterotoxin, and *B. cereus*, *C. perfringens* and VTEC toxins, although only tests for *C. difficile* are used in routine microbiology laboratories.

SELF-CHECK 10.7

What are the advantages and disadvantages of using cell culture for detection of *C. difficile* toxins?

Typing techniques

Although the major bacterial species responsible for infectious intestinal disease have been known for many years, it was quickly realized that it would be helpful to find ways of differentiating apparently identical organisms in order to track their **epidemiology** and to clarify their role in human illness.

epidemiology
Epidemiology is the study of disease origin and spread.

Serotyping

All bacteria possess a variety of antigenic compounds on the cell surface, to which specific antisera can be raised in animals. In Gram-negative bacteria, it is usually the polysaccharide **'O' antigen**, forming part of the cell wall that has properties characteristic of the particular strain. By reacting bacterial suspensions with a series of specific antisera, many organisms can be attributed to serogroups and specific serotypes.

If the bacterium possesses **flagellae,** antisera to these 'H' antigens can also be used further to subdivide the strains. Serotyping is used in the differentiation of *Campylobacter* spp. (Lior and Penner types), *Salmonella* spp., *E. coli*, *Shigella* spp., *Yersinia* spp., *V. cholerae*, *B. cereus* and *C. perfringens.*

flagellae
These are hair-like projections which project from the cell body. Their primary function is locomotion.

Unfortunately, related bacteria share many of these antigens, for example *Shigella* spp. share many common antigens with non-pathogenic *E. coli*, so it is vitally important to ensure full biochemical identification is secure before proceeding to serotyping.

For salmonellae, a complex serotyping scheme has been developed over the years, which is known as the **Kauffmann-White scheme**. Originally developed by Kauffmann in Copenhagen and White in London, this scheme depended upon the reactions of suspensions of salmonellae with crude antisera raised against other strains. By painstaking comparisons, different antigenic groups were identified and designated as *Salmonella* 'species'. It was discovered later that the *Salmonella* antigens were derived from three sources: the 'O' polysaccharide component of the bacterial cell wall, the 'H' flagellar antigens present in most strains and the 'Vi' virulence antigen which is a specialized polysaccharide, capsular antigen, present mainly, but not exclusively, in *S. typhi*.

Kauffmann-White's original scheme developed as specific antisera were raised against these three antigenic groups. It was found that each strain could possess several 'O' determinants and, on this basis, they could be attributed to a particular 'O' group; for example, group A salmonellae such as *S. paratyphi* possess 'O' antigen 2 as their major determinant. Further subdivision can be achieved on the basis of the flagellar 'H' antigens.

Although initially treated as separate species, of which there are currently well over 1500, these strains are now recognized as serovars of *Salmonella enterica* and are named as such, for example *S. enterica* var. *typhimurium*. Most clinical laboratories stock only a limited number of specific antisera with which to identify the most commonly isolated strains. Those unable to be identified are routinely sent to a reference laboratory where a wider range of antisera can be applied.

Phage typing

Bacteriophages are viruses that attack bacteria, causing them to burst or lyse. They will usually attack only a certain range of bacteria, so, if a selection of lytic bacteriophages is applied to a culture of a particular bacterium, the pattern of lysis is fairly specific to that strain. Phage typing can be used further to subdivide bacterial serotypes or to type those species which are too homogeneous for serotyping to be of much use. This technique is used to differentiate strains of *S. aureus* and VTEC and to further subdivide *S. sonnei*, *S. enterica* serovars *enteritidis*, *typhimurium*, *virchow* and *hadar*.

PCR ribotyping

This technique analyses strain-to-strain variations in particular regions of bacterial ribosomal RNA. It can be used to try to differentiate many bacteria but is particularly useful in identifying distinct strains of *C. difficile*.

SELF-CHECK 10.8

Which would you consider to be the most definitive for the identification of a *Salmonella* or *Shigella* - serotyping or biochemical identification?

10.6 Patient management and control of infection

Any patient with diarrhoea and/or vomiting should be treated as potentially infectious until the cause of the illness is determined. Person-to-person spread occurs via the faeco-oral route, that is faecal contamination of hands, food, water or the environment, followed by ingestion.

For the toxin-mediated food poisonings associated with *S. aureus*, *B. cereus* and *C. perfringens*, person-to-person spread does not occur. With the exception of *Campylobacter* spp., where person-to-person spread is rare, for most other bacterial causes the likelihood of spread depends on the infectious dose and the susceptibility of the person exposed. VTEC and *Shigella* spp. are highly infectious due to their low infective dose (Table 10.1).

C. difficile is a special case, as the patient's environment quickly becomes heavily contaminated with resistant spores, exposing vulnerable people to infection via inanimate objects.

It is important that, in health-care settings, patients are isolated as quickly as possible. They should have separate toilet facilities and all those attending them should observe scrupulous hand hygiene and wear disposable gloves and aprons to protect themselves. Enhanced cleaning of the environment, especially 'touch points' (e.g. door handles) should be carried out using an appropriate disinfectant. Visiting must be restricted to the minimum. Patients must remain isolated for at least 48 h after cessation of all symptoms.

Although many may continue to excrete the causative organism for some time after clinical recovery, with good personal hygiene there is much less risk of transmission at this time.

SELF-CHECK 10.9

Would you wait for the culture results to be positive before you would isolate the patient? If not, why not?

10.7 Therapeutic regimes

For most common gastrointestinal pathogens, no specific treatment is required, other than supportive measures to ensure that the patient remains well hydrated and in electrolyte balance. Simple pain relief for abdominal cramps and sponging with tepid water to reduce fever are usually all that is necessary.

Antibiotic treatment is not usually needed and may, in fact, prolong carriage of the pathogen. Widespread overuse of antibiotics for such infections has resulted in major problems with bacterial resistance and, for seriously ill patients, it is now essential to check antibiotic susceptibilities. Antibiotics are indicated for the enteric fevers, severe cases of *Salmonella* food poisoning and shigellosis, although their use takes second place to the correction of fluid and electrolyte depletion. For VTEC, the role of antibiotics is unclear but there is evidence that they may actually make the illness worse.

10.8 Non-bacterial causes of gastrointestinal disease

Apart from the bacterial causes outlined above, three other groups of organisms cause infectious intestinal diseases, that is viruses, protozoa and worms (**helminths**). These will be discussed later in this chapter; however, it is important to be aware that symptoms similar to those of infectious intestinal diseases may be caused by a number of other factors.

Structural or functional abnormalities of the gut, including surgical procedures, cancers, **malabsorptive states** and diverticular disease can all cause diarrhoea, abdominal pain and/or vomiting. Many medicines can have similar effects, as can dietary factors; for example, patients receiving **nasogastric feeds** often develop diarrhoea. It is important that patients with unexplained and persistent symptoms are fully investigated by a specialist.

Viral pathogens

Most cases of gastroenteritis caused by viruses are mild and short-lived, lasting only 1 to 2 days and rarely cause the person to seek medical attention. Transmission is principally person-to-person, but can also be via the inhalation of airborne droplets, which are produced during explosive vomiting. Food-borne outbreaks occur and usually result from contamination during food preparation by an infected food handler, or from food contamination by sewage.

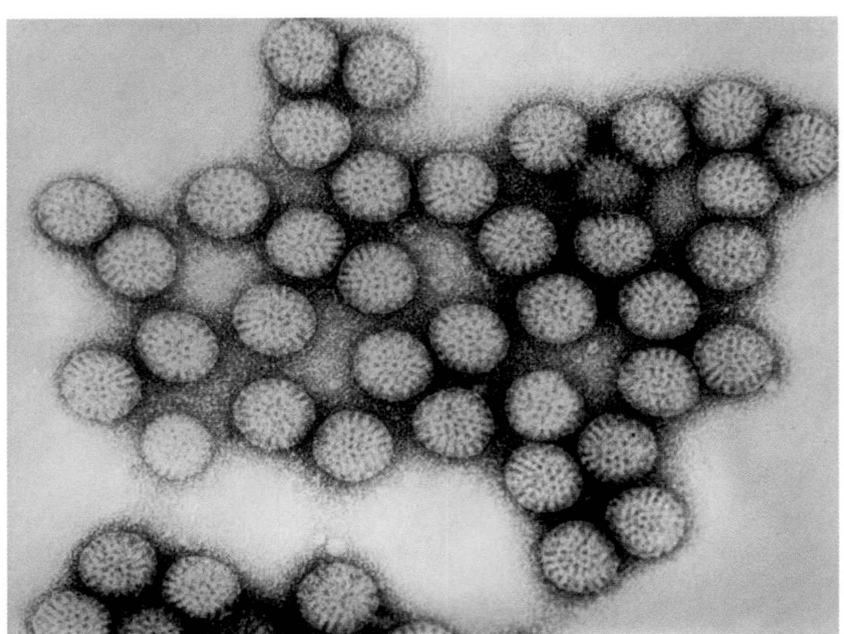

FIGURE 10.5
Electron micrograph of a rotavirus. (Courtesy of CDC/Dr Erskine Palmer.)

Two of the most important enteric viruses are rotavirus and norovirus. Both are highly infectious and occur most frequently in the winter months, although cases can occur at any time.

- Rotaviruses (Figure 10.5) affect mainly children under 2 years of age, spreading rapidly in nurseries and playgroups. Older children are less often affected as they have usually acquired immunity by the age of 3.
- Norovirus affects all age groups and, due to its very low infective dose (<100 particles), its capacity to spread via aerosols and ability to survive for long periods in the environment, causes many thousands of infections each year.
- Numerous serotypes are recognized and immunity after infection lasts only 3 to 4 months for each, so repeated infection is common. Outbreaks are particularly common where people are in close proximity, e.g. schools, hospitals, nursing homes, hotels and even cruise liners.
- Caliciviruses are similar, yet distinct, and sometimes cause outbreaks of diarrhoea in young children. They are not a common cause of diarrhoea.
- Adenoviruses usually cause respiratory infections, but certain strains of subgenus F cause gastrointestinal symptoms. Infections tend to occur in nursery and primary school children and occur throughout the year.
- Astroviruses are infrequent causes of gastroenteritis, the peak incidence being in winter and early spring. Children are mainly affected. A number of different serotypes have been described.

Laboratory detection

Electron microscopy was once the only method of detecting enteric viruses. Although it is expensive, slow and relatively insensitive, it allowed visualization of the viral particles, hence the names of some of the first to be detected, for example rota (wheel-like) and astro (star-shaped) viruses.

Latex agglutination and enzyme immunoassay techniques were then developed for the more common viruses, although only selected serotypes could be detected.

More recently, a variety of PCR-based methods have largely superseded these tests. However, in most laboratories, viral detection is only used in specific circumstances, such as outbreaks or during recognized seasonal increases.

SELF-CHECK 10.10

Why do most laboratories only look for viral pathogens in specific circumstances?

Patient management and infection control

There is no specific treatment for any of the viral infections discussed and, for the most part, due to the relatively mild and short-lived nature of the illness, only hydration, rest and simple analgesia are required.

Infection control is the main problem as, even in previously fit people, the onset of symptoms can be surprisingly abrupt, with vomiting a prominent feature, resulting in infectious aerosol formation and contamination of the environment. Patients should stay at home, or, in the health-care setting, be isolated until 48 h after the symptoms have completely resolved. Rigorous cleaning with a chlorine-based disinfectant is essential, concentrating on 'touch points'. As alcohol hand rubs are not **virucidal**, thorough hand washing with soap and water is most important.

Parasitic infections

Protozoa

Protozoa are single-celled organisms which, when ingested, can, in some cases, cause gastrointestinal disease. Worldwide, they are major causes of morbidity and mortality but in the UK they are much less common than bacterial or viral infections. The three most important are *Cryptosporidium parvum* (Figure 10.6), *Giardia intestinalis* and *Entamoeba histolytica*. They can survive well in the environment – *Giardia* and *Entamoeba* forming resilient cysts – and often contaminate surface waters. They are resistant to chlorine and so may survive even in treated tap water.

Cryptosporidium affects animals as well as humans and can be acquired by direct contact as well as via contaminated water. It causes an unpleasant, sometimes prolonged, but self-limiting diarrhoea. However, in immunocompromised people, especially those with HIV infection, it may be severe and life threatening. There is no specific treatment, although several drugs,

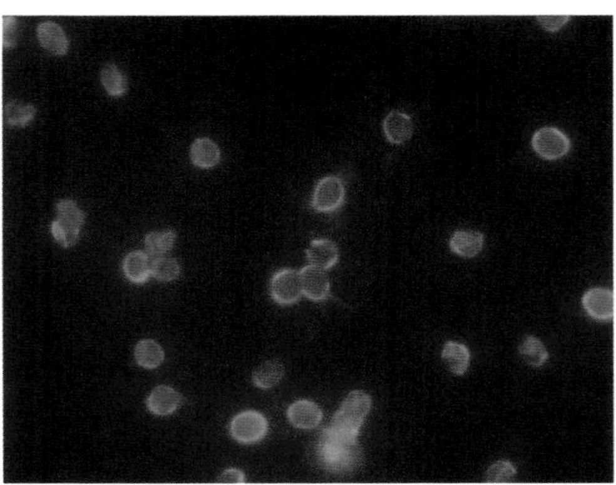

FIGURE 10.6
Immunofluorescence of *Cryptosporidium parvum* oocysts from a stool sample. (Courtesy of EPA/ H.D.A. Lindquist.)

including gamma interferon, have been tried in an attempt to enhance the immune response. For *Cryptosporidium parvum* a smear of faecal material is made on a glass slide and, after allowing it to dry at room temperature, it is generally stained using either a fluorescent stain or a modified **Ziehl–Neelsen** technique. For all other parasites, a concentration method using differential centrifugation (formol–ether separation) is used in order to increase the chances of finding cysts and ova. A wet preparation of the supernatant is placed under a cover slip and examined by low-power light microscopy. A 10% solution of iodine can be used to enhance the features of the organisms. Great expertise and experience is needed to make an accurate examination as many non-pathogenic protozoa and food debris such as plant cells can be mistaken for clinically important parasites.

asymptomatic
Asymptomatic indicates the presence of the organism but without symptoms developing.

- Giardia is usually acquired from water contaminated by sewage, but also spreads person-to-person. It may be entirely **asymptomatic** but, in some cases, diarrhoea may persist and damage to the gut mucosa leads to malabsorption and weight loss. Treatment, if needed, is usually with metronidazole or tinidazole.
- Several species of *Entamoeba* live as part of the normal human gut microflora, but only *E. histolytica* causes disease. Some strains appear to be more virulent than others, but in its full-blown form it causes amoebic dysentery, in which amoebae invade the colonic mucosa forming shallow ulcers. Diarrhoea is severe, with blood and mucus. Amoebae may spread to other organs, usually the liver, via the bloodstream and form abscesses. Treatment is with **metronidazole**, but patients should also receive a course of diloxanide furoate to eradicate cysts, which may remain. Cysts of *E. histolytica* are identical to those of the non-pathogenic *Entamoeba dispar*. The diagnosis of amoebiasis therefore rests on the examination of freshly passed, 'hot' stool, in order to see motile trophozoites, which, characteristically, ingest red blood cells.

Although faeco-oral spread may occur, good sanitation and clean water are, as usual, the best preventive measures.

Worms (helminths)

Helminths are classified into three main groups, the nematodes (roundworms), cestodes (tapeworms) and trematodes (flukes). Although extremely common in less developed parts of the world, where sanitation is poor, most are rare in the UK.

The thread or pin worm, *Enterobius vermicularis*, is the only commonly occurring helminth in this country and causes few symptoms other than perianal itching. In fact, most helminths do not cause symptoms of gastrointestinal disease, but instead their presence leads to a variety of conditions, ranging from anaemia to pneumonitis. In less developed countries, the worm's demand for nutrients from the host may exacerbate malnutrition.

With the exception of pin worm infestation, which is diagnosed by the examination of a strip of sellotape applied to the perianal skin to remove ova, laboratory detection of both protozoa and helminths is by means of light microscopy of faeces, where protozoal cysts, small helminths and the ova of larger helminths can be identified by their typical appearance. Because the excretion of some parasites in faeces is intermittent, it is usual to examine at least three faecal samples obtained over a period of 7 to 10 days. For *Giardia intestinalis*, samples of duodenal aspirate or biopsy material may also be examined.

One of the most commonly described nematodes is *Trichuris trichiura*. Infection with this parasite is more common in warm climates and can often be seen as a mixed infection with *Ascaris* spp. Human infection is acquired through ingestion of eggs present in contaminated soil. These eggs hatch in the small intestine and eventually attach in the large intestine. Diagnosis is by the detection of eggs in the stool (Figure 10.7).

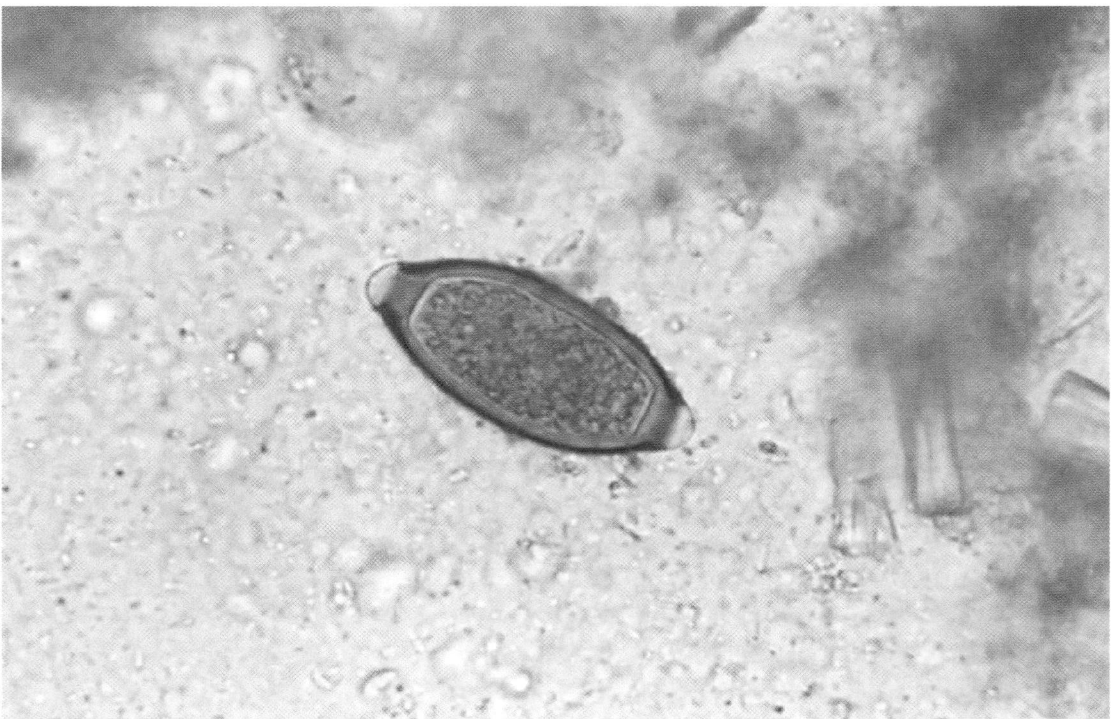

FIGURE 10.7
Photomicrograph of *Trichuris trichiura* as seen in a stool sample following sample concentration. (Courtesy of CDC/Dr Mae Melvin.)

A number of methods are available for the detection of parasites in stool samples, the most common of which is the parasite concentration method. Briefly, the sample is mixed with formalin to kill any harmful organisms and it is then mixed with an organic solvent such as ethyl acetate. This solvent extracts fat and debris from the faeces sample in order to ensure the 'cleanest' possible preparation is produced. Following centrifugation the parasites are concentrated at the bottom of the tube and a drop is examined by light microscopy.

A variety of treatments are available for all the helminthic infections, although some are more successful than others. Apart from the simplest infections, patients are generally referred to specialist infectious disease physicians for treatment.

SELF-CHECK 10.11

What are the common treatments for diarrhoea? Is it appropriate to treat gastrointestinal infections with antibiotics? If not, why not?

10.9 The role of the reference laboratory

Most developed countries have one or two highly specialized laboratories which have the capability to give a definitive verdict on the identification and nature of microorganisms implicated in disease. In order to be able to do this work, a large proportion of their effort is spent in developing and evaluating new methods, some of which may, eventually, find their way into routine use.

There are four situations in which samples or isolates are sent to the reference laboratory:

- To confirm the result obtained by the initial testing laboratory
- To give a definitive identification where preliminary results are inconclusive
- To characterize an organism fully, including typing which is unavailable routinely
- To perform tests that are either too complex for routine use or so infrequently performed in individual laboratories that staff have insufficient expertise to render the test reliable.

In addition, the reference laboratory can collect data on all isolates referred, in order to detect regional and national patterns of disease, which may not always be obvious at the local level. Finally, experts are able to act in an advisory capacity to assist staff in local laboratories in their management of individual patients and outbreaks.

CASE STUDY 10.1 An outbreak of gastroenteritis

The initial report

The manager of a residential home for elderly people telephones the environmental health department of the local authority to report a suspected outbreak of food poisoning.

Over a 24-h period, 18 out of 80 residents and three members of staff have become ill, the main symptoms being vomiting, diarrhoea, headache and feeling feverish. Two residents have had to be admitted to hospital.

The investigation

An environmental health officer (EHO) visits the home in order to question staff and residents, examine the catering and hygiene arrangements and to provide sample pots and request forms so that faecal samples can be collected for submission to the local laboratory.

It is quickly established that all the affected residents and staff had attended a birthday celebration at the home around 18 h before the first resident developed symptoms. No one who was not at the party had become ill.

Food for the party was prepared in the kitchen of the home by a catering assistant with help from an untrained member of staff, as the catering manager was off sick with a broken leg. Egg mayonnaise, ham and cheese sandwiches had been served as well as a homemade trifle. Although no sandwiches were left, a portion of the trifle had been taken home by a member of staff and frozen. The remainder of the egg mayonnaise was kept in a plastic box in the refrigerator.

The kitchens were noticeably untidy, with poor separation of preparation areas and utensils for raw and cooked food. There was only one sink, which was used for washing large utensils and food as well as hand washing.

Sampling

The EHO arranges with the home manager to obtain faecal samples from all affected residents and staff. He also requests faecal samples from the two food handlers, although they have not shown any symptoms. Samples of the egg mayonnaise and frozen trifle are obtained using aseptic technique, carefully labelled and packed in a cool box with a continuous temperature monitoring system to ensure that the food reaches the laboratory in the same condition it was in at the point of sampling.

The EHO calls ahead to a senior member of staff at the laboratory to inform them of the outbreak and to discuss the most likely causes so that testing can be prioritized appropriately.

Infection control and public health

All cases of suspected food poisoning must, by law, be notified to the 'proper officer', who is usually the local consultant in communicable disease control, based at the area Health Protection Unit (HPU).

The EHO calls to make the notification and the HPU sends an infection control nurse to the home to advise on measures needed to limit the spread of infection. The EHO also points out the shortcomings in the catering arrangements and instructs that the kitchen be closed until such time as thorough cleaning has been carried out and trained staff recruited to provide the service.

The infection control nurse advises that, as far as possible, all residents should keep to their own rooms, and that affected residents should use designated toilets and bathrooms. All bathrooms, toilets and touch points must be cleaned several times a day with a chlorine-based disinfectant in order to limit transmission via hands. The hospitals where the most severely ill patients have been admitted are notified of the outbreak so that the patients can be isolated and spread of infection to other hospital patients avoided.

Laboratory investigations

Over the next 3 days, 10 faecal samples are received from residents and three from staff - two being from the food handlers. Given the incubation period and symptomatology, investigations for *Salmonella* and *Campylobacter* are prioritized. The food samples are frozen until the results of faecal investigations are known so that the samples are used to best advantage.

After 48 h, samples on two residents yield presumptive *Salmonella*, which is confirmed on further testing. Initial serotyping indicates it to be *S. enterica* var. *typhimurium*, a serovar often associated with poultry meat and eggs.

Food samples are then thawed and cultures set up for *Salmonella*. Two days later, presumptive salmonella of an identical serotype is detected in both food samples. Cultures from the food handlers prove negative.

All isolates are submitted, without delay, to the national reference laboratory, which confirms that the human and food isolates are identical.

At all stages, the laboratory keeps the EHOs and HPU fully informed of the results and provides expert advice on their interpretation.

Follow-up and discussion

On re-questioning the food handlers, it was admitted that, due to only one sink being available, it had been used, on the day of the party, for washing raw chicken pieces prior to cooking for the evening meal, as well as for cooling the boiled eggs and washing hands. There was no colour coding to indicate which chopping boards and other utensils were to be used for raw and cooked foods, so that eggs may have been chopped on a surface previously used for cutting raw meat.

The most probable cause of the outbreak was cross-contamination from raw chicken to the cooked food, either directly or via hands - all the foods served at the party required a lot of handling.

The importance of scrupulous hygiene in the kitchen, especially when serving a high-risk population, cannot be overstated. Catering staff should have formal food hygiene training, commensurate with their seniority. The EHO has legal authority to close the kitchen until satisfied that it is safe to reopen and hygiene offences may incur a fine of £20,000.

- What were the conditions which contributed most to the outbreak?
- Why did the EHO use aseptic techniques for sampling and send the food sample to the laboratory in a cool box?
- What measures would have been implemented if a food handler proved to be positive also?

CHAPTER SUMMARY

After reading this chapter you should know and understand:

- How common gastrointestinal diseases are throughout the world
- The importance of the various host defences to ensure gastrointestinal pathogens do not initiate infection
- The wide range of bacterial gastrointestinal pathogens and why some are associated more with developing nations

- Culture is still the gold standard for detection of most common bacterial pathogens and the range of culture media available for their isolation
- The basic principles of such culture media and how they are used to isolate and detect bacterial pathogens
- The complex mechanisms by which bacterial pathogens can evade the host defences and initiate infection
- The role of toxins A and B in the promotion of *C. difficile* infection and the national significance of this organism
- Important viral pathogens and the significance of norovirus outbreaks
- How the reference laboratory can be used to identify suspect isolates and confirm the presence of toxin genes, etc., and the role this plays in outbreak analysis
- The potential control of infection issues for both hospitalized and community patients when a gastrointestinal pathogen is detected
- The role of environmental health officers in the investigation of outbreaks.

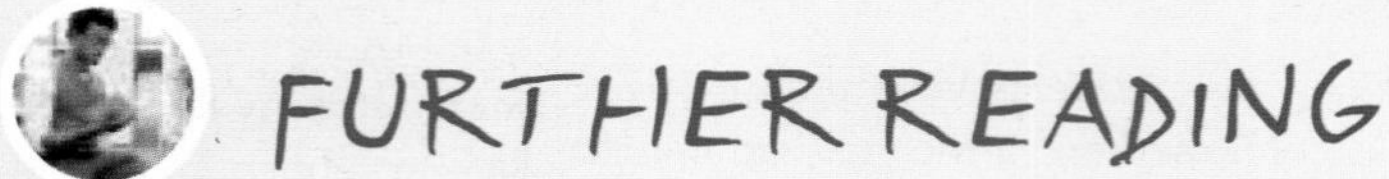

- **World Health Organization. http://www.who.int**

  Numerous publications on infectious diseases and current global programmes.

- **Centers for Disease Control (CDC). http://www.cdc.gov.**

  Excellent resource for all aspects of health care and an excellent image bank.

- **Health Protection Agency. Infectious diseases. http://www.hpa.org.uk/webw/HPAweb&Page&HPAwebAutoListName/Page/1153846673347?p=1153846673347.**

  Extensive library of information on infectious diseases and many other health-care related topics.

- **American Society for Microbiology. http://www.asm.org.**

  One of the largest and oldest microbiology worldwide resource.

10.1 Are all cases of bacterial gastroenteritis confirmed by culture? If not why not?

10.2 Do you believe that commonly used culture media for the isolation of bacterial pathogens from stool samples are sensitive and specific enough?

10.3 What measures would you use to ensure gastrointestinal infections are eradicated worldwide?

11

Clinical mycology

Derek Law

Mycology is the study of fungi and clinical mycology is the study of human infections caused by fungi. Mycology is a small but growing part of the world of clinical microbiology and the laboratory practitioner must be aware of fungal diseases that are commonly encountered in the laboratory. This chapter describes the clinical symptoms and manifestations of fungal infections, the methods used in laboratory diagnosis of infections and treatment of such infections.

Learning objectives

After studying this chapter you should be able to:

- Describe the difference between bacteria and fungi
- List the fungi that commonly cause disease in man
- Describe the different types of fungal infection and the types of patient that are infected
- Describe the signs and symptoms of fungal infections
- Describe the isolation and identification of common fungi
- Discuss antifungal drugs and methods of susceptibility testing of fungi

11.1 What are fungi?

Before discussing fungi and their diseases it is necessary to first of all consider what fungi are and how they differ from bacteria. It is difficult to produce a simple definition of what a fungus is as there are very many species of fungi and they vary immensely in terms of structure and morphology.

Bacteria are simple prokaryotic cells; fungi are more complex and are classed as eukaryotic, like human cells.

- They have a nucleus, and membrane-bound organelles such as mitochondria and endoplasmic reticulum.
- They exhibit mitosis.
- They have a rigid cell wall (like plants).
- Unlike plants they do not have chlorophyll and therefore do not carry out photosynthesis.
- Fungi are saprophytic (living on dead organic matter) or parasitic (utilizing living tissue).

Fungi are found throughout nature in many different environments, most are invisible to the naked eye and live in soil and decaying matter, where they carry out an essential role in decomposing organic material and cycling of nutrients in the environment.

There are estimated to be 1.5 million different fungal species of which about only 70 000 have been described and characterized. Of these only about 200 are known to cause diseases in man and the vast majority of human infections are caused by about 20 species.

> Key Point
>
> **Fungi play an important role in nature in helping break down organic material such as plants. They also have uses in the baking and brewing industries, in the production of bread, beers and wines. Fungi also provide us with food - such as mushrooms.**

Fungi exist in two main forms: firstly, as a unicellular yeast form that reproduces by simple budding. The most familiar of the yeasts encountered in the microbiology laboratory is *Candida albicans* which we will learn about in detail in a later section. The second form is a filamentous or mould form composed of long thread-like-filaments; these filaments are often known as **hyphae** or mycelia.

> **hypha**
>
> **Hypha (plural hyphae) is a long, branching filamentous cell of a fungus. Hyphae are the main mode of vegetative growth, and are collectively called a mycelium.**

Fungi such as mushrooms are just large masses of hyphae packed tightly together. Reproduction in filamentous fungi is by the production of spores or conidia which are produced by specialized structures known as conidiophores.

Look at Figure 11.1a; this shows a stained preparation of *C. albicans*, a common yeast pathogen of man. Note the typical oval shape of the yeast cells, which are about 5 μm in length. Also visible are yeast cells which are reproducing by budding. In Figure 11.1b a filamentous fungus, *Aspergillus fumigatus*, is shown - note the filamentous hyphae and a conidiophore covered in conidia, with numerous free conidia. The conidia disseminate in the same way as plant seeds. Diseases caused by *A. fumigatus* will be described in detail later in this chapter.

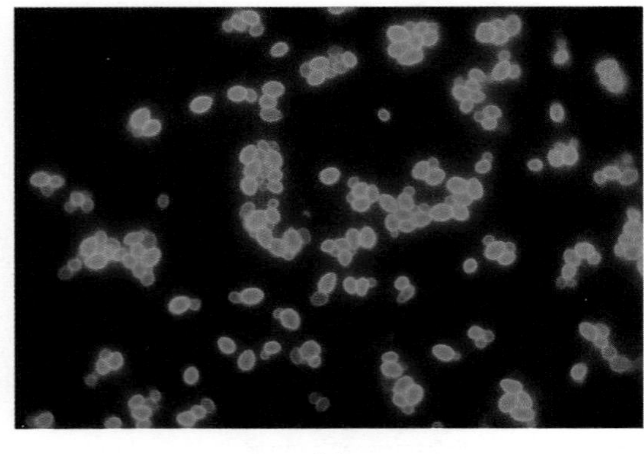

(a)

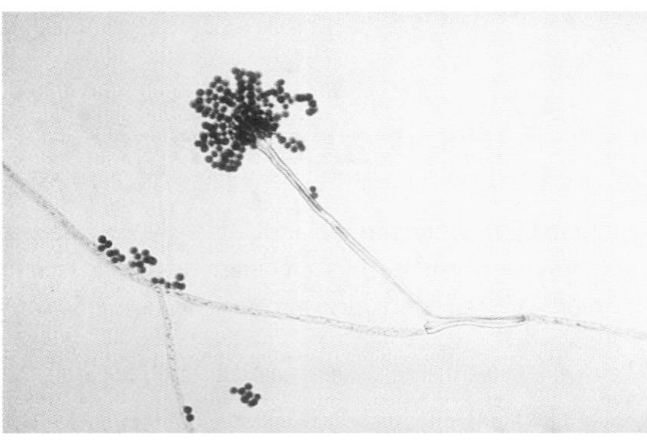

(b)

FIGURE 11.1

(a) Photomicrograph of a yeast stained with a fluorescent antibody. Typical ovoid yeast cells and some budding yeasts are highlighted. (Figure kindly provided by CDC/Maxine Jalbert, Dr Leo Kaufman.) (b) Photomicrograph of *Aspergillus fumigatus*, showing hyphae, conidiophore with attached conidia and some free conidia. (Figure kindly provided by CDC.)

As we will learn later, fungal spores are important as a source of infection; their shape and arrangement are also useful identification aids for various types of fungi. Spores are able to resist heat and drying in a similar way to bacterial spores. They are present in air, soil, dust and water.

The classification of fungi into yeasts and filamentous fungi is not quite so simple, however. Some fungi are known as **dimorphic** (dimorphic - two forms); this means that they can exist in both a yeast form and a hyphal form. In most cases the environmental conditions dictate the form of the fungus. For example *Histoplasma capsulatum* grows as a hyphal form at 25°C but grows as a yeast at 37°C.

Key Point

Most fungi exist in one of two forms—a yeast form or a filamentous or hyphal form. Certain fungi are dimorphic, i.e. they can exist in either a yeast form or a filamentous form depending on environmental conditions.

Although fungi will grow on many types of culture media used in bacteriology, specialized fungal media are available. The most common medium used is Sabouraud's agar; this is a glucose rich medium, which preferentially stimulates growth of fungi. In many fungal media the pH is low (e.g. pH 5); this favours fungal growth whilst suppressing bacterial growth. To make the media even more selective antibacterial antibiotics such as chloramphenicol, tetracycline or gentamicin are frequently added.

SELF-CHECK 11.1

What are the main differences between bacteria and fungi?

Cross reference
Culture media are described in Chapter 3.

11.2 Clinical mycology in the 21st century

Twenty or thirty years ago laboratory mycologists dealt primarily with two forms of infection which we consider in detail later, namely infections of skin, nails and hair caused by **dermatophytes** and infections of the mucosa caused by the yeast *C. albicans* and other *Candida* spp.

dermatophytes
Dermatophytes are types of fungi that cause infections of skin, nails and hair; they are described in more detail later in this chapter.

Such infections were and still are very common but rarely life threatening. Indeed life-threatening infections caused by fungi were extremely rare, poorly understood and in many cases often only recognized at post mortem. Therapy was also limited by the paucity of available agents for treatment. It is now widely recognized that the scope of fungal infections goes beyond mild, non-life-threatening infections to encompass a range of more serious infections where fungi invade the tissues of the body, producing infections which carry a high mortality rate despite treatment with antifungal antibiotics. Fortunately, such infections are still rare, occurring primarily in patients with some underlying illness which reduces or compromises the normal host defences. Examples of conditions and illnesses predisposing to systemic fungal infection will be described in detail in later sections of this chapter.

Although severe fungal infections are rare and much less common than severe bacterial infections, their occurrence in selected populations of patients such as transplant patients, cancer patients undergoing chemotherapy, and in patients with HIV infection and AIDS is sufficient that laboratories must be capable of isolating and characterizing fungi from such infections. In addition there are still many cases of minor fungal infections and most laboratories receive samples from such infections on a daily basis. It is therefore essential that anyone working in a microbiology laboratory has a basic understanding of the fungal infections likely to be encountered in the laboratory, how to process the relevant samples, and the organisms likely to be isolated and methods used to identify such organisms. It is also important to have an understanding of the underlying conditions that predispose to fungal infections, and the antifungal drugs that are available to treat both localized and invasive fungal infections.

As mentioned previously, the incidence of serious invasive fungal infections is increasing. There are four main reasons for this:

immunocompromised, immunodeficient and immunosuppressed
A state where the patient's immune function is absent or impaired so the ability to fight infection is reduced.

1. The majority of invasive fungal infections occur in patients who are **immunocompromised/immunosuppressed** or have other underlying conditions that weaken host defences. The number of such patients is rising because of the increasing number of organ transplants carried out, the increase in patients with malignancy undergoing chemotherapy and the increase in numbers of severely ill patients with underlying conditions or treatments that predispose to reduced immunity and compromised host defences. This can be as a result of drug action (e.g. immunosuppressive drugs used for organ transplantation) or disease (e.g. HIV infection or congenital immunodeficiencies).
2. There is now greater awareness among clinicians regarding the occurrence of invasive fungal infections in at-risk patients. This allows suspicion of an infection in high-risk patients to prompt the correct diagnostic procedures to be instituted, allowing early diagnosis of infection.
3. There are now several different safer antibiotic therapies available for treating severe fungal infections - this has spurred on the need for early recognition of fungal diseases and better diagnostic procedures. Antifungal drugs are discussed later in this chapter.
4. In recent years, improvements in diagnostic tools for invasive fungal diseases have occurred. These include clinical procedures such as **computed tomography scans** (**CT scans**) and radiography, which can in some cases identify signs specific for fungal infections, and bronchoalveolar lavage (BAL), which can provide better samples for analysis. There are now various commercially available, laboratory-based diagnostic tools which detect fungal cell wall components released during fungal growth and which can be used for early detection of infection.

computed tomography (CT) scan
CT scans are a medical imaging technique which generates a three-dimensional image of the inside of the body.

Cross reference
Processing of bronchoalveolar lavages is described in Chapter 9.

SELF-CHECK 11.2

What are the main reasons for the increase in fungal infections?

CLINICAL CORRELATION

Serious fungal infections

Most serious fungal infections occur in patients with some other underlying illness which reduces immunity to infection. Common underlying conditions include AIDS, cytotoxic therapy for malignancy, the use of steroid therapy, immunosuppression for transplantation, diabetes and following abdominal surgery.

HEALTH & SAFETY

The vast majority of fungi are safe to work with, and yeasts can be handled quite safely on an open bench. Spore-producing fungi are more problematic; although most fungi have low virulence, inhalation of spores should be avoided. Whenever cultures start to grow filamentous moulds they

should be handled and examined in a safety cabinet, and plates should be taped up to prevent spores being released.

Certain fungi are classed as high risk such as *Penicillium marneffii* and the organisms of the endemic mycoses. If these organisms are suspected (patients usually have a history of foreign travel) then specimens should be referred to a reference laboratory.

11.3 Common fungal infections of man

There are many fungi capable of causing infection in man; however, the vast majority of fungal infections are caused by a small number of fungi. Only the fungi seen with any degree of regularity and clinical significance will be discussed in this chapter. For details on other fungi the reader is urged to consult specialized textbooks or websites (for example the website www.doctorfungus.org gives an indepth account of fungi encountered in human infections, including photographs of cultures to assist in identification). We will learn about three main forms of fungal disease in this chapter; they differ in the types of fungi involved, how they are acquired, the types of patient involved and their severity and outcome. These infections are:

1. Infections of the skin, nails and hair
    - These infections are caused mainly by dermatophyte fungi.
    - Infection is usually confined to **keratinized** outer layers of skin.
    - Infection can occur in any age group and in healthy people, although infections are more common in children.
    - Sources of infection are the environment, animals (pets and farm animals) and infected people.
2. Infections of the mucosa
    - These infections are commonly caused by yeasts such as *C. albicans* which occur in small numbers as part of the **normal flora**.
    - Infections can occur in healthy individuals but commonly require some predisposing factor.
    - Common predisposing factors include diabetes, extremes of age, abrasion or occlusion of skin, and use of broad-spectrum antibacterial antibiotics.

keratin
Keratin is a protein found in the outer layers of skin. It serves to protect skin.

normal flora
Normal flora is the collective term for organisms colonizing a given body site.

Key Point

The use of broad-spectrum antibiotics removes much of the normal bacterial flora, allowing overgrowth of yeasts at various body sites. This will predispose to infection with yeasts. Antibacterial antibiotics have no effects on fungi.

CLINICAL CORRELATION

Fungal infections and diabetes

Fungal infections are common in diabetic patients and are one of the presenting features of diabetes. Patients with recurring fungal infections should be screened for diabetes.

3. Invasive fungal infections

- Infection can occur in any organ in the body but the lung is the main site of infection.
- They typically occur in patients with underlying illness that reduces immunity or host defences.
- They carry a high mortality rate even when treated with antifungal drugs.
- A wide range of fungi can be the cause, many of which are acquired from the environment; however, *Candida* spp. and *Aspergillus* spp. are the most commonly encountered organisms.
- Often the fungi have low **virulence** and are only capable of establishing infection when the immune system of the host is compromised.

virulence
Virulence is the ability of an organism to produce disease – highly virulent organisms are able to produce disease easily.

SELF-CHECK 11.3

What factors predispose to fungal infections? Are these factors different for bacterial infections?

11.4 Infections of the skin, nails and hair

The first types of fungal infection that we will consider in this chapter are infections of the skin, nails and hair. Such infections are common all over the world and, although they are rarely severe, they may cause considerable discomfort to the sufferer. Infections of the skin, nails and hair are caused by a group of fungi known as the dermatophytes. There are three genera of fungi that comprise the dermatophytes: *Trichophyton*, *Microsporum* and *Epidermophyton*. Look at Table 11.1 for individual species and the main sites of infection. All of the dermatophyte fungi are filamentous fungi and produce spores which can withstand adverse conditions, such as heat and drying, and can remain infective in soil, on floors and in the environment for long periods of time.

TABLE 11.1 **Dermatophyte species and main sites of infection.**

Species	Main sites of infection	Comment
Epidermophyton floccosum	Skin and nails	Common cause of tinea cruris
Trichophyton rubrum	Skin, nails and hair	Most common dermatophyte in UK
Trichophyton mentagrophytes	Skin, nails and hair	Common cause of nail infections
Trichophyton schoenleinii	Skin, nails and hair	
Trichophyton violaceum	Skin, nails and hair	
Trichophyton tonsurans	Skin, nails and hair	
Microsporum canis	Hair and skin	Often acquired from cats and dogs
Microsporum gypseum	Hair and skin	
Microsporum audouinii	Hair and skin	

Sources of dermatophyte infection

There are three main sources of dermatophyte infections:

- The environment; some species live in soil where they live on skin and feathers shed by animals and birds
- From animals such as dogs, cats and farm animals
- From other humans; infection is acquired through contact with spores.

There are various terms used to describe fungal infections of skin, nails and hair. The term **tinea** followed by the part of the body is often used to describe an infection, for example tinea pedis is used to describe infection of the foot, tinea capitis, infection of the scalp.

tinea
Tinea is the name given to a fungal infection of the body and is usually followed by a term to describe the body site, e.g. tinea capitis.

The main symptoms of dermatophyte infections of the skin are scaling of the skin with or without redness and itching. Infection is limited to the outer layers of keratinized skin but can occur on any body site.

CLINICAL CORRELATION

Dermatophyte infections of the skin

Only the outer layers of skin contain keratin. Dermatophytes are able to break down keratin and use it as a nutrient. That is why infection is restricted to the outer layer of skin - the epidermis.

The most common site of dermatophyte infection is the foot, leading to the condition athlete's foot. The infection may spread from person to person in communal showers and bathrooms or in other moist areas where infected people walk barefoot. Infection occurs most frequently in the warm, moist areas between the toes, but can spread to involve the entire sole of the foot. Sometimes scaling is severe and cracks or fissures appear, which can become infected by bacteria. Severe cases can be very painful, and the toe nails may also become involved. Infection of the nails is termed **onychomycosis**; in this condition the fungus attacks and grows in the nail which becomes discoloured, thickened and deformed.

onychomycosis
Onychomycosis is the term given to a fungal infection of the toe or finger nails.

Look at Figure 11.2 - it shows the foot of a patient with athlete's foot caused by *T. rubrum*. Notice the flaking skin between the toes, the most common site of infection. Infection can spread to the sole of the feet and the toe nails. Secondary bacterial infections can also occur.

The finger nails can also become infected following infection of the skin of the hands, and can be unsightly.

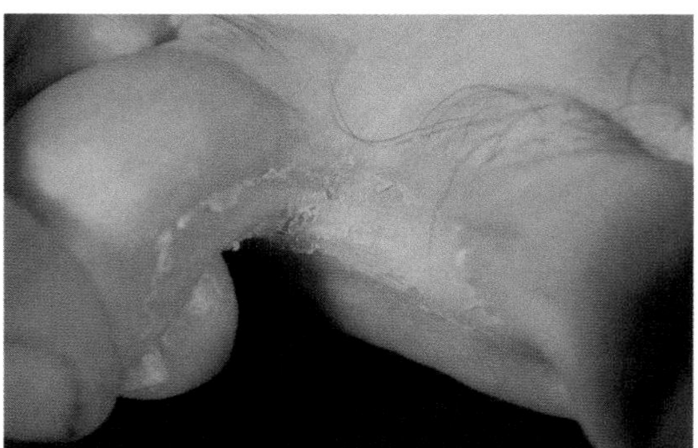

FIGURE 11.2
Photograph of the foot of a patient with athlete's foot showing typical lesions between the toes. (Figure kindly provided by CDC/Dr Lucille K. Georg.)

Infection of the scalp (known as tinea capitis) is primarily caused by *Trichophyton* spp. It is very common among children and can be highly contagious, and outbreaks of infection may occur in schools. The commonest source of infection is often close contact with pet animals or close contact with an infected individual. The symptoms of tinea capitis can vary but usually involve several discrete area of redness, scaling and hair loss. In chronic cases the infection can spread to cover the entire scalp. Some species of fungi are able to penetrate and infect the hair itself. With certain organisms the hair shows fluorescence under **Wood's light**.

Wood's light
Wood's light is a long-wave ultraviolet light used as a diagnostic tool by dermatologists to distinguish dermatophyte infections from other causes of scalp irritation and hair loss.

Infection of the body is known as tinea corporis - infection can occur anywhere on the body and usually presents as a round itchy area with pink scaly borders. The infection site gets larger if untreated and can spread to other parts of the body. The lesions may heal in the centre whilst the edge spreads outwards. Eventually several concentric rings may become apparent on the skin; this leads to the term **ringworm** which is another common name for tinea corporis. When infection occurs in the groin or genital area it is known as tinea cruris. Infection may be transferred from one area of the body to another by scratching or contact.

ringworm
Ringworm is the common name for fungal infections of the skin and hair.

Figure 11.3 shows a typical skin lesion caused by a dermatophyte fungus. The lesion is circular and the outer edges of the lesion show the greatest degree of inflammation.

SELF-CHECK 11.4

Are dermatophytes likely to cause severe infections such as septicaemia? If not why not?

Diagnosis of dermatophyte infections

The symptoms of dermatophyte infections are non-specific and similar symptoms can be caused by other non-infectious dermatological conditions such as eczema, psoriasis or infections caused by other organisms. It is important to have a confirmed diagnosis of a dermatophyte infection for three reasons:

- Specific antifungal agents can be prescribed to cure dermatophyte infections - but treatment can be prolonged and drugs may have side effects and it is therefore important that only true dermatophyte infections are treated with antifungal drugs.
- It important to have a correct diagnosis for epidemiological purposes so that sources of infection can be identified and cross infection can be eliminated.
- A negative diagnosis of a dermatophyte infection allows other causes of the complaint to be investigated.

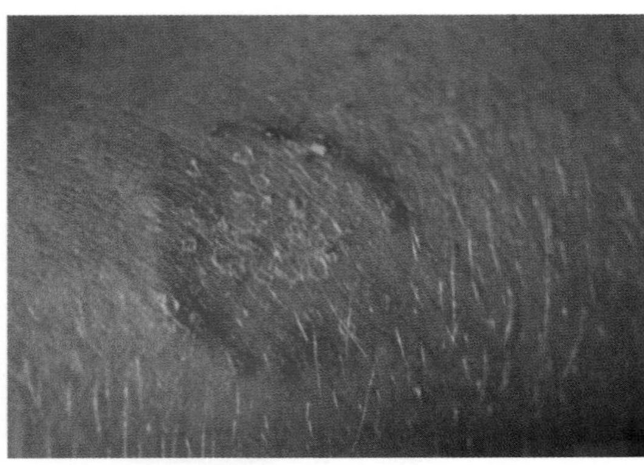

FIGURE 11.3
Photograph of a skin lesion from a patient with dermatophyte infection, showing a typical circular lesion with inflammation on the periphery. (Figure kindly provided by CDC/Dr Lucille K. Georg.)

Key Point

Fungal infections of the skin are often indistinguishable from other skin conditions. Laboratory testing is required to confirm infection and to ensure appropriate therapy is given, and to allow further investigations in the case of a negative fungal diagnosis.

The first stage in diagnosis of dermatophyte infections is examination of infected skin and hair sites using a Wood's light; fluorescence is a strong indicator of a fungal infection. Dermatologists will often commence treatment on the basis of clinical signs and a positive Wood's light sign. However, not all dermatophyte infections show fluorescence and therefore its absence cannot rule out a dermatophyte infection.

Collection of samples for diagnosis of dermatophyte infection

As we have seen in previous chapters, good quality samples are necessary for diagnosis, and diagnosis of dermatophytes is no exception; poorly taken samples may prevent an accurate diagnosis. A brief description of the type of samples and how they are taken is given below.

Collection of samples requires skill and experience to select the best sites for sampling and to ensure that sufficient quality material is taken. First of all, skin and nail sites should be cleaned with 70% alcohol; this reduces bacterial contamination and helps remove creams and ointments which can interfere with diagnosis. Skin samples are taken from the edges of lesion where the hyphae are usually most prolific. If multiple lesions are present then several of these should be sampled; newer lesions are likely to be more productive than old lesions. When nails are infected, friable, discoloured material can easily be scraped from the surface of the nail with a scalpel.

Potentially infected hairs which are broken or distorted should be removed using forceps. The use of a Wood's light to identify fluorescent hairs, which can then be removed, is useful. All samples should be placed in black absorbent paper - this serves two purposes: the samples dry out which reduces bacterial contamination during culture and also the samples are much easier to visualize and work with when on a black background. Fungi remain viable within skin samples for many weeks, and there is therefore no requirement for urgent transport to the laboratory; however, any undue delays should be avoided as this will delay diagnosis.

Key Point

Good quality skin, nail and hair material is required to enable the correct diagnostic procedures to be carried out.

Laboratory procedures for diagnosis of dermatophyte infection

When received in the laboratory samples should be opened carefully to ensure that the contents remain intact. It is very easy for paper to flex and bend creating a shower of skin flakes. The use of gloves by the operator is useful to prevent the possibility of cross infection. Two laboratory methods are used for diagnosis – namely microscopy and culture. Microscopy serves as an initial diagnostic tool and, although not sensitive, can provide a rapid diagnosis when fungal hyphae are seen in samples. Culture is the second method and is generally more sensitive than microscopy; however, growth of dermatophytes can be slow and culture and identification can take several weeks. It may therefore take up to a month before results are available. However, negative culture results are usually reported within 2–3 weeks. Both methods are important and the procedures involved in both will be described in detail. If there is insufficient sample for both microscopy and culture then culture should be set up.

Key Point

Laboratory diagnosis involves two separate techniques: microscopy to identify fungal hyphae in the sample and culture to isolate the infecting fungus. Both methods are important.

Microscopy

The purpose of microscopy is to detect the presence of fungal hyphae in the skin, nail or hair samples. This is a rapid technique, providing a result within a day of sample receipt, allowing early initiation of antifungal therapy when fungi are detected. However, microscopy is not a straightforward technique for several reasons. The quantity of fungal elements present in a sample may be low in relation to the number of skin cells or nail debris and is very dependent on how well the sample was taken. Fungal hyphae may also be difficult to distinguish from cellular material on microscopy. In the Method Box is a brief description of how microscopy is carried out.

Recently, Calcoflour white has been shown to improve the sensitivity of microscopic examination. Calcoflour white is a dye that fluoresces when bound to cellulose or chitin; both are present in fungal cell walls but absent from human cells. Addition of this to KOH preparations and examination under UV light shows up fungal hyphae as bright fluorescing elements. The microscopy results should be reported as either '*fungal elements seen – culture results to follow*' or '*No fungal elements seen – culture results to follow*'.

Figure 11.4 shows a Calcoflour white preparation of a skin sample from a patient with a dermatophyte infection. Compare the upper picture which shows the view under traditional phase contrast microscopy – fungal hyphae are present but difficult to see because of debris. The lower picture is the same field of view but using UV microscopy – fungal hyphae which

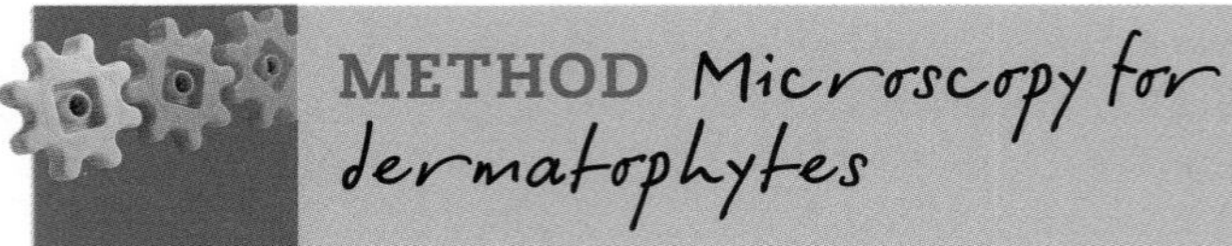

- **Place a drop of 30% KOH on a slide.**
- **Place fragments of skin, nail or hair in the drop and apply a coverslip.**
- **Allow time (15–60 min) for KOH to dissolve skin.**
- **Examine under ×200–400 magnification for fungal hyphae. These show up as slightly greenish, branching threads which contrast with the colourless material of skin or nails.**
- **Experience is required to distinguish hyphae from skin and nail debris.**

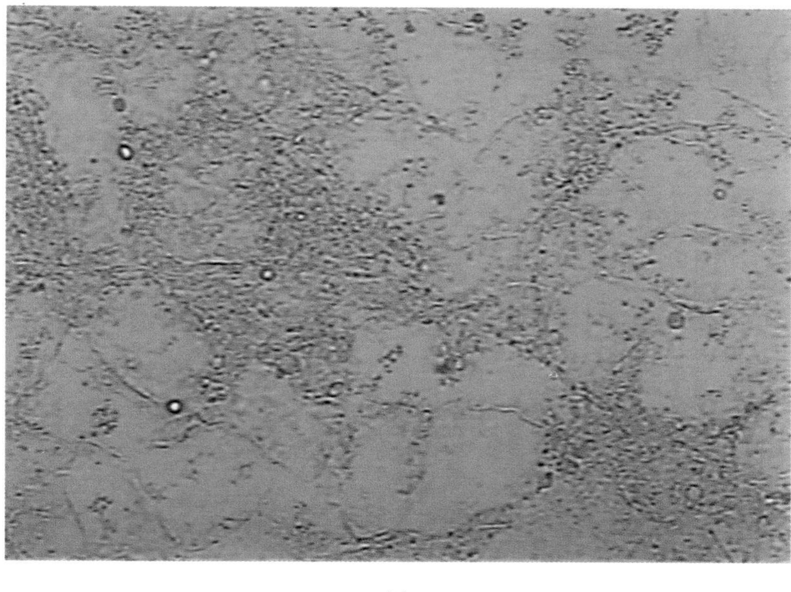

(a)

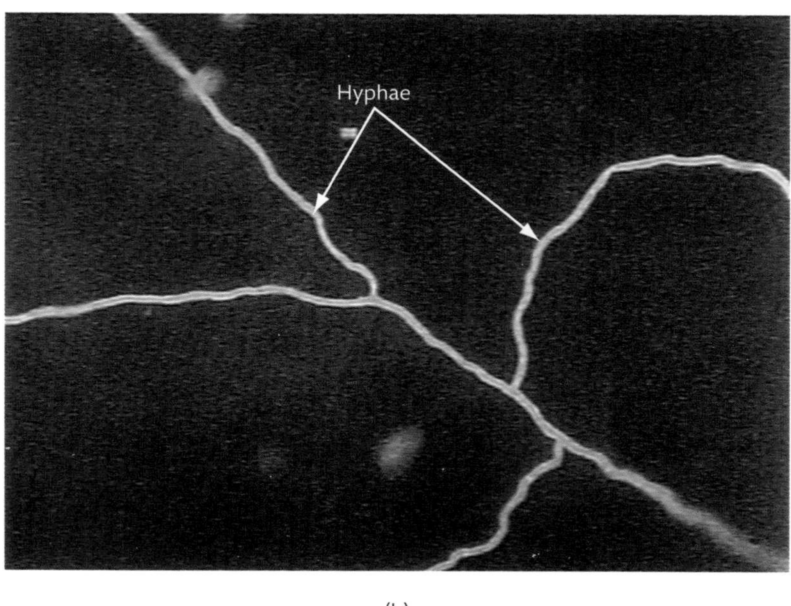

(b)

FIGURE 11.4

Phase contrast (a) and fluorescent microscopy (b) of same optical field of a skin scraping from a patient with a dermatophyte infection. Hyphae are difficult to see in the phase contrast slide because of skin debris, but are visualized very clearly in the fluorescent slide after staining with Calcoflour white which binds to fungal chitin. (Figure kindly provided by Liz Hamer.)

have taken up the Calcoflour stain fluoresce white against the dark background of epithelial cells and are clearly much more visible than in the phase contrast view.

Culture for dermatophytes

Culture is the most reliable method for identifying and confirming a fungal skin or nail infection and allows the isolation of the causative organism. However, culture is not infallible and may be negative when microscopy is positive; this may be because of uneven distribution of fungal hyphae throughout the sample or the effects of prior antifungal treatment.

Culture media for dermatophyte isolation are usually prepared in bottles rather than in Petri dishes. Cultures are often kept for up to 4 weeks and the use of bottles prevents dehydration of the media. Also, because samples may contain fungi that produce spores in profusion, contamination may be problematic; using closed bottles prevents spores escaping from the bottle. Specimens should be cultured onto two different mycological media; typically one of the media will contain the antibiotic **cycloheximide** (this antibiotic inhibits the growth of **non-dermatophyte fungi**), and the other will contain no cycloheximide. Media also contain antibacterial antibiotics to suppress growth of bacteria found in skin and nail samples. Typical media that are used include Sabouraud's agar or variations of this medium.

cycloheximide
Cycloheximide is a useful additive to culture media for dermatophytes as it suppresses growth of fungi such as *Aspergillus* and *Penicillium* which are common in the environment and may contaminate skin and nail samples.

non-dermatophyte fungi
Non-dermatophyte fungi are other filamentous fungi that can cause infections. There are several species that can cause infections of nails.

SELF-CHECK 11.5

What advantage does microscopy for dermatophytes have over culture as a diagnostic tool?

Key Point

Use of media with and without cycloheximide helps in the presumptive identification of a dermatophyte isolate, as other fungi are often inhibited by it.

Confirmation of dermatophyte isolation

When fungal growth occurs on both media and fungal elements are seen on microscopy there is a high probability that a dermatophyte is present in the sample. Confirmation of this is

METHOD *Culture for dermatophytes*

- **Moisten the end of a right-angled inoculating wire; use the moistened tip to pick up samples of the material.**
- **Press the material into the surface of the agar medium to ensure that the fungal elements come into contact with the nutrients and are able to grow. If there is sufficient sample it can be inoculated at multiple sites on the slope.**
- **Incubate for 2–3 weeks at 30°C. The optimal temperature for dermatophytes is below 37°C.**
- **Examine cultures twice weekly for growth - dermatophytes grow as white velvet/fluffy colonies, and growth appears usually after 3–4 days of incubation although this may be much later in partially treated cases or where the amount of fungus in the sample is low. Eventually the colonies grow in size and may become slightly pigmented; some organisms such as *Trichophyton rubrum* produce a red pigment on the underside of the colony.**
- **If fungal growth occurs on both cycloheximide-containing and cycloheximide-free media, the organism is probably a dermatophyte; if there is only growth on the cycloheximide medium it is likely that the fungus is a non-dermatophyte. Non-dermatophyte fungi may be contaminants from the sample or in some cases may be significant.**

METHOD Identification of dermatophytes

- Remove some growth from the culture and emulsify in **lactophenol blue.**
- Examine microscopically for the presence of microconidia and macroconidia.
- If results are inconclusive inoculate isolate onto a medium to encourage sporulation, such as potato dextrose agar.
- Repeat the process when growth occurs.
- The characteristic appearance of macroconidia and microconidia should help allocate the organism to one of the three dermatophyte genera.

lactophenol blue
Lactophenol blue acts as a good background to visualize fungal structures.

needed and this is carried out by examining the colony for distinct morphological structures. Dermatophytes produce two types of conidia, **microconidia** and **macroconidia**; the structures of these are distinct for each genus of dermatophyte. By examining colonies for the presence and shape of both microconidia and macroconidia it is possible to assign a dermatophyte to one of the three genera described earlier in this chapter.

microconidia, macroconidia
Microconidia and macroconidia are the names given to the spores produced by dermatophyte fungi. They differ in size: microconidia are small, often spherical or tear-drop shaped, 2–4 μM in diameter; macroconidia are much larger – their morphology varies between the three genera with slight variations between species in each genus.

Figure 11.5 shows the different morphology of the dermatophyte macroconidia. Note the distinctive shapes which are characteristic of each dermatophyte genus. Also present in the *Trichophyton* photomicrograph (Figure 11.5b) are some smaller microconidia.

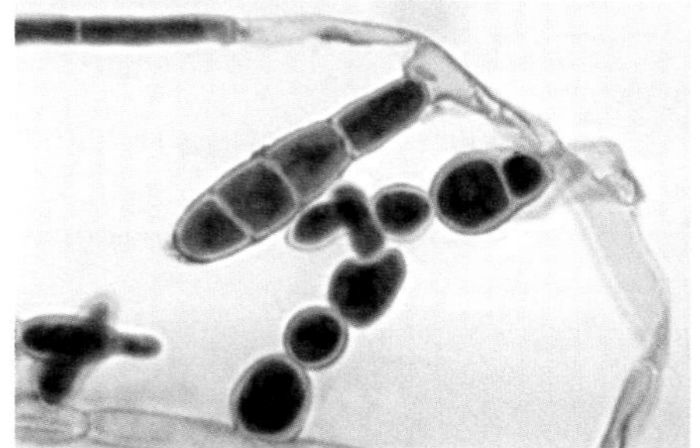

(a)

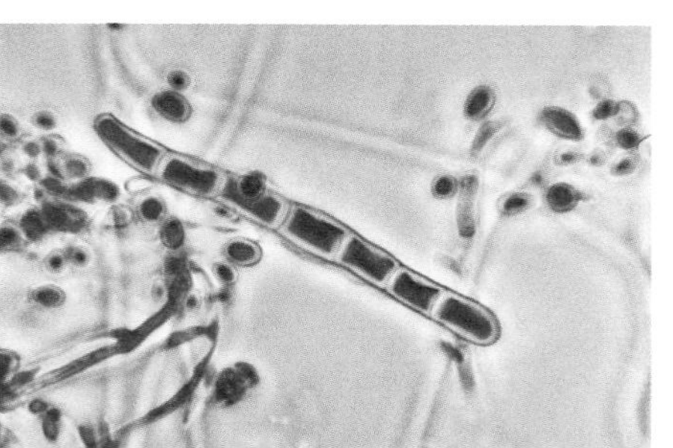

(b)

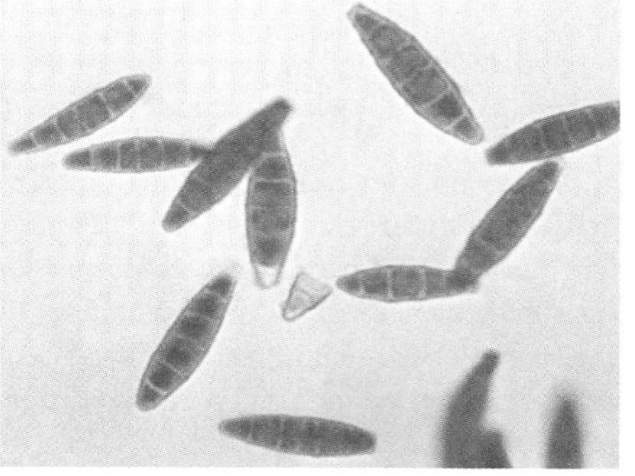

(c)

FIGURE 11.5
Photomicrographs of lactophenol blue preparations of three genera of dermatophyte—(a) *Epidermophyton*, (b) *Trichophyton* and (c) *Microsporum*—showing typical macroconidial morphology. (Figures kindly provided by CDC/Dr Libero Ajello (a, b) and Dr Lucille K. Georg (c).)

Final confirmation of the species identify is complex, requiring various biochemical tests including the ability to grow on various media, colonial characteristics and microscopic morphology. Specific mycology text books or websites should be consulted for identification methods and characteristics of common isolates. For laboratories handling few isolates, dermatophyte isolates should be referred to a reference laboratory for identification. It is easy for the inexperienced to misidentify isolates and preparation and maintenance of stocks of specialized media is rarely cost effective.

If a colony is isolated with typical dermatophyte morphology which grows on cycloheximide-containing media, a preliminary report should be issued stating that *'Probable dermatophyte isolated – further identification to follow'*.

Susceptibility tests are rarely carried out on dermatophytes but may be appropriate in cases of infection refractory to treatment. Such tests are best carried out in a reference laboratory. Repeat sampling should always be considered in cases of suspected dermatophyte infection with negative laboratory reports.

Key Point

Identification of dermatophytes is by cultural characteristics of the colony, the ability to grow on cycloheximide-containing media, and the presence and arrangement of conidia.

Non-dermatophyte fungi as a cause of skin infections

lipophilic
Lipophilic means lipid loving – these organisms require a supply of lipid in the growth medium.

Not all skin and nail infections are caused by dermatophyte fungi; there are several infections that can be encountered that the laboratory mycologist should be aware of, such as infection caused by the **lipophilic** yeast *Malasezzia furfur*. This organism causes scaling and darker patches of the skin, particularly on the chest. Under Wood's light the lesions show fluorescence in various colours, leading to the terminology tinea versicolor. Diagnosis is typically carried out by microscopy. Examination of skin scrapings from tinea versicolor typically shows large numbers of yeasts and hyphal fragments, which is characteristic of this condition and is very different from the microscopic picture from lesions caused by dermatophyte fungi. *M. furfur* will not grow on Sabouraud's agar unless it is supplemented with lipids such as olive oil.

Look at Figure 11.6 – it shows a photomicrograph of a stained skin sample from a patient with tinea versicolor. Note the abundant yeasts and hyphae characteristic of *M. furfur*. This is clearly different from the pattern seen in dermatophyte infections (Figure 11.4).

Although many infections of the nail are caused by dermatophytes, several other fungi are capable of causing infection with a similar clinical picture. These fungi include *Scopulariopsis brevicaulis* and *Scytalidium hyalium*. Microscopically such infections are indistinguishable from those of dermatophyte infections. Diagnosis is by culturing the organism and identification. Because such organisms will often fail to grow on cycloheximide-containing media, cycloheximide-free media must always be inoculated. Identification is by examining colony appearance and microscopic structures.

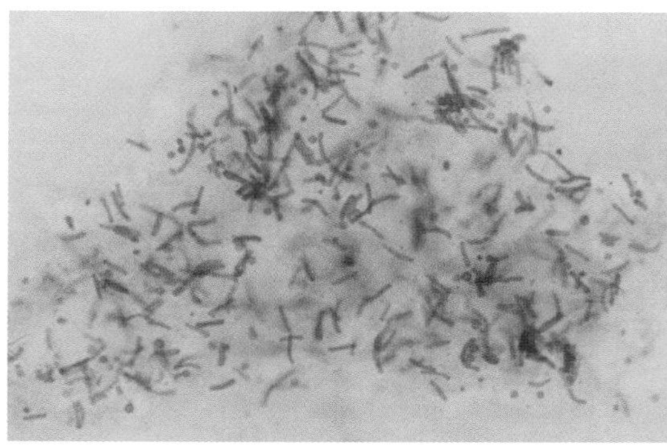

FIGURE 11.6
Photomicrograph of a stained skin sample from a patient with tinea versicolor showing abundant yeast and hyphal forms characteristic of infection with *M. furfur*. (Figure kindly provided by CDC/Dr Lucille K. Georg.)

Candida spp. can also cause infections of nails; it is common in people whose hands are continually immersed in water. Such infections will often be evident by growth of large numbers of yeast on culture. These should be identified as described below, and their presence reported to clinicians.

Treatment of skin, nail and hair infections

Infections of small areas of skin such as athlete's foot can usually be treated with the application of topical antifungal creams or powders. Treatment of more extensive lesions and where nail or hair is infected usually requires oral therapy given for weeks or months. Drugs such as griseofulvin, terbinafine or itraconazole are usually given for periods of 2 weeks to 6 months, depending on the severity and site of infection.

11.5 *Candida* infections of the mucosa

Yeasts of the genus *Candida* form part of the normal flora of the body and can be found on the skin, throughout the alimentary tract and in the female genitalia. Infections can occur in normal healthy individuals. In such cases infection is usually due to an impaired epithelial barrier function, for example occlusion or abrasions of skin, excessive sweating, or disturbance of the normal flora through the use of antibacterial antibiotics which result in overgrowth of yeasts.

candidiasis
Candidiasis is the name given to infections caused by *Candida* spp.

Candidiasis of the mouth and throat are rarely seen in normal individuals but are common in the newborn and elderly; in the latter group they are often related to the presence of dentures. Infections in other individuals are often associated with some form of immunological impairment, for example leukaemia, neutropenia, HIV infection and diabetes. Prior therapy with cytotoxic drugs, radiation or broad-spectrum antibiotics also predispose to infection. Infection presents as discrete, white plaques on the mucosal surfaces of the mouth and throat. Symptoms include pain and burning on swallowing. A severe form of infection, oesophageal candidiasis, is frequently seen in severely immunosuppressed patients. It is a frequent complication of patients receiving cytotoxic therapy and occurs commonly in AIDS patients. Infection involves the oesophagus and the mouth; the main symptoms are difficulty in and a burning pain on swallowing. Clinical diagnosis often requires the use of an endoscope to visualize the plaques in the oesophagus. Figure 11.7 shows the mouth of an HIV patient with

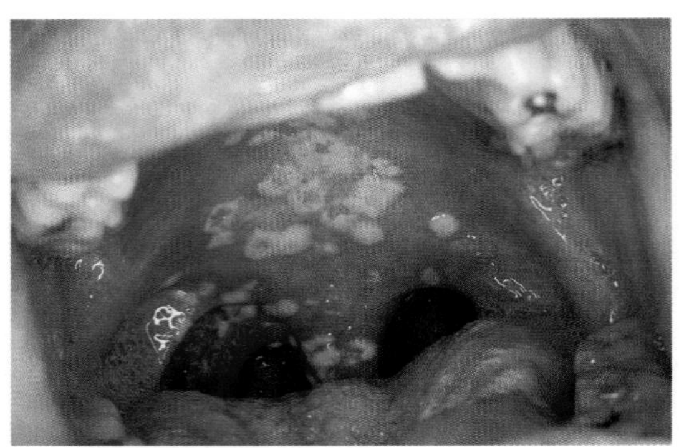

FIGURE 11.7
Photograph of oral cavity of an AIDS patient showing white candidal plaques. (Figure kindly provided by CDC/Sol Silverman, Jr, DDS.)

oral candidiasis. The discrete white plaques are evident on the mucosa. This is characteristic of *Candida* infections.

Candidiasis of the skin is most frequently seen in the groin, underneath breasts, in the axillae and the umbilicus. Such infections require some predisposing factors such as occlusion or abrasion of skin, or excessive sweating; such infections are common in the obese, patients with diabetes and with the use of broad-spectrum antibiotics. Nappy rash seen in infants is commonly caused by *Candida* spp.

Vaginal *Candida* infection is a common infection in women. *Candida* spp. occur in the vagina in low numbers in women of child bearing age; however, the use of broad-spectrum antibiotics can remove the normal bacterial flora allowing *Candida* spp. to proliferate. Diabetes and the use of oral contraceptives may be contributing factors and infection is common in the third trimester of pregnancy. The most common symptoms include itching and burning with a creamy white vaginal discharge. Infection can also extend to the external genitalia. In a small proportion of women repeated episodes are common, although the reason for this is unclear.

Key Point

Vaginal infections are one of the commonest infections caused by *Candida* spp. and affect a large proportion of women at some point during their life.

CLINICAL CORRELATION

Infection of the skin caused by *Candida* usually requires some underlying condition—e.g. immunosuppression, diabetes, use of broad-spectrum antibiotics, cytotoxic therapy, intravascular lines or excessive moisture.

Diagnosis of mucosal *Candida* infections

Mucosal infections caused by *Candida* spp. are the most frequently encountered fungal infections in clinical laboratories. Each day most laboratories receive many samples requiring culture for *Candida*. Fortunately, diagnosis of such infections is relatively simple.

Many samples will be swabs from infected skin sites, or samples of vaginal discharge. Along with any bacteriological investigation required, samples for the diagnosis of *Candida* infection are routinely plated onto Sabouraud's agar. Plates are incubated at 35–37°C for 24–48 h and plates examined for growth of *Candida*. *Candida* spp. grow on Sabouraud's agar as creamy white, **butyrous** colonies, with a typical sweet yeasty odour. Colony size after 24 h is typically 2–3 mm diameter but after 48 h colonies are 4–5 mm diameter. The amount of growth is typically recorded as +, ++ or +++; this will help in the interpretation of the result. True *Candida* infections will normally show ++ or +++ growth if samples have been taken correctly. Very small numbers of organisms will often represent colonization of a site rather than infection. Interpretation of results requires both culture and clinical data.

Microscopy is useful for vaginal swabs. The presence of large numbers of yeasts and white blood cells (pus cells) is often indicative of *Candida* infection. This can often be carried out alongside microscopy for other pathogens such as *Trichomonas vaginalis* and *Gardnerella vaginalis*. Various stains, such as a Gram and acridine orange, may be used to visualize yeasts.

Key Point

***Candida* spp. grow readily on Sabouraud's agar, and the inclusion of antibiotics helps suppress bacterial growth. *Candida* spp. grow as creamy white colonies with a characteristic odour. Quantitation of the growth helps in interpretation of significance.**

The majority of human *Candida* infections are caused by the species *Candida albicans*, although other species will occasionally cause disease in similar situations. The other species commonly found are *C. glabrata*, *C. tropicalis*, *C. krusei*, *C. parapsilosis* and *C. dubliniensis*. Disease caused by these species is indistinguishable from disease caused by *C. albicans*. Some of these species show resistance to various antifungal drugs. In most laboratories identification of yeasts isolated from skin and vaginal swabs is not carried out and the isolation of a yeast with the quantity of growth is reported.

METHOD The germ tube test

- **Place a small amount of the colony to test into a small volume of human or horse serum to produce a faintly turbid suspension.**
- **Incubate at 35–37°C for 2.5–3 h but do not exceed this.**
- **Remove a small sample of the fluid and examine microscopically.**
- **Germ tube production is indicated by the presence of a hyphal structure emerging from the cell.**
- **True germ tubes of *Candida albicans* have parallel walls where they leave the yeast cell; occasionally other species of yeast may show similar structures, but have constrictions at the site where they leave the yeast cell.**
- **The quantity of germ tubes produced by different *C. albicans* strains varies; some produce many germ tubes, in other strains the number of cells producing germ tubes is low.**

Key Point

Candida glabrata is increasingly isolated from clinical samples and is often resistant to azole antifungals such as fluconazole.

Rapid identification of *C. albicans* is possible by carrying out a germ tube test. The germ tube test is an important identification characteristic and is specific for *C. albicans*, the most common yeast in human infections.

Figure 11.8 shows a preparation of *C. albicans* that has been incubated in human serum for 2 h. A yeast cell is producing a structure that protrudes from the cell – this is a germ tube. It is clearly distinct from the other yeast cell which is not producing a germ tube.

Germ tube negative isolates are typically often reported as *Candida* species not *albicans*. Further identification of germ tube negative species can be carried out using commercially available kits or by using chromogenic culture media. Chromogenic media can also be used as a primary isolation medium and are useful in that they can identify mixed infection, that is infections caused by two or more different species of *Candida*. This can be useful as the colonial appearance of *Candida* spp. on Sabouraud's agar, with the exception of *C. krusei*, is very similar. Figure 11.9 shows the typical colonial appearance of various species of *Candida* on a *Candida* chromogenic medium. The common species of *Candida* are easily distinguished and differentiated on such media. Mixtures of two species of *Candida* are relatively common; however, it is very unlikely that all five species will be seen in a clinical sample, as seen on this plate.

Cross reference

See Chapter 3 for information on chromogenic media.

Treatment of mucosal *Candida* infections

Many mucosal infections are self-limiting and can often be cured by removal of the predisposing condition, e.g. reducing obesity or controlling diabetes. Treatment with locally applied antifungal drugs can often speed up recovery, but unless the underlying conditions are tackled infection may recur. Treatment with topical antifungal agents is used in cases of vaginal candidiasis; however, oral therapy is also effective and has better compliance. Oesophageal candidiasis is a more serious condition and treatment often requires hospitalization and use of intravenous antifungal drugs.

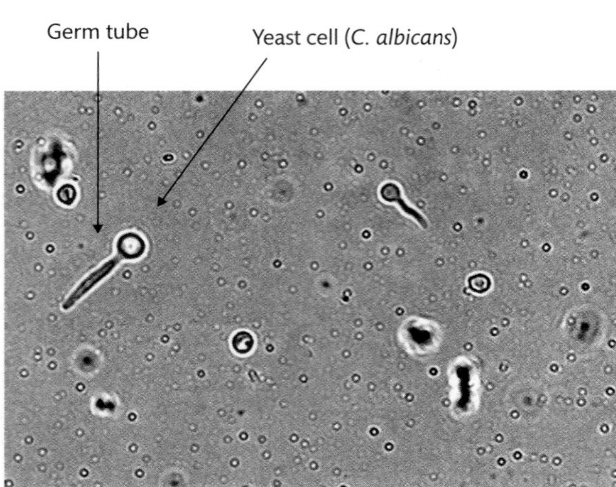

FIGURE 11.8

Photomicrograph of *C. albicans* after 2 h incubation at 37°C in human serum showing production of germ tubes.

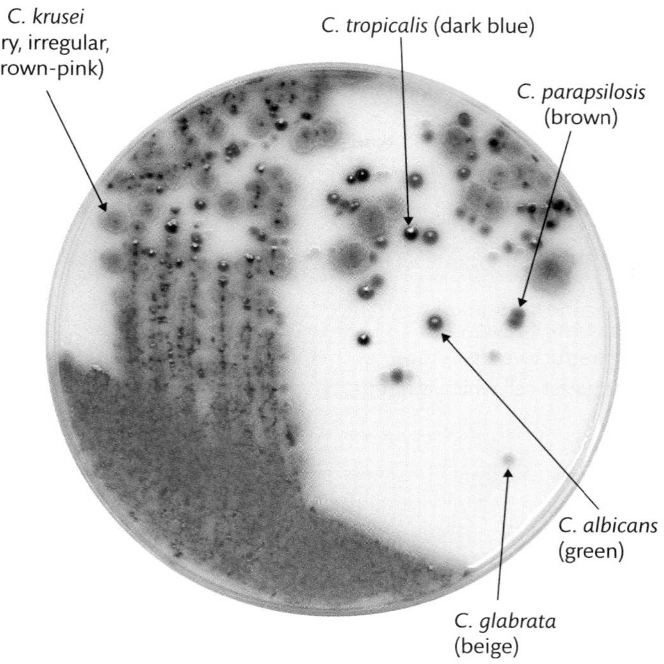

FIGURE 11.9
Photograph of chromogenic *Candida* medium showing colonial appearance of five common species of *Candida* after 48 h incubation. (Photograph copyright of Oxoid, reproduced here with their permission.)

Key Point

***Candida* infections frequently recur after treatment, unless the underlying condition predisposing to infection is also treated, e.g. stabilization of diabetes or weight loss in the obese.**

11.6 Invasive fungal infections

Invasive fungal infections are caused principally by *Candida* spp. and *Aspergillus* spp. but can also be caused by a variety of other fungi. These are fungal infections of the body which occur almost exclusively in debilitated patients whose normal defence mechanisms are impaired. The commonest fungal organisms causing such infections are *Candida* spp. and infections with these organisms are commonly derived from the patient's own flora.

Invasive infections caused by filamentous fungi such as *Aspergillus* spp. are commonly derived from the environment. Infection is often acquired through the inhalation of spores and the lung is the commonest site of infection, although infection can spread (disseminate) to other body sites, including the brain. The organisms that cause such infections often have low intrinsic virulence. With very rare exceptions they are unable to cause infection in the normal healthy host. The increased incidence of these infections is due to the emergence of AIDS, the use of aggressive cytotoxic chemotherapy for malignancy, increased incidence of transplantation requiring immunosuppressive regimes, the widespread use of broad-spectrum antibacterial antibiotics, **cytotoxic agents**, corticosteroids, and surgical and other procedures that result in lowered resistance of the host.

There are several other types of infection caused by fungi, but these are beyond the scope of this book; they include allergic-type reactions to fungal products such as farmer's lung and intoxications caused by fungal-produced toxins found in food. In the USA and other parts of the world there are a group of infections known as the endemic mycoses, caused by several different genera of dimorphic fungi. Although these infections can be severe, they will not be considered here as they are extremely rare in the UK. The types of infection considered in this chapter are described below.

SELF-CHECK 11.6

Why is the respiratory tract the most common site of invasive *Aspergillus* infections?

Key Point

Invasive fungal infections occur mainly in ill patients, particularly those who are immunosuppressed. Such infections carry a high mortality rate—partly because infections are difficult to recognize and because of the underlying status of the patient. Dissemination from an initial infection site is common.

Invasive *Candida* infections

The most common invasive form of *Candida* infection is candidaemia, defined as the presence of *Candida* in the blood. *Candida* is one of the ten most common pathogens isolated from blood cultures; in some centres it may be the fourth or fifth most common pathogen. Predisposing factors include the use of intravenous catheters, surgical procedures (particularly abdominal surgery), burns, parenteral nutrition, chemotherapy-induced damage to the oral or intestinal mucosa and the use of antibacterial antibiotics. Patients are usually ill but not necessarily immunosuppressed. Infection is initiated on a mucosal surface or following infection of an IV catheter; infection then spreads to the blood stream and can disseminate to various body organs, causing abscess formation.

There are various other forms of invasive candidiasis, such as gastrointestinal candidiasis; this occurs in patients receiving cytotoxic therapy which damages the mucosa of the intestine. Perforation of the mucosa can lead to spread of organisms to the peritoneum, blood stream and other organs such as the liver and spleen.

Invasive candidiasis can occur in babies and neonates, partially because of the immaturity of the immune system, but low birth weight and prematurity are predisposing factors. Organisms often gain access to the blood stream via intravascular catheters and *C. parapsilosis* is isolated frequently from neonates. Infection may disseminate to cause meningitis and infection in other organs.

One of the most problematic *Candida* infections to diagnose is infection of the urinary tract. Asymptomatic *Candida* infection of the urinary tract occurs mainly in women and occurs following antibiotic or steroid treatment that promotes colonization and growth of *Candida* in the gastrointestinal and genital tracts. Infections of the bladder are also common in diabetic patients and patients with urinary tract abnormalities, especially if the patient is catheterized or receiving antibacterial antibiotics. Infection may spread to the kidney either via the ureters or through haematogenous spread. Diagnosis of renal infection is difficult as isolation of *Candida* from a urine sample may be due to a genuine bladder infection or contamination of a urine sample due to colonization of the genital tract. Samples of urine taken directly from the bladder (supra-pubic urine) and blood cultures may help confirm the diagnosis of renal candidiasis.

Cross reference

See Chapter 6 for information on taking urine samples and interpretation of urine cultures.

Diagnosis of invasive candidiasis

Blood cultures are frequently taken to diagnose invasive candidiasis, although they may not always yield *Candida* spp.

Cross reference

Chapter 5 deals with blood culture samples.

Samples can also be taken from other potentially infected body sites, such as the peritoneal cavity, liver and lung, to assist with diagnosis. Isolation of *Candida* spp. from a normally sterile site is indicative of invasive candidiasis. However, isolation of *Candida* from sputum, urine or skin is not conclusive as these sites are frequently colonized in hospitalized patients, particularly if they have been treated with antibacterial antibiotics. Repeated isolation of large numbers of yeasts from a site may be a more reliable diagnostic marker for infection. As yet, there are no reliable markers for diagnosis of systemic infection although PCR tests may become more widespread for detecting and quantitating *Candida* in clinical samples.

Following diagnosis of invasive candidiasis by isolation of yeasts from blood cultures or from another sterile body site in an ill patient, urgent treatment is required. *C. albicans* is usually susceptible to fluconazole, whereas germ tube-negative yeasts such as *C. glabrata* and *C. krusei* are often fluconazole resistant. An **echinocandin** antifungal agent may also be used as they are highly active against most species of *Candida*. In many situations therapy will be started before an organism has been isolated; this is known as **empiric therapy.** It is essential to initiate treatment early because delays can increase mortality. Antibiotic therapy may change when the identity of an organism is known or a susceptibility test is performed.

empiric therapy

Empiric therapy is the commencement of therapy before it is known what the infective organism is or what its sensitivity is.

Cryptococcal meningitis

Most serious yeast infections are caused by organisms of the *Candida* genus. However, there is one other yeast which can cause severe systemic infection; this is *Cryptococcus neoformans*.

CASE STUDY 11.1 *Candida septicaemia*

A 60-year-old male diabetic was admitted for colonic surgery. He was given fluconazole prophylaxis as well as antibacterial antibiotic prophylaxis prior to surgery. After surgery he was transferred to the ICU and 3 days later the patient spiked a temperature and felt unwell. Blood cultures were taken and the patient's antibiotics were changed although fluconazole was continued. Then, 24 h later, the patient's condition began to deteriorate, further blood cultures were taken and IV catheters were removed and sent for culture. A further 24 h later, the patient was still febrile, culture results showed that one of the catheters grew +++ yeast, and a germ tube test was carried out and was negative, indicating a non-albicans *Candida*. Because of this fluconazole was discontinued, and an echinocandin drug was given to the patient. Within 12 h the patient's temperature started to fall and his condition improved. Eventually several of the blood cultures became positive and microscopy showed the presence of a yeast. Culture on chromogenic media showed colonies typical of *Candida glabrata*. The yeast isolated from the catheter tip was also identified as *C. glabrata*. The blood culture and catheter isolates were sent to a reference laboratory for susceptibility tests and the MIC to fluconazole was >128 mg/L and to caspofungin was 0.05 mg/L.

- **What are the risk factors in this patient for a *Candida* infection?**
- **Why do you think *C. glabrata* was isolated from this patient?**
- **What test led to the change in antifungal therapy and why?**

capsule
A capsule is a polysaccharide coat that covers some organisms such as *C. neoformans*. Capsules help resist ingestion by white blood cells.

C. neoformans is a **capsulate** yeast and causes infections of the human respiratory tract and nervous system. Infection is usually acquired by inhalation of desiccated yeast forms or spores and the primary site of infection is the respiratory tract. Infection at this site is often asymptomatic. In patients who are immunosuppressed the organism can disseminate, typically to the nervous system where it causes meningitis. *C. neoformans* is the commonest cause of fungal meningitis and affects up to 10% of AIDS patients. The main symptoms are headache and fever; occasionally, skin lesions may be evident. If left untreated the infection is invariably fatal. Cryptococcal infection should be suspected in any immunosuppressed patient with symptoms of meningitis. Rapid and accurate diagnosis is required so that antifungal therapy can be instituted as soon as possible.

Cross reference
Methods for processing CSF samples are covered in Chapter 8.

The main samples for diagnosis are CSF samples. Three techniques can be used for diagnosis: microscopy, culture and antigen detection.

The capsule is essential for *C. neoformans* to cause disease, and can be detected and quantified in CSF samples using commercially available kits. Antigen detection can be performed on serial CSF samples to monitor the response to therapy as levels decline with successful therapy.

Samples for *C. neoformans* should be cultured on Sabouraud's agar at 30°C and 37°C. Any yeast isolated from a CSF sample should be examined for the presence of a capsule using an India ink stain, identified to species level, and should have a susceptibility test performed to ensure the correct therapy is given. Treatment of cryptococcal infection often involves combinations of antifungal drugs to maximize the chances of eradication of the organism. Patients often have to remain on therapy for the remainder of their lives to prevent relapse of disease.

Cross reference
Further details on *Cryptococcus* are included in Chapter 8.

Key Point

Cryptococcal meningitis is a frequent complication of AIDS. When meningitis-like symptoms occur in an AIDS patient, the laboratory must be alert for cryptococcal infections.

Aspergillus infections

Aspergillus fumigatus and some other species of *Aspergillus*, such as *A. terreus*, *A. flavus* and *A. niger*, can cause a wide range of infections in man, ranging from mild allergic disease to life-threatening, invasive illness. Below is a list of the common conditions associated with *Aspergillus* spp. Aspergillosis is the term given to an infection caused by an *Aspergillus* spp.

1. Allergic bronchopulmonary aspergillosis (ABPA). This is primarily an allergic reaction to *Aspergillus* antigens. It follows *Aspergillus* colonization of damaged lung tissue although there is no tissue invasion. The main symptoms are wheezing, chest pain, fever and coughing up brown mucus plugs. It typically occurs in immunocompetent people, but some previous lung damage, for example through cystic fibrosis, heavy smoking or emphysema, predisposes to this condition.
2. Non-invasive aspergillosis or aspergilloma. This is caused by growth of aspergilli in preformed lung cavities, usually secondary to tuberculosis. An aspergilloma may persist for many years without symptoms. The usual symptoms when they occur are a cough with blood-stained sputum.
3. Acute invasive aspergillosis. This is the most serious infection and carries a high mortality rate despite antifungal therapy. The main site of infection is the lung, and it is caused by inhalation of spores from the air. It is a disease of immunosuppressed patients, particularly

those who are neutropenic (reduced number of white blood cells). It is most common in leukaemic patients, bone marrow transplant patients, and patients on immunosuppressive or cytotoxic therapy. Symptoms include fever, cough, haemoptysis and pleuritic chest pain and are often non-specific. Lack of response to antibacterial antibiotics often raises suspicion of a fungal infection.

Figure 11.10 is a photomicrograph of a section of a lung infected with *A. fumigatus*. Using a silver stain aspergilli appear black and are easily distinguished against the pink tissue background. A colony of *A. fumigatus* is seen spreading through tissue as the hyphae grow. The spreading nature of the hyphae is responsible for the organism invading adjacent tissue.

In patients who are severely immunosuppressed, the organism may disseminate from the lung to other body sites, including the brain, kidney and bones. Cerebral aspergillosis carries a very high mortality rate. Occasionally, infection will start at a different body site. The nasal sinuses are a common source of infection, again due to inhalation of spores. Infections have frequently been associated with building work in hospitals as this produces dust and disturbs soil, producing large numbers of airborne spores. Other primary sites of infection include the skin; such infections are most commonly seen in patients with severe burns or severe trauma.

For detailed information concerning *Aspergillus* infections (clinical aspects, diagnosis and treatment) the *Aspergillus* website has a wealth of information (http://www.aspergillus.org.uk).

Key Point

Aspergilloma and ABPA occur in patients who are immunocompetent, disease is localized and there is no invasion of tissue by *Aspergillus*. Invasive aspergillosis typically occurs in patients with compromised immune systems and there is invasion of tissue and dissemination of infection around the body.

Key Point

***Aspergillus* spp. can cause three main conditions in man: aspergilloma, allergic bronchopulmonary aspergillosis and finally invasive aspergillosis, which carries a high mortality rate even with treatment.**

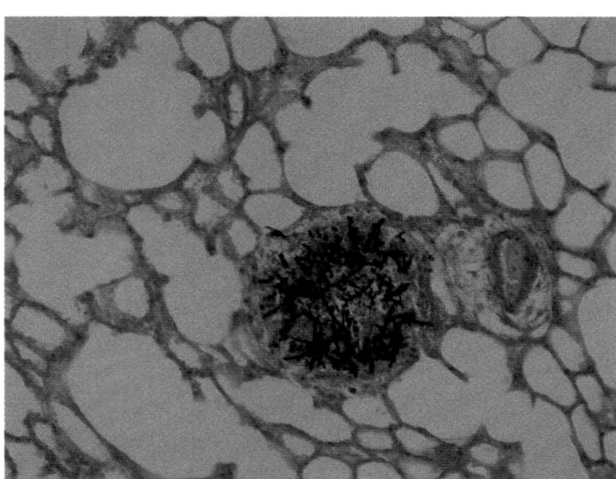

FIGURE 11.10

Photomicrograph of lung infected with *Aspergillus fumigatus*. The hyphae (stained black) can be seen spreading throughout the tissue. (Thanks to P. Warn for providing the sample.)

Diagnosis of ABPA and aspergilloma

In cases of allergic pulmonary aspergillosis and aspergilloma diagnosis is based mainly on clinical findings such as symptoms, underlying disease and chest X-rays. The laboratory can play an important role as isolation of *Aspergillus* from sputum can help confirm the diagnosis. Sputum samples may be sent to the laboratory and these should be cultured on standard bacterial media for respiratory bacteriological analysis (including TB if haemoptysis is present) as well as on Sabouraud's agar and incubated for up to 7 days. *Aspergillus* spp. grow as white fluffy colonies 1–2 mm in diameter after 24 h. As the culture is incubated further, colonies grow to 5–10 mm in diameter and change colour as the culture starts to sporulate. *A. fumigatus* typically produces dark green colonies after 2–4 days' incubation; other species of aspergilli may produce colonies of different colours. Figure 11.11 shows the typical colonial appearance of the main *Aspergillus* spp. growing on Sabouraud's agar for 3–4 days. Note the different coloration of the colonies which help in the differentiation of species. The coloration is due to different pigments in the spores or conidia. Some other non-*Aspergillus* fungi can produce colonies that look like aspergilli, and therefore caution is required in identifying a colony solely on the basis of it coloration.

The quantity of growth should be recorded in a semiquantitative fashion. It should be noted that because the air contains *Aspergillus* spores, aspergilli are common laboratory contaminants and may be found in the mouth, nose and sputum of healthy people. The isolation of very small numbers of *Aspergillus* from sputa may not be significant. If there is doubt then repeat samples should be cultured. In the case of aspergilloma sputum samples may yield *Aspergillus* in only 50% of cases.

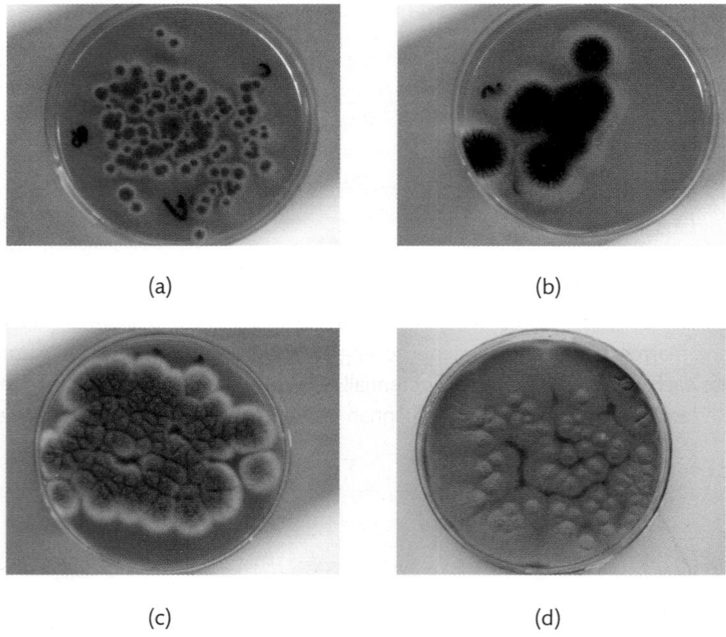

(a) (b) (c) (d)

FIGURE 11.11

Appearance of four main species of *Aspergillus* after 3–4 days' growth on Sabouraud's agar; the differences in coloration of the different species are clear. (a) *Aspergillus fumigatus*, (b) *Aspergillus niger*, (c) *Aspergillus flavus*, (d) *Aspergillus terreus*.

Cross reference

Chapter 9 deals with methods of processing respiratory samples.

Blood samples may also be taken to detect antibodies to *Aspergillus* and to measure levels of IgE antibodies. These tests are usually carried out by immunology or mycology reference laboratories.

Treatment for aspergilloma is in most cases conservative - attempts to remove the aspergilloma are dangerous. The main form of treatment of allergic pulmonary aspergillosis is the administration of steroids to suppress the immune response. However, in some instances this can result in invasive aspergillosis. The use of antifungal therapy to reduce the fungal load is being explored and itraconazole has shown promising results.

CLINICAL CORRELATION

Antibodies to *Aspergillus*

High levels of IgE antibodies are indicative of an allergic response. In cases of allergic pulmonary aspergillosis serum contains high levels of total IgE antibodies and high levels of both IgG and IgE to *Aspergillus* antigens. In cases of aspergilloma antibodies to *Aspergillus* are primarily of the IgG class. In cases of invasive aspergillosis because of the poor immune status of the patient the antibody response is very poor.

Key Point

Isolation of *Aspergillus* spp. from a sputum sample is not conclusive evidence of an *Aspergillus* infection. Because *Aspergillus* spores occur in the atmosphere, *Aspergillus* may occur as a contaminant in sputum cultures.

Diagnosis of invasive aspergillosis

Invasive aspergillosis carries a high mortality rate despite treatment with antifungal drugs and numerous studies have shown that the outcome of treatment is influenced by how quickly therapy is administered. Early intervention with antifungal drugs can increase survival rates whereas delayed treatment results in poorer survival rates. Therefore early diagnosis of infection is essential. Clinical diagnosis can be difficult as symptoms are non-specific and may mimic other conditions, including bacterial infections. The main symptoms include fever, cough and pleuritic chest pain. Suspicion of a possible *Aspergillus* infection is raised in patients who are immunosuppressed and where symptoms are not resolved by antibacterial antibiotics.

Chest X-rays and CT scans may be helpful in identifying lung lesions. In high-risk patient groups, such as bone marrow transplant recipients, testing of blood samples by **galactomannan** assay several times each week can help identify potentially infected patients early, allowing further diagnostic procedures to be instituted (galactomannan tests are described below in this section).

Although sputum samples may be produced by patients with invasive aspergillosis, such samples do not always yield *Aspergillus*. Bronchoalveolar lavages and lung biopsies are used to obtain samples deep in the lung and may provide better yields of organisms.

Laboratory diagnosis involves microscopy and culture of samples. BAL samples should be concentrated by centrifugation and examined by microscopy with Gram staining and using KOH preparations similar to dermatophyte microscopy. The presence of branched hyphae in a sample from a patient with supporting clinical symptoms is considered significant. Sputum samples should also be examined microscopically and tissue samples should be sent to a histology laboratory for appropriate processing and staining with special stains such as PAS or methanamine silver to stain fungal structures.

Cross reference

See Bancroft D, Gamble M. *Theory and Practice of Histological Techniques*, listed in the Reference section, for details of stains suitable for diagnosing fungal infections.

Cultures should also be set up on Sabouraud's agar; colonial appearances are described above. The identity of any *Aspergillus* isolate should be determined as this can help in selection of appropriate therapy, for example *A. terreus* and *A. flavus* are commonly resistant to **amphotericin B**. Other samples (e.g. skin, pus or tissue from other sites) should be cultured on standard bacteriological media and Sabouraud's agar and microscopy attempted to identify fungal hyphae in the sample (see relevant sections above for how to process the samples).

There is currently a lot of interest in the use of galactomannan and β-glucan detection kits. Galactomannan and β-glucan are fungal cell wall components released into the blood or tissue during fungal growth. Both tests are available as ELISA tests with colorimetric endpoints which measure the levels of the particular component. These can be used as specific diagnostic tests on blood or BAL samples from patients suspected of invasive aspergillosis. Galactomannan and β-glucan have also found utility in monitoring at-risk patients, as they can give a very early indication of infection, often before symptoms develop.

Key Point

Fungal cell walls are composed of an array of carbohydrate polymers such as galactomannan, β-glucan and chitin. These are released from the cell as the fungus grows and galactomannan and β-glucan have found utility in diagnosing fungal disease. The presence of these substances in body fluids is often indicative of a systemic fungal infection.

Treatment of invasive aspergillosis

Early recognition of infection and early treatment are key to successful therapy of invasive aspergillosis (IA). Because safer, less-toxic drugs are available treatment may be instituted before there is a definite diagnosis - this is known as empiric therapy. At the present time voriconazole is the drug of choice for IA, although alternatives such as amphotericin B and the echinocandins are available in cases of voriconazole toxicity, intolerance or treatment failure. There is growing interest in the use of combination therapies involving two or more drugs.

In some units at-risk patients may be given prophylactic antibiotics to prevent infections developing. However, fungal infections do occur in such patients and are more difficult to treat as prophylactic antibiotics may compromise the results of galactomannan and β-glucan tests and therapy of breakthrough infections often requires the use of a different class of drug in case resistance has developed to the prophylactic agent.

Although resistance to antifungal drugs is rare, resistance can develop during therapy, and certain species such as *A. terreus* and *A. flavus* are intrinsically resistant to amphotericin B. Any *Aspergillus* isolate from a case of IA should have a susceptibility test carried out to confirm susceptibility to the drug used initially and to identify alternatives in case a change of therapy is required.

SELF-CHECK 11.7

What are the most common clinical conditions associated with *Aspergillus*?

Other filamentous fungi as a cause of invasive disease

Although the majority of invasive fungal infections are caused by *Candida* spp. and *Aspergillus* spp., a small number of infections are caused by other filamentous fungi. The most common non-*Aspergillus* fungi encountered are listed below. These organisms have low virulence in humans and infections typically occur in severely ill, highly immunocompromised patients. The respiratory tract is the most common site of infection as infection is usually acquired through inhalation of spores, but the organisms can disseminate and infect any body organ. Diagnosis of these infections is based on clinical signs and demonstration of fungi in relevant samples by microscopy and culture. Any fungus isolated from a BAL sample or a tissue sample must be considered potentially significant and should be fully identified and a susceptibility test carried out as many non-*Aspergillus* organisms that are resistant to many of the antifungal agents used.

Organisms that are associated with invasive fungal infections include:

- *Fusarium oxysporum*
- *Fusarium solani*
- *Scedosporium prolificans* and *Scedosporium apiospermum*
- *Absidia corymbifera*
- *Rhizopus* spp.
- *Mucor* spp.

For details of these organisms you should refer to specialized books or websites such as http://www.doctorfungus.org.

Fungal infections of the eye

Fungi can cause infections of the eye - fungal **keratitis**. The cornea is the site of infection and typically filamentous fungi such as *Fusarium* spp. are involved. Infection usually follows some trauma to the cornea, although in recent years infections have been seen in contact lens wearers.

Scrapings from the cornea are required to culture the causative organism. Cultures for bacteria and fungi should be set up together and virology may be considered.

Treatment is usually with antibiotic eye drops although in severe cases of infection involving other parts of the eye intravenous antibiotics may also be given.

11.7 Antifungal susceptibility testing

Unlike susceptibility testing of bacteria, susceptibility testing of fungi is a relatively new phenomenon. This is partly due to the small number of available antifungal agents, the complexity of developing a standardized method and the small numbers of tests required. In this section we will look at the common antifungal agents available, methods of susceptibility testing and reasons for susceptibility testing.

Antifungal agents

There are many antibacterial antibiotics available but only about 10 clinically useful antifungal drugs, and even less when considering drugs for systemic use for serious infections. The drugs fall into five main classes. The main points on each drug are given in Table 11.2. For more information on antifungal drugs, have a look at the Further Reading resources at the end of this chapter.

Methods of susceptibility testing

Susceptibility testing of fungi is more difficult than susceptibility testing of bacteria. It is only in the last 10 years that standardized methods for susceptibility testing have existed. Methods in use include MIC tests, usually in microtitre plates, disc tests, E-tests and various commercially available kits, some of which are automated. At the present time there are few data to correlate fungal MICs with clinical outcome for certain drugs and breakpoint are only established for several of the azole drugs. However, efforts to develop better susceptibility testing methods and accurate breakpoints are ongoing.

The most important question we are faced with in the laboratory is, 'When should susceptibility testing be carried out on fungi?'

Because of the complex nature and difficulty in carrying out susceptibility tests on fungi, testing is carried out only in limited cases. For dermatophytes and *Candida* spp. isolated from localized infections, susceptibility testing is rarely carried out except in cases of treatment failure. For isolates of fungi from cases of invasive infection or where there is a strong suspicion of invasive infection susceptibility tests should be carried out. In most circumstances clinicians will start therapy before waiting for the susceptibility test result. The test results when available confirm the correct choice of drug or will on occasion result in a change of antifungal agent.

Because of their specialized nature and the relatively low numbers of fungal susceptibility tests required, it is common practice to refer isolates to specialized centres for testing. This adds extra delays before results are available. In-house testing will give earlier results but if laboratories choose to carry out their own testing, then the methodology chosen must be rigorously tested and controlled. The methods chosen will reflect the number of tests carried out per annum and budgetary considerations. Tests can be time consuming and expensive.

Key Point

Fungal susceptibility tests are complex and difficult to perform and are carried out mainly on isolates from invasive infection or in cases of recurrent infection or where there is the possibility of the development of resistance.

Cross reference

E-tests are discussed in Chapter 4 covering antibiotic susceptibility testing.

SELF-CHECK 11.8

Why are antifungal susceptibility tests carried out less frequently than susceptibility tests on bacteria?

TABLE 11.2 Antifungals and clinical uses.

Class	Drugs	Route of administration	Spectrum of activity	Indications	Adverse effects/problems
Polyenes	Nystatin	Oral and topical	Broad-spectrum yeast and moulds	Oral infections and superficial infections	Too toxic for systemic use
	Amphotericin B	IV	Broad-spectrum yeast and moulds	Invasive yeast and mould infections	Nephrotoxic, causes reactions on IV injection
	Lipid formulations of amphotericin B	IV	Broad-spectrum yeast and moulds	Invasive yeast and mould infections	Less toxic than conventional amphotericin B Better tolerated, but expensive
Nucleotide analogues	Flucytosine	Oral and IV	Yeasts only	In combination with other drugs for treatment of cryptococcal meningitis or candidiasis	Toxic at high doses Levels must be monitored regularly
Allylamines	Terbinafine	Oral and topical	Dermatophytes	Treatment of dermatophyte infections Topical treatment for simple local skin infections Oral therapy for nail and widespread skin/hair infections	Long-term treatment required especially for nail infections (6–12 weeks).
Azoles	Fluconazole	Oral and IV	Yeasts	Treatment of vaginal candidiasis Treatment of systemic candidiasis	Well tolerated. *C. krusei* and *C. glabrata* may be resistant.
	Itraconazole	Oral and IV	Yeast and moulds	Treatment for dermatophyte infections Treatment for invasive yeast and mould infections	Has some effects on liver, interacts with many drugs, oral absorption may be variable Requirement to measure levels on oral therapy
	Voriconazole	Oral and IV	Yeast and moulds, wider spectrum than itraconazole	Treatment for invasive yeast and moulds	Has side effects, interacts with many drugs Oral absorption variable Requirement to measure levels.
	Posaconazole	Oral only	Yeast and moulds, wider spectrum than voriconazole	Prophylaxis in high-risk patients Treatment of invasive yeast and mould infections	Can only be given orally, some side effects
Echinocandins	Caspofungin Micafungin Anidulafungin	IV only	Yeast and moulds	Treatment of systemic candidiasis (most species/strains susceptible) Treatment of invasive mould infections	Can only be given by IV route Generally well tolerated, few interactions with other drugs

11.8 The role of the reference laboratory

Mycology reference laboratories have an important role in mycology, providing a range of useful services. These include:

- Identification of fungi
- Susceptibility testing of fungi
- Measurement of antifungal drug levels in blood
- Other specialized tests such as serological investigations
- A source of advice for mycological queries, such as significance of isolates, interpretation of test results and advice on therapy.

The extent to which a laboratory uses a reference laboratory is dependent on the amount of mycology work carried out in the laboratory and the expertise available.

Most laboratories should be able to isolate and identify *Candida* spp. using commercially available kits or chromogenic agars. Whether susceptibility tests should be carried out in house is dependent on the number of strains requiring such tests; where there are sufficient numbers a case could be made for investing and training in some form of testing method such as a commercialized system or E-test. Where only small numbers of isolates are seen it is probably better to refer these to a reference laboratory.

Filamentous fungi isolated from patients with suspected or proven invasive fungal infection should be fully identified and have a susceptibility test performed. As the number of these encountered in a routine laboratory is small and the tests are both complex and time consuming they should be referred to a reference laboratory for testing. Tests for *Aspergillus* antibodies which are useful in the diagnosis of ABPA and aspergilloma are best carried out in a reference laboratory.

For dermatophytes most laboratories should be capable of identifying to genus level the common dermatophytes. Further identification is usually unnecessary; susceptibility testing is not routinely performed on dermatophytes except in cases of treatment failure or development of resistance during therapy.

One important function of the reference laboratory is in the measurement of antifungal drug levels in blood. Unlike antibacterial antibiotics, commercially available automated methods are not available for antifungal drugs. Measurements are carried out using bioassays or HPLC methods; both methods are complex and time consuming, and are probably beyond the capabilities of the average microbiology laboratory. It is therefore common practice to refer antifungal drug level samples to a reference laboratory.

There are several reasons to measure the levels of antifungal drugs:

- To ensure that adequate levels of drugs are achieved; this is particularly important with drugs given orally as absorption may be poor and other medications can both increase and decrease the levels of antifungal drugs.
- In patients receiving long-term therapy measurement ensures that patients are taking the drug and that levels are maintained in the therapeutic range.
- Certain drugs such as flucytosine can be toxic if recommended levels are exceeded; measurement of levels allows alteration of dose to ensure toxic levels are not achieved.

Reference laboratories serve a further function in that they can provide advice across a range of mycological areas and assist both laboratory staff and clinicians in the interpretation of both culture results and test results, and they can suggest alterations in therapy based on both susceptibility test results and drug levels.

Key Point

Reference laboratories serve an important function in identifying and testing unusual fungi, carrying out susceptibility tests on fungi and acting as a source of advice on fungi, fungal diseases and therapy.

CHAPTER SUMMARY

After reading this chapter you should know and understand:

- Numerous fungi exist in nature but only a very limited number cause human infections.
- Fungal infections are increasing mainly due to more patients who are immunocompromised.
- Infections of the skin, hair and nails are caused by dermatophyte fungi and are of low severity.
- The main site of invasive fungal infection in the body is the lung.
- Invasive fungal infections are very serious and are associated with high mortality rates.
- Antifungal susceptibility testing is increasing as better and more standardized methods are used.
- The role of the reference laboratory is very important and is used for expert advice and for a wide range of specialized services such as antifungal drug monitoring.

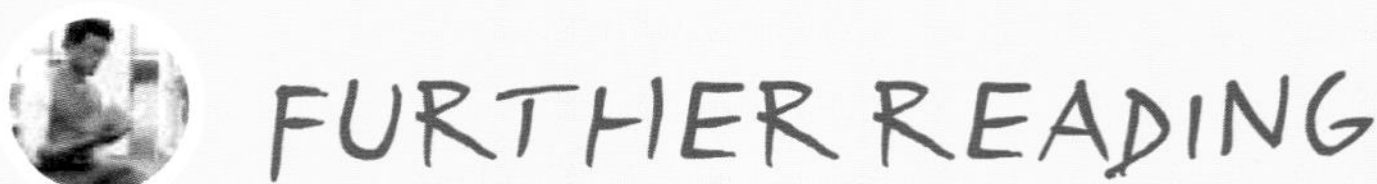

FURTHER READING

The following websites carry much useful information on antifungal drugs, microscopic and macroscopic appearance of fungi, protocols for identification and differentiation of fungi.

- **Aspergillus website. http://www.aspergillus.org.uk.**

  Excellent resource for material regarding all aspect of *Aspergillus* diagnosis, therapy and laboratory investigations.

- **University of Adelaide Mycology Department. http://www.mycology.adelaide.edu.au.**

  All round resource for information and photographs related to clinical mycology.

- **Centers for Disease Control, USA. http://phil.cdc.gov/phil/home.asp.**

  Many images of fungi and fungal infections.

- **Rex JH, Pfaller MA, Walsh TJ *et al*. Antifungal susceptibility testing: practical aspects and current challenges. *Clinical Microbiology Reviews* 2001; 14: 643–658.**

  Useful text which provides further information on all aspects of antifungal susceptibility testing.

11.1 Why do you think that most laboratories don't provide a full range of fungal testing services?

11.2 Why do you think it is critical for the laboratory to be as free as possible from *Aspergillus* spores?

11.3 Why do you think fungal infections are on the increase? Is there anything we can do to reverse this trend?

12

Infection prevention and control

Sheila Morgan and Michael Ford

Infection prevention and control has always been a major priority within any healthcare environment. As an increasing range of organisms are causing outbreaks of infection, the importance of Infection Prevention and Control has never been higher. Whilst several high-profile outbreaks of methicillin-resistant *Staphylococcus aureus* (MRSA) and *C. difficile* have focused attention of the governments and media, other organisms can be equally as important. As a result, the infection prevention and control team has a very varied role in both preventing and controlling the spread of infection.

Learning objectives

Upon reading this chapter the reader will have an understanding of the vital role of the laboratory as part of the infection prevention and control team. This will include:

- Procedures for the isolation and detection of MRSA
- Procedures for the isolation and detection of vancomycin-resistant enterococci (VRE) strains
- Pathogens requiring the laboratory to instigate a screening protocol
- Types of isolation media used for infection prevention and control samples
- Presumptive identification of these pathogens
- Use of typing systems
- Importance of *C. difficile* in hospital outbreaks
- Use of rapid techniques, e.g. rapid PCR
- Future problem organisms

Upon reading this chapter the student will have an understanding of why infection prevention and control is important and the various strategies involved in its implementation. These include:

- Ward-based protocols - hand washing etc.
- Barrier nursing, management of cohorts
- Current guidelines governing air quality in theatres
- Disinfection policies used in a hospital

- Disposal policy used in the hospital
- Monitoring of air quality in theatres
- Techniques to check the microbial cleanliness of the hospital environment and equipment
- Policies for eradication

12.1 What is Infection Prevention and Control?

This can be simply defined as the attempt to stop the spread of an infectious agent (bacterial, fungal or viral) from contaminating or infecting another individual. Most commonly, this takes place in a hospital environment.

Infection prevention and control has now become very high profile with headlines appearing in national newspapers almost on a daily basis. At present this is due mainly to methicillin-resistant *Staphylococcus aureus*, more commonly known as MRSA. This is a strain of *S. aureus* that has developed the ability to break down a very commonly used antibiotic - methicillin. In addition, most MRSA strains express resistance to a number of other commonly used antibiotics. This has made MRSA a particular problem, as infection with MRSA causes the same life threatening illnesses as *S. aureus* but it is very difficult to treat—especially if it enters the patient's bloodstream. Hospitals care for patients whose immunity is compromised, or who have IV lines *in situ* or other breaks to the skin—so preventing MRSA getting into such sites is of paramount importance. Once the patient develops an initial focus of infection, or becomes **colonized**, it is not a huge step for the organism to spread into the bloodstream and produce life threatening septicaemia. Typically, hospitals examine staff and patients as part of an overall screening strategy to detect the presence of MRSA.

Cross reference
Chapter 5.

colonization
Colonization is where the patient carries an organism on or in a particular body site but it is not doing them any harm. With MRSA it is often the nose that is colonized.

12.2 MRSA screening

MRSA screening is very different from most other samples sent to microbiology laboratories. This is because the patient may have no signs of infection. Indeed it is not just patients who are examined but staff also. The reason for screening is to detect the MRSA and hopefully eradicate it before it spreads to other individuals.

Consequences of MRSA colonization

Often MRSA colonization can go undetected. The MRSA is often not doing the patient or staff member any harm. It is the consequences of its spread that are the most apparent - particularly in a hospital environment.

Historically, the various approaches to MRSA screening have been somewhat haphazard and at times poorly organized. This has allowed some infected/colonized patients onto the wards with their MRSA status unknown. Similarly, staff who are colonized may have commenced work before having swabs taken and this can inadvertently lead to infection of patients on the unit.

A new screening regime called Search and Destroy is becoming popular. This is where all staff and patients are screened on admission and discharge, and in some larger hospitals results in nearly half a million swabs sent to the laboratory per annum.

Justification for MRSA screening

Overall MRSA screening is linked to many government targets to reduce the number of cases of MRSA bacteraemia (bloodstream infection). The main consequence of MRSA bacteraemia is the very high associated patient morbidity and mortality. The second reason is that failing to reduce the numbers of MRSA bacteraemia to within government targets can result in so-called hit squads being called into the hospital from the Department of Health, for example to make recommendations and insist standards of hygiene and techniques are improved. Naturally, this is seen as a failure of the infection prevention and control process and even the management of the hospital in general.

Key Point

Would you want to go or send a relative into a hospital knowing that it was failing in its hygiene standards?

One of the problems most hospitals face in complying with MRSA bacteraemia targets is the transfer of patients from other hospitals and nursing homes. These people may already be infected or colonized with MRSA when they arrive in the hospital. If they develop an MRSA bacteraemia where does the accountability lie? This would count as a bacteraemia for the hospital they are currently in which is argued as unfair. Hence, Search and Destroy programmes have arisen. By screening all patients on admission and discharge it can be proved that they either arrived with MRSA or acquired it during their hospital stay. This type of screening programme helps determine the possible cause when the number of cases of MRSA bacteraemia is examined.

Sample types and collection

The sample types and their collection can vary between hospitals but, generally, swabs are taken from inside the nose and the throat of staff and patients, plus swabs from any lesions (e.g. wounds). The commonest source of carriage of MRSA is the nose. The practice, however, may differ slightly in hospitals depending upon local policy; perineum (near the anus) and axilla (armpit) swabs are taken in some hospitals in addition to the above. The swabs are then correctly labelled and sent to the laboratory for processing.

SELF-CHECK 12.1

Why are nasal swabs almost always sent for MRSA screening?

Sample processing

There are several different strategies for how MRSA is detected from screening swabs. The most common are enrichment culture, selective agars, chromogenic agar and molecular methods (PCR). We will examine each of these in turn.

Enrichment culture

Since there may only be a few cells of MRSA on a screening swab, finding them can be as difficult as finding a needle in a haystack. Some experts argue that a few cells are hardly going

transient colonization
Studies have shown that nurses at the start of a shift can be free of MRSA but the organism may be detected in nasal swabs in very low numbers later on. At the start of the next shift the MRSA strain may not be detected. This is an example of transient colonization.

to be responsible for a hospital outbreak and may represent **transient colonization**. Opinion is still divided on how important small numbers of organisms are; however, several different enrichment methods have developed. These generally take the form of a broth culture containing antibiotics which suppresses other competing organisms, allowing the MRSA strains to multiply to high numbers so they are more easily detected.

MRSA selective broth is an example of a selective enrichment medium that is commercially available. It is a medium containing several antibiotics to inhibit methicillin-sensitive *S. aureus* as well as other organisms. The medium is based on the ability of the surviving MRSA strains to ferment mannitol and change the pH indicator from red to yellow due to acid being produced from the fermentation of mannitol.

It is important to understand how the microbiology department receives, processes and reports results from MRSA screening tests and Figure 12.1 illustrates an example of the process which could be used from receipt of swab to issuing a report. Note how many stages are involved and how labour intensive the processes can be.

Whilst showing excellent sensitivity and specificity from nasal swabs, false-positive results from other sites are common. This is because swabs may contain high numbers of other organisms that may not be totally inhibited by the antibiotics present. As a result the carbohydrate can be fermented and a false-positive result obtained. Perineum swabs may contain high numbers of 'faecal organisms' and it is common for false-positive results to be obtained from such swabs using the broth enrichment method.

The major disadvantage of broth enrichment is that confirmed results may take 48 h to become available. As you can imagine this will have an impact on controlling the spread of MRSA if the patient or staff member is infected/colonized for 2 days before it is detected.

As a result there is a demand for more rapid methods, and newer chromogenic methods in particular can achieve excellent results in 18–24 h. A reduction from 48 to 24 h for the detection of MRSA may be hugely important in limiting its spread and reducing the chances of a patient subsequently developing MRSA bacteraemia. Despite broth enrichment having an advantage in detecting small numbers of MRSA in screening swabs, the method is being superseded because of a demand for more rapid methods.

Key Point

Can you imagine how far an infectious organism such as MRSA can spread if a result is not obtained for 48 h.

Selective media (non-chromogenic)

Most commonly this involves the use of a solid medium (similar principle to MRSA broth) which utilizes the fermentation of mannitol as the method of MRSA colony detection. The medium, mannitol salt agar, contains the antibiotic oxacillin (structurally similar to methicillin) and has a high salt concentration to suppress other bacteria which may interfere with the recognition of MRSA colonies. (MRSA strains survive well in high salt concentrations.) These agents give the medium its selectivity for MRSA. Unfortunately, like so many selective agars which contain carbohydrates, the acidity released from fermentation diffuses through the agar, making the individual colonies of MRSA difficult to detect,

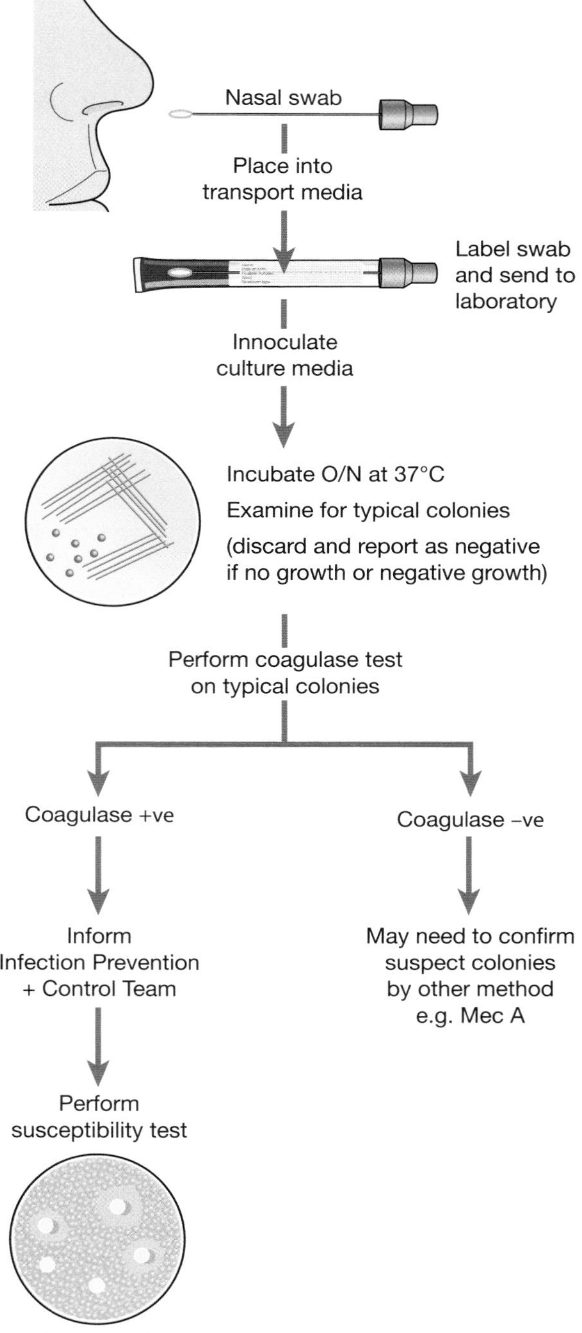

FIGURE 12.1

Example of an algorithm for processing MRSA screening swabs. (Courtesy of M. Ford.)

particularly in mixed culture. This is shown in Figure 12.2. Some authors have demonstrated its limited sensitivity and selectivity for the direct isolation of MRSA from clinical samples and screening swabs. Despite these drawbacks the medium is still widely used for isolation of MRSA.

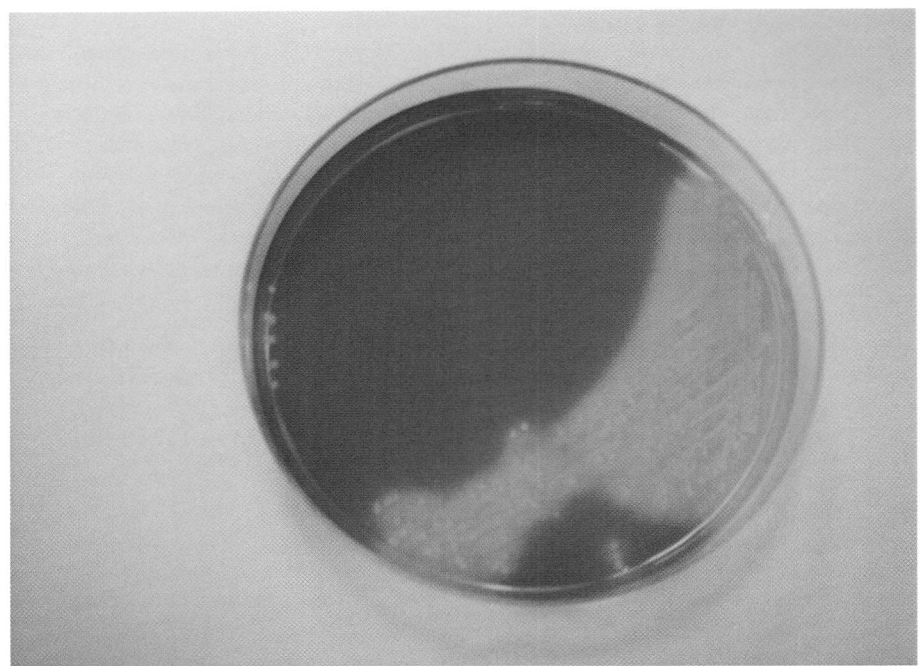

FIGURE 12.2
Growth of MRSA on mannitol salt agar. Note the diffusion of acidity away from the colonies. (Courtesy of M. Ford.)

Another commonly used, non-chromogenic, selective MRSA agar is Baird Parker agar containing ciprofloxacin and potassium tellurite as the selective agents. MRSA strains are seen on the medium as black colonies due to the ability of *S. aureus* (and MRSA strains) to reduce the potassium tellurite and produce black coloration. As with any medium the colonies must be confirmed by a positive coagulase test.

Cross reference
Chapter 3, Section 3.3.

Use of the medium is now dwindling due to several MRSA strains being susceptible to ciprofloxacin. This means the MRSA which is being sought may not grow on the medium, limiting its use.

Selective media (chromogenic)

These media are becoming very popular due to both high sensitivity and selectivity when compared to broth enrichment and other selective MRSA media. The results are available in 18–24 h, thus aiding the infection prevention and control infrastructure in its fight to contain and eradicate MRSA as quickly as possible. These media contain cefoxitin, to which MRSA strains are resistant. You may ask why methicillin is not used as the selective antibiotic in MRSA media. This is difficult to explain within the context of this chapter but it appears the structurally similar cefoxitin is better at inducing resistance in MRSA strains than methicillin.

Detection of the MRSA colonies is through hydrolysis of a chromogenic substrate. The principles of these substrates have been described in Chapter 3.

Cross reference
Chapter 3, Section 3.6.

Key Point

Cefoxitin is a cephalosporin-type antibiotic which is structurally similar to penicillin. Studies have shown that this antibiotic induces the *S. aureus* strains to express methicillin resistance better than methicillin itself.

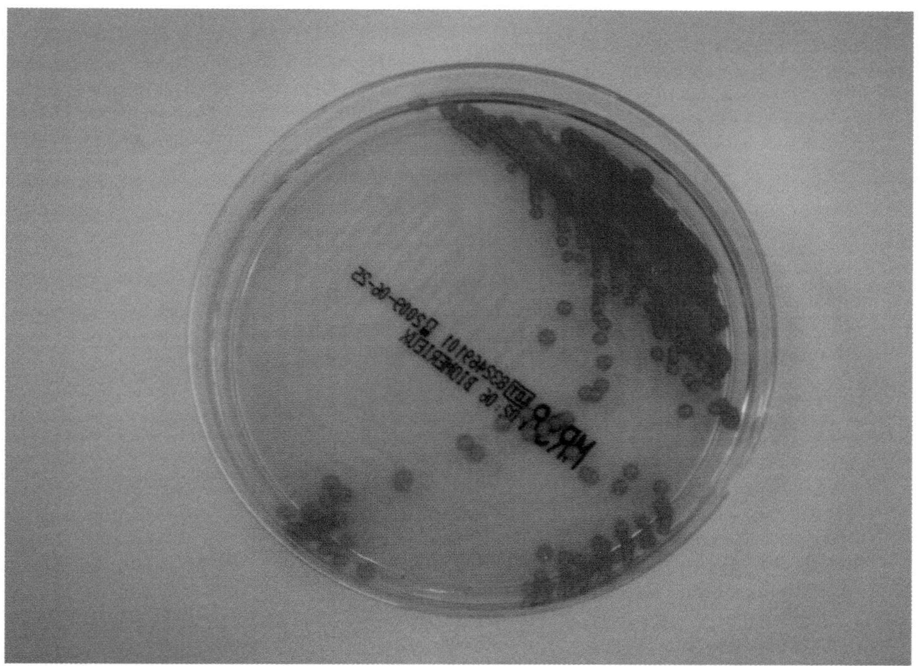

FIGURE 12.3
Photograph of MRSA colonies growing on MRSA ID agar (bioMérieux). Note the difference between this medium and the mannitol salt agar (Figure 12.2) in the appearance of the colonies and the colour of the surrounding medium. (Courtesy of M. Ford.)

Briefly, the colonies of MRSA which grow on the medium produce enzyme(s) which hydrolyse or 'break down' the colourless chromogenic substrate in the medium. Upon substrate hydrolysis, a rapid reaction with atmospheric oxygen occurs and a coloured compound is produced, as you can see in Figure 12.3. These coloured products are hydrophobic and stick to the surface of the colony. Each MRSA colony is therefore easily recognized, even in mixed culture. Several types of media are available from a variety of commercial companies.

SELF-CHECK 12.2

Why are individual MRSA colonies in mixed culture more easy to spot on a chromogenic plate than on mannitol salt agar?

Molecular methods

Molecular methods would be expected to be ideal in the ever increasing demand for rapid MRSA results as well as giving even higher sensitivity and specificity for MRSA screening when compared to culture methods. However, whilst promising to revolutionize MRSA screening, molecular methods have still not been adopted widely. The possible reasons are high cost, reduced sensitivity for samples other than nasal swabs, non-validation for swabs other than nasal and low throughput compared with conventional culture methods. In addition, it is generally perceived that specialist staff are required to perform molecular assays. However, systems such as the Cepheid Xpert™ (Cepheid, Sunnyvale, CA, USA) for MRSA and other assays (e.g. for *C. difficile* and *M. tuberculosis*) are extremely easy to use and can even be taken out of the laboratory setting and used as a point of care test (POCT) by medical and nursing staff at ward level. The assay can generate results in 1 h, which may be highly advantageous in certain clinical situations.

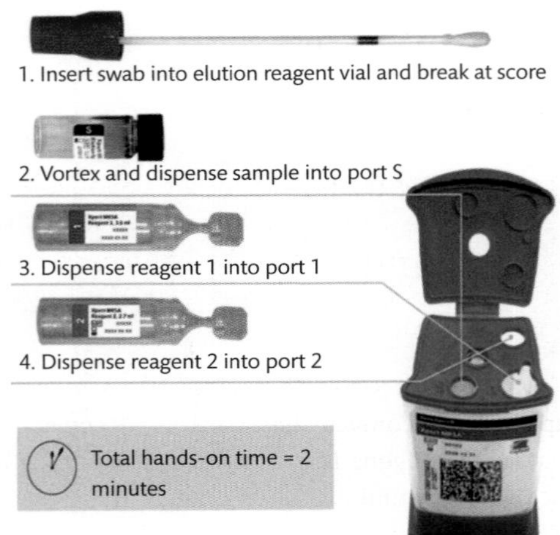

FIGURE 12.4
Sample protocol for detection of MRSA in nasal swabs using the Cepheid Xpert™ MRSA system. (Courtesy of Cepheid.)

The Cepheid Xpert™ for MRSA combines a user friendly sample preparation system with real-time PCR amplification, with the process being fully automated (Figure 12.4). In trials, the positive predictive value was shown to be 80.5% and the negative predictive value (NPV) 96.6% for nasal swabs. This is higher than results achieved using chromogenic agars.

The cost of MRSA screening using molecular methods is, however, significantly higher in comparison to the costs of selective agars but it is important to always consider the 'bigger picture' in that significant savings may be achievable elsewhere due to operations not being cancelled because of the rapidity of such results, antibiotics may be avoided, spread of MRSA may not occur, etc. In the US, over $2.5 billion excess health-care costs have been attributed to MRSA infection.

Key Point

Remember that health care is a business and everything has to be paid for.

Certainly, reducing the screening time to an hour instead of overnight incubation would be expected to have significant benefits. This could save the patient requiring to be barrier nursed or having an operation cancelled. However, there are studies which have shown that following introduction of molecular methods costs rose significantly but there was little reduction in the number of MRSA cases.

SELF-CHECK 12.3

What are the benefits of molecular over cultural methods? What are the disadvantages of these methods and, having considered this, what single method would you recommend for MRSA screening in your institution?

SELF-CHECK 12.4

Laboratories have been screening for MRSA for a number of years now and yet cases of MRSA bacteraemia are still high. Is screening the only answer?

12.3 Vancomycin-resistant enterococci (VRE)

Enterococci are part of the normal flora of the gastrointestinal tract. As you have seen in previous chapters they cause a wide variety of infections, most commonly urinary tract infections, which can develop into more serious conditions such as septicaemia. Generally, treatment of these infections is relatively straightforward using antimicrobials such as amoxycillin, which generally give a high concentration in the urine under standard dosage regimes.

One of the most important groups of antibiotics are the glycopeptides, particularly vancomycin. As you can see in Chapter 4, vancomycin has very good activity against all commonly encountered Gram-positive pathogens (e.g. *S. aureus*, β-haemolytic streptococci and enterococci) as resistance to the antibiotic is very rare. Therefore, if a patient is admitted with a suspected Gram-positive infection, clinicians can generally be safe in the knowledge that the patient should respond to vancomycin. This may not be true in all institutions as different geographical regions can have very different patterns of resistance. However, whilst resistance is still extremely rare in staphylococci, more and more infections with VRE strains are being reported, particularly in *Enterococcus faecium*, as you can see in Tables 12.1 and 12.2.

Cross reference
Chapter 4.

TABLE 12.1 **EARSS data on the percentage of invasive infections (blood culture isolates) caused by glycopeptides-resistant *E. faecium* in the EU in 2007.**

Country	Invasive infections (%)
Greece	37
Ireland	33
Portugal	29
Cyprus	25
Israel	24
United Kingdom	21
Germany	15
Italy	11
Turkey	8
Czech Republic	6
Slovenia	5
Spain	2
Croatia	2
Austria	2

(Continued)

TABLE 12.1 (*Continued*)

Country	Invasive infections (%)
France	1
Denmark	1
Finland	1
The Netherlands	0

EARSS: European Antimicrobial Resistance Surveillance System.

TABLE 12.2 EARSS data on the percentage of invasive infections (blood culture isolates) caused by glycopeptides-resistant *E. faecalis* in the EU in 2007.

Country	Invasive infections (%)
Greece	6.7
Portugal	3.6
Ireland	2.6
United Kingdom	2.5
Italy	1.8
Czech Republic	1.4
Belgium	1.0
Germany	0.8
Denmark	0.5
Israel	0.5
Austria	0.3
Turkey	0.3
Spain	0.1

phenotype

A phenotype is a typical characteristic of an organism. This could be seen as its colonial appearance or biochemical properties. A typical *E. coli*, for example, may show different biochemical properties but it is still an *E. coli*.

VRE strains show various levels of resistance. Those expressing a Van A **phenotype** are classed as high level resistant to both vancomycin and teicoplanin, another commonly used glycopeptide. Such VRE strains have an MIC of >128 mg/L and would show little or no response to treatment with vancomycin. In addition, VRE strains also carry resistance mechanisms to other antimicrobials making them truly multidrug resistant. Naturally, having VRE strains commonly isolated in a hospital environment would have a huge impact on the management of patients. For example if glycopeptides could no longer be relied upon for treating severely ill patients suspected of having Gram-positive bacteraemia or other serious conditions what else could be used? Some VRE strains can be resistant to all penicillins, aminoglycosides and now glycopeptides, thus severely limiting therapeutic options.

Key Point

It is thought that VRE first developed through the misuse of antibiotics in animal feedstuffs and these resistant organisms transferred to humans through the food chain.

Sample processing

Because resistance to such key groups of antibiotics is extremely important, screening strategies have been developed to detect these in the hospital environment. Similar to MRSA screening there are several different methodologies: selective media, enrichment broth and molecular methods.

SELF-CHECK 12.5

Why is controlling the spread of VRE so important?

Selective media

Since enterococci are part of the gastrointestinal tract it is sensible to obtain samples from sites most likely to detect carriage of VRE in patients. Commonly, rectal swabs are sent, although faeces samples can be sent. The former is generally more acceptable to patients and staff, particularly those on Intensive Care Units (ICUs). These samples may contain high numbers of other bacterial types so culture media contain additional selective agents to remove these competing organisms.

There are several different media that have been used to isolate VRE strains from both clinical and environmental samples, all of which contain vancomycin as the selective agent. These media have varied from simple blood agar plus 4 mg/L vancomycin to bile esculin azide vancomycin (BEAV) agar which utilizes sodium azide and vancomycin as the selective agents (Figure 12.5). The latter has the advantage of utilizing esculin which is hydrolysed by enterococci to produce black colonies making detecting VRE strains easier. This medium, however, does not differentiate between species of enterococci and further tests are required to identify the vancomycin resistant strain.

Key Point

Remember other Gram-positive bacteria can be resistant to vancomycin – so just because an organism grows on the medium doesn't mean it is always a VRE!

Chromogenic VRE media

Media such as BEAV are now being superseded by chromogenic media (e.g. chromID VRE agar from bioMérieux). This medium has the advantage of allowing differentiation between the commonest enterococcal strain types, *E. faecalis*, and *E. faecium*, based on the colour of the colony obtained (Figure 12.6). Studies have shown improved recovery of VRE from stool samples over BEAV agar in 24 h. Chromogenic media are gradually superseding

FIGURE 12.5
Growth of *E. faecium* on BEAV agar showing typical black colonies due to the organism's ability to hydrolyse esculin.

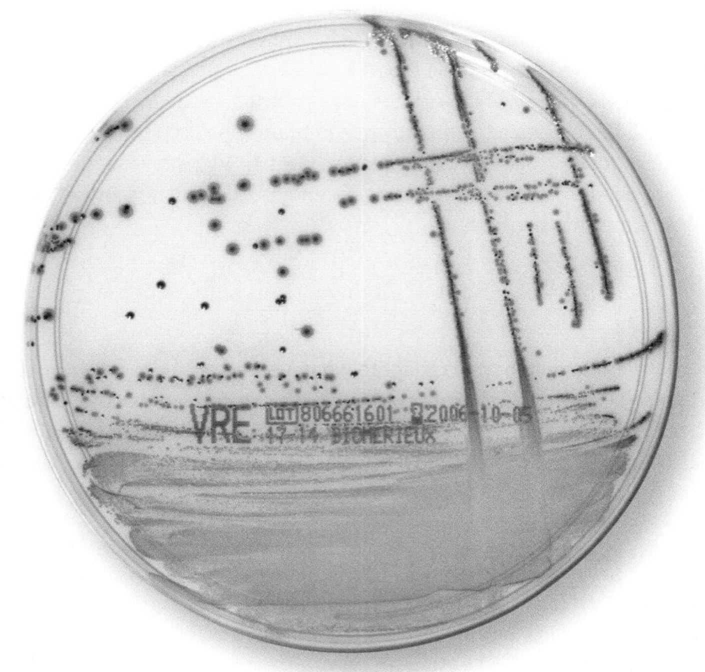

FIGURE 12.6
Growth of *E. faecium* (violet) and *E. faecalis* on chromID VRE agar. Note both the differential aspects of the medium and the localization of the chromogen on the colonies, and lack of diffusion into the surrounding medium. (Courtesy of bioMérieux.)

conventional media for the isolation of VRE strains, such as those based on esculin. These allow differentiation of *E. faecium* and *E. faecalis* directly. In addition, these organisms grow as discreet colonies with no diffusion of the reaction products into the surrounding medium.

Enrichment culture

Again similar to some MRSA screening strategies, low numbers of VRE in a sample can be enriched to high numbers by the use of broth enrichment. Bile esculin broth (BEB) plus vancomycin at a concentration of 15 mg/L has been compared with BEAV and led to significantly more VRE strains being detected than direct culture. Whilst detection is enhanced it takes a further 24 h for a result to be available compared with direct plating. The enrichment broth is also differential in that the broth turning black indicates the presence of a VRE.

Key Point

Hydrolysis of esculin by bacteria results in the formation of a diffusible black pigment.

12.4 Detection of pathogens that require initiation of a screening programme

In theory the patient-to-patient transmission of any pathogen could prompt the laboratory to initiate a screening programme. Altematively, this could be from staff to patients or *vice versa* or even amongst a group of staff. In general, a screening programme would be initiated following detection of a multidrug-resistant bacterium or a significant organism such as a Group A *Streptococcus*. It is important that pathogenic organisms are monitored in highly important clinical areas, such as ICUs, special care baby units (SCBU) or sterile theatres, as the transmission of pathogenic organisms in such environments is likely to have serious consequences and outbreaks could occur. In addition, fungal pathogens can cause significant morbidity and mortality and finding them may also trigger a screening programme.

The laboratory would have to adapt to whatever the problem organism is and develop a plan for its isolation and identification. Good communication with the infection prevention and control team is essential to plan which samples are to arrive and when. In addition, it is imperative to relay the laboratory results back to the team as soon as possible.

Technically, you would need to determine the resistance pattern of the potential outbreak strain as this would be important in determining which media could be used for its selective isolation. You would also need to consider the organism's growth pattern and the atmospheric conditions it is best suited to for optimal isolation. In addition, you will need to consider which enrichment media to use should the situation warrant its use. As you can imagine there are hundreds of possibilities based on the nature of the outbreak strain. Finally, consider the staff resources which may be necessary.

12.5 Importance of *Clostridium difficile* (*C. difficile*) in hospital environments

C. difficile is the major cause of antibiotic associated diarrhoea (AAD) and *Pseudomonas* colitis throughout the world. In many instances the organism is carried harmlessly as part of the normal human gastrointestinal flora. Indeed it is estimated that over half of all healthy neonates carry the organism asymptomatically. So how does an organism which exists harmlessly in the gastrointestinal tract initiate the disease process?

C. difficile is a Gram-positive anaerobic rod which is carried as a harmless commensal in the gastrointestinal tract of many individuals. Strains can produce one or both of two toxins, A and B. These are large protein toxins. Toxin A is an **enterotoxin** and toxin B is a **cytotoxin**. Toxin A appears to be most important in **pseudomembranous colitis**. The organism becomes a problem when it can multiply in the gastrointestinal tract and is no longer kept in check by the high numbers of other bacteria in the gastrointestinal tract. Such a proliferation of *C. difficile* can occur when the patient has had, or is undergoing, a course of antibiotics. The antibiotics wipe out the other competing bacterial flora and allow *C. difficile* to multiply. Patients in hospital are often on antibiotics and this tends to increase in the winter months when colds and other viruses trigger off respiratory infections. Since *C. difficile* strains produce a highly cell-damaging toxin, pseudomembranous colitis (Figure 12.7) can develop which is a life-threatening condition, especially in the elderly.

enterotoxin
An enterotoxin is a toxin released by an organism in the intestine. It is a protein and frequently cytotoxic, killing the host cells.

cytotoxin
A cytotoxin is a toxin that kills cells.

pseudomembranous colitis
Pseudomembranous colitis is an infection of the colon, the symptoms of which are severe diarrhoea, abdominal pain and fever. The disease can often be fatal, particularly in elderly patients.

C. difficile has developed a very high profile as an organism capable of causing significant morbidity and mortality, especially in the elderly. Reduction in the number of *C. difficile* infections is now a high government priority. Outbreaks have occurred worldwide, which have resulted in numerous patient deaths and raised the awareness of standards of cleanliness in hospitals generally. Therefore the rapid detection, containment and treatment of patients with *C. difficile* infection are important targets for both governments and hospitals. In the USA, it is estimated that each case of *C. difficile* infection results in costs of over $15 000.

Whilst *C. difficile* infection rates in the vulnerable over 65-year-old group are starting to fall, the UK Health Protection Agency reported that there were over 13 000 cases between April and June 2007. It is in this age group that most of the deaths due to *C. difficile* occur, with an increase in cases noted in the winter months.

SELF-CHECK 12.6

Why do *C. difficile* infection rates increase in the winter months?

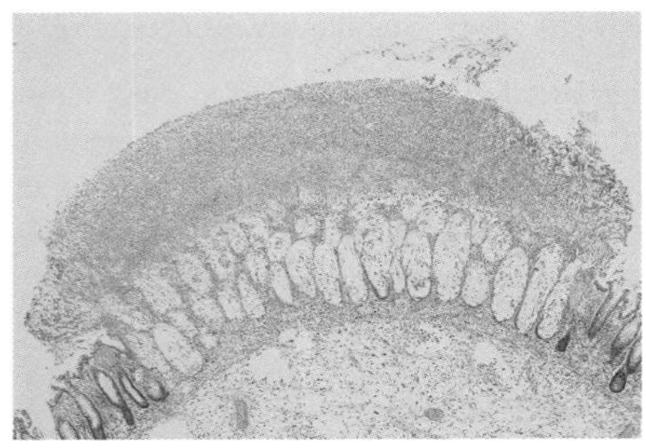

FIGURE 12.7
Section of the large intestine showing C. difficile infection. Pseudomembrane (comprising fibrin, red blood cells, epithelial debris and mucus) overlies dilated crypts in which epithelium is largely lost. (Reproduced by kind permission of Professor Alan Patterson).

Laboratory diagnosis of *C. difficile*

The accepted gold standard for the detection of *C. difficile* is a cytotoxin assay which utilizes cell culture. Essentially, this is to demonstrate that a faeces supernatant can kill a monolayer of cells, and this effect is lost if an antibody is added which reacts with the toxin and thus protects the cells from damage. Some laboratories use an alternative gold standard (cytotoxigenic culture) where *C. difficile* strains are grown from the sample followed by cell culture to determine toxin production. The latter is used to detect *C. difficile* strains which have the capacity to produce toxin, rather than detecting the presence of toxin in a sample. These assays, however, suffer from the drawbacks of maintenance of cell lines, high turnaround times and being quite labour intensive to perform. As a result the detection of toxins using automated platforms has become widespread.

Routinely, most laboratories detect the toxin(s) present in the stool sample by ELISA methods. These are now relatively simple to perform as they are usually on automated platforms. These have superseded the use of cell culture methods which are quite labour intensive with contamination of cell lines a particular problem. The ELISA method takes around 90 min to produce a result, which is then given to both clinicians and infection prevention and control staff so a patient positive for *C. difficile* toxin can be managed appropriately. These patients are often **barrier nursed** in an attempt to contain the organism and stop any potential spread of infection.

barrier nursing

Barrier nursing involves putting the patient in a side room or with other individuals with exactly the same infection in the same room. The nursing staff wear disposable aprons and gloves when dealing with these patients to minimize any spread of the infection to other patients.

The organism can be cultured on solid media like any other organism. Unfortunately, the colonies smell of horse manure so it is not a particularly pleasant job! The organism has a typical colonial appearance, under anaerobic conditions. It generally takes 48 h to grow the organism, and only then can it be tested to determine if it produces toxin(s). This is a more lengthy process than simply testing the stool sample for toxin. However, culture can be useful to determine if the organism causing the infection is exactly the same as that infecting other patients on the same ward or environment. The process used is that of **ribotyping** which involves extraction of RNA from isolated *C. difficile* cells and comparison with that of other *C. difficile* ribotypes. In the UK, the common ribotypes are 001, 016 and 019; however, an outbreak strain which caused significant mortality in Canada was shown to be a rare ribotype 027. The significance of this strain is high due to its ability to produce much more toxin than other *C. difficile* ribotypes. This 027 strain is known as a hyper toxin producing strain.

ribotyping

Ribotyping is the process of trying to 'fingerprint' an organism by detection of specific genomic DNA.

In severe cases of *C. difficile* infection, the patient would be started on either metronidazole or oral vancomycin for 10 to 14 days. In mild cases, due to antibiotic use, it is usually enough to withdraw the offending drug, allowing diarrhoea to settle within a few days. Since many patients may be frail it is particularly important to correct dietary deficiencies, especially where they are low in protein. In severe cases, the bowel may distend greatly and perforate, leading to peritonitis, in which case urgent surgery is needed.

Other detection methods

Many workers have questioned the use of ELISA methods alone for the detection of *C. difficile* disease as some have a low sensitivity and specificity for the toxin. Some studies have highlighted false-negative rates as high as 50%. As a result, a number of new techniques are being used to detect other markers associated with *C. difficile* infection to complement the detection of toxins A+B. The first of these is the use of faecal lactoferrin which is a marker for intestinal inflammation. This can be caused by a number of conditions including *C. difficile* disease.

Cross reference

Chapter 13.

Whilst non-specific it may be useful as part of an algorithm combined with other tests for the detection of *C. difficile* infection.

Perhaps the most promising is the use of glutamate dehydrogenase (GDH) which is produced by both toxigenic and non-toxigenic strains of *C. difficile*. It is produced in large amounts and is therefore a useful antigenic marker for the organism. GDH in some studies has a negative predictive value of >99%; thus samples can be screened using a GDH assay and if negative results could be sent out in 2 h. Since GDH does not distinguish between toxin-positive and toxin-negative strains, positive samples would then be tested for toxin A+B by ELISA or other method. Since *C. difficile* is a life-threatening disease the faster the disease is diagnosed the better for both management of the patient and reduction of spread of the disease. Currently, algorithms are being prepared in a number of institutions to incorporate GDH testing in particular.

There are now rapid molecular methods being developed which can detect the presence of toxin-producing *C. difficile* in less than 1 h. Compared with the gold standard of cell culture the sensitivity of molecular assays was shown to be 100% with a specificity of 93%. These assays are being developed so they are ward based which removes any potential delays in transportation to the laboratory. It is vital, however, that the infection prevention and control team knows the results generated at ward level from such assays.

Key Point

No single laboratory method is optimal for the detection of C. *difficile* infection.

SELF-CHECK 12.7

What would you consider to be the best algorithm for the laboratory detection of *C. difficile* infection? What are your reasons?

12.6 Future problem organisms

It is very difficult to predict what future problem organisms are likely to be encountered, as new pathogens may be discovered of which we currently have no knowledge. It is only 25 years since the causative organism of Legionnaire's disease was first discovered and even less for *Helicobacter pylori*. However, if we look over the last few years then there is a trend developing which shows multiantibiotic resistant bacteria being isolated more frequently. Examples of this include the extended spectrum β-lactamases (ESBL) seen in some Gram-negative bacteria such as *E. coli* and *Klebsiella* spp. These strains have caused outbreaks in some hospitals and in the community. It is reported that their general prevalence is increasing, leading to significant morbidity and mortality. One only has to look at the initial impact of swine flu to see how within a few weeks a global pandemic was predicted.

Highly pathogenic organisms can even require a worldwide strategy for dealing with their spread. This could include the stockpiling of antiviral drugs, involvement of local and national organizations and dissemination of information to both the public and health-care workers. The infection prevention and control teams need to assess each situation on an individual basis and develop policies and procedures to manage a multitude of different scenarios. Close links therefore with microbiology staff and departments are essential.

Panton-Valentine leukocidin-producing *S. aureus*

Recently, *S. aureus* strains producing a toxin called Panton-Valentine leukocidin (PVL) have been implicated in outbreaks of severe disease in the community. PVL is a toxin produced by *S. aureus* which can kill white cells. Fortunately, it is only produced by a minority of strains at present but infections with such strains are severe, leading to significant morbidity and mortality. Most notably the toxin can be produced by MRSA strains which show susceptibility to ciprofloxacin.

Think back to what you have read so far and if it could be problematic to isolate these strains from screening swabs.

Carbapenem-resistant Enterobacteriaceae

Resistance to the carbapenem antibiotics is of serious concern as they are one of the most important antibiotics of choice, particularly for patients with a serious, systemic Gram-negative infection. Such resistance is starting to emerge in ESBL-producing Enterobacteriaceae and is associated with a cell change which hinders the antibiotic entering the cell. Resistance by production of an enzyme (**carbapenemase**) can also be seen in some bacteria. It is likely that as such strains will be encountered more frequently and strategies will need to be developed for their detection.

carbapenemase
Carbapenemase is an enzyme which hydrolyses not only penicillins but also cephalosporins, monobactams and carbapenems. Strains that produce these enzymes are highly drug resistant.

12.7 Other resistance screening and media

New media are becoming commercially available every year and are targeted at resistant organisms or organisms that are currently difficult to isolate using currently available media. Most of the new media arriving on the market contain chromogenic substrates as these make target colonies much easier to detect. It is important that the biomedical scientist is aware of any new developments in microbiology so that the best possible media and techniques can be applied for the benefit of patients.

12.8 Infection Prevention and Control

It is estimated that at any one time, over 1.4 million people worldwide are suffering from infections acquired in hospital and between 5 and 10% of patients admitted to modern hospitals in the developed world acquire one or more infections. As a result, the emphasis on infection prevention and control has never been greater.

Infection prevention and control has always been at the cornerstone of all clinical practices. Since 2000, however, it has assumed new political significance and taken centre stage in government policy. National Audit Office Reports in 2000 and 2004 identified that there needed to be a cultural shift towards infection prevention as well as control as the cost of hospital acquired infection (HAI) to the UK NHS was stated to be as much as £1 billion per year, with 9% of patients acquiring an infection as a result of an inpatient episode, resulting in 5000 deaths per year. It was estimated that 15% of HAIs could be prevented, releasing circa £150 million back into the NHS.

Since then there have been many national strategies put in place to reduce the incidence of HAIs. These include:

- Audit and financing improvements in decontamination facilities
- **Mandatory reporting** of MRSA bacteraemia
- Mandatory reporting of VRE bacteraemia
- Nationally and locally agreed targets for MRSA bacteraemia
- Recording incidences of *Clostridium difficile*-associated diarrhoea
- New post of Director for infection prevention and control

mandatory reporting
Mandatory reporting is the obligatory reporting of a particular condition to local or national authorities. This often applies to communicable disease. Reporting of MRSA bacteraemia and *C. difficile* infection is mandatory in the UK.

All these strategies have culminated in The Health Act 2006: Code of Practice for the Prevention and Control of Health Care Associated Infections in the UK. This document is statute and therefore compliance with the elements of the Act is mandatory. The Health Act sets out the criteria for senior managers of each NHS organization 'to ensure that patients are cared for in a clean environment, and the risk of Health Care Associated Infections are kept as low as possible'. The Care Quality Commission is responsible for auditing compliance with these criteria. Failure to comply could result in significant consequences for the NHS organization involved. Some of the topics discussed below are encompassed within The Health Act 2006.

Hand hygiene

If hand hygiene is carried out properly then it is the single most important and cost effective method of preventing the transfer of organisms from the environment or patient to another patient and thereby preventing potential infections occurring. However, it is often performed ineffectively and infrequently. Various campaigns and initiative have been tried in a bid to improve hand washing compliance. One of these, the '5 moments for Hand Hygiene', has recently been launched to show ward staff in particular that hands must be washed five times in the whole period of patient contact (Figure 12.8).

Hand hygiene consists of two main methods: (a) hand washing and (b) application of alcohol-based hand disinfectants. Both methods disinfect the hands; that is, the processes involved *reduce* the number of potentially harmful organisms on the skin. Organisms found on the skin can be **resident** or transient.

resident organisms
Resident organisms are also known as normal flora or commensals. These are deeply seated within the epidermis in skin crevices, hair follicles, sebaceous glands and beneath fingernails. Their purpose is protecting the skin against colonization by transient organisms. They are not readily removed by hand washing.

The type of soap used will depend upon the nature of the task or procedure that is to be undertaken and the susceptibility of the patient. There are three methods of rendering hands safe: social, disinfectant and surgical hand wash.

Social hand washing involves the use of soap and water and is used in most activities such as washing hands upon leaving a toilet facility, handling soiled nappies and preparing food. In general, washing hands for 10–15 seconds with simple liquid soap will suffice for these tasks. This process removes transient organisms.

Liquid disinfectant soaps with greater antimicrobial activity are used for many hand washing practices in hospitals, as patients are more vulnerable to infection. These products remove and destroy transient organisms and remove some detachable resident organisms. Some products, such as chlorhexidine gluconate, have a residual effect.

Surgical hand washes are used prior to theatre procedures and their purpose is to further reduce the number of resident organisms on the skin of the operator/clinician to a safe level to prevent potential infection of the surgical site.

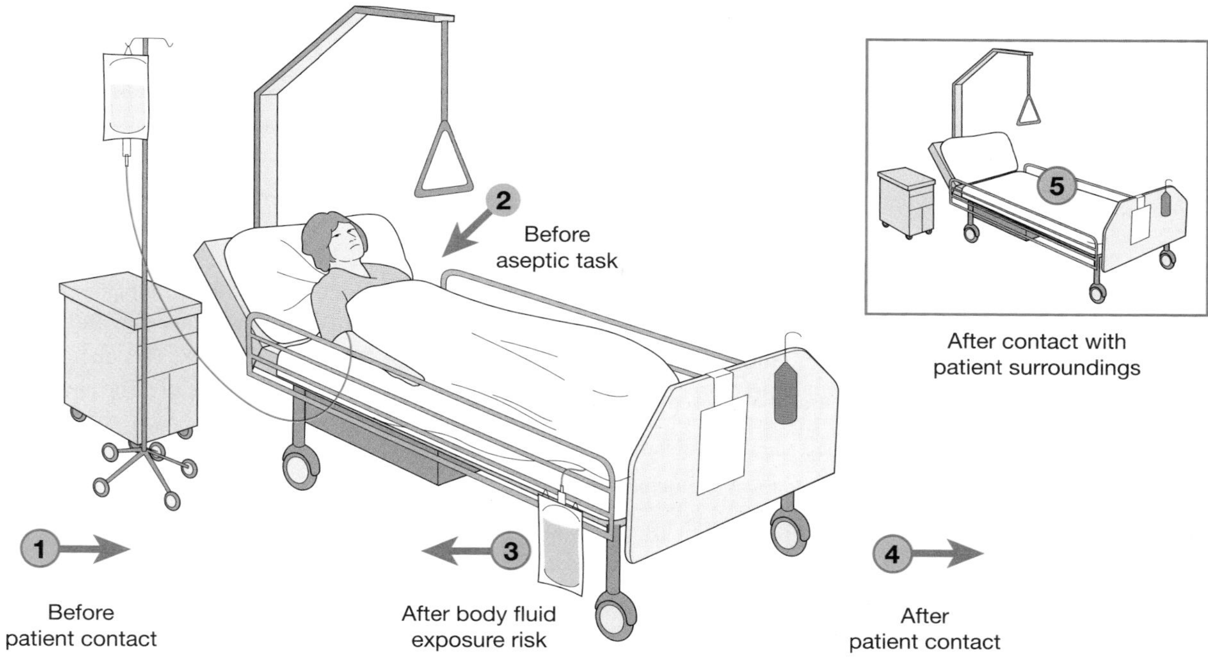

FIGURE 12.8
Image showing the five moments of hand hygiene, used as an educational aid to health-care workers.

In September 2004, the National Patient Safety Agency issued a Patient Safety Alert to all acute trusts in England instructing that alcohol-based hand gels (AHBG) should be installed at all points of use, for example at ends of beds and bedside lockers, by April 2005 in order to improve hand hygiene compliance to reduce the incidence of **hospital-acquired infections**.

hospital-acquired infection
Hospital-acquired infection is also called nosocomial infection. It is an infection which appears more than 48 h after a patient has been admitted to hospital or other health-care facility.

Alcohol-based hand disinfectants have good activity against most organisms save spores. These products should only be used on clean hands, as they are known to be less effective in the presence of soiling and dirt. They do not have any residual antimicrobial activity.

SELF-CHECK 12.8

Do you think ABHG would be useful to eliminate *C. difficile* from the ward environment?

Standard Precautions

Universal Precautions were a set of clinical actions initially introduced in the USA during the 1980s by the Centers for Disease Control (CDC) mainly to address the rising incidence of blood-borne viral infections (e.g. hepatitis B and human immunodeficiency virus) in health-care workers; it was noted that many of these individuals had suffered skin contamination with blood from patients or received sharps injuries.

The main principle of these precautions was to prevent parenteral, mucous membrane and non-intact skin exposures of health-care workers to blood-borne pathogens through sharps injuries and body fluid splashes/exposure whilst protecting patients from cross-infection. Since then, these principles have been adopted by other national health-care providers and are referred to as Standard Precautions.

Key Point

Sharps injuries often occur from contaminated needles which have been used to take blood and not discarded properly. This situation is totally avoidable!

Standard Precautions involve identifying high-risk procedures rather than high-risk individuals and also have the benefit of preventing patient discrimination. Health-care staff should consider all blood and body fluids (including semen, vaginal secretions, cerebrospinal fluid, amniotic fluid, peritoneal fluid, pericardial fluid, pleural and synovial fluids) as high risk, that is potentially infective, whether visibly blood-stained or not and wear the appropriate personal protective equipment (PPE) - masks, goggles, gloves, aprons, gowns, etc.

Other essential elements apart from PPE that comprise Standard Precautions are:

- Good standards of hand hygiene
- Use of gloves (sterile or non-sterile as appropriate)
- Safe disposal of sharps and single use/disposable equipment
- Management of used and soiled laundry
- Decontamination of equipment

Isolation nursing

Patients who are known to have, or are suspected of having, certain serious and potentially communicable (transmissible) infections require isolation precautions and management in order to prevent cross infection. Equally, patients who are particularly vulnerable to infection may also require isolation accommodation.

Ordinarily, the patient is managed in single room accommodation, with an ante-room/lobby, and ideally with en-suite facilities. All doors are kept shut at all times. Sometimes it is necessary to restrict the number of relatives and visitors to the patient depending upon their clinical condition and in some circumstances this can last for several weeks. As you can imagine these facilities are expensive to build and therefore their numbers are limited.

There are two main classes of isolation:

- Source - for known or suspected infectious patients
- Protective - for patients who are rendered highly susceptible to infection by their disease or therapy

The common means of transmission of infection are by:

- Airborne route
- Faecal-oral route
- Direct contact

Source isolation

This class of isolation can be divided into three main categories:

- **Strict** - for highly transmissible or dangerous diseases such as viral haemorrhagic fevers (VHF), e.g. Lassa, Ebola and Marburg. Currently there are only two of these highly specialized Category 4 units in the United Kingdom sited in Newcastle and London

- **Respiratory** – for diseases where the main route of transmission is airborne, e.g. pulmonary tuberculosis, measles, mumps, rubella
- **Standard** – for most other communicable disease, e.g. MRSA, viral gastroenteritis, *Clostridium difficile*-associated diarrhoea

Protective isolation

As patient susceptibility to infection varies considerably, depending upon the patient's disease and/or therapy, so does the isolation requirement for these particular patients. Patients with extensive burns, neutropenia and those receiving immunosuppressive therapy such as transplant recipients would require protective isolation nursing and management.

Management of cohorts

Due to diseases such as seasonal viral gastroenteritis, for example, it is not always possible to isolate all affected and symptomatic patients due to the shortages and greater demand made upon single room accommodation during such outbreaks. However, whenever possible it is best practice to find appropriate single room accommodation. Where and when this is not possible patients are managed in 'cohorts'. In circumstances such as these, patients who have had contact with the **index patient** are kept together and managed as a group. This prevents potential spreading of the causative organism to patients who have not been exposed to the index carrier thereby minimizing the numbers of patients affected.

index patient
An index patient is the individual who first shows signs of the disease.

SELF-CHECK 12.9

Why are cohorts managed together, generally in the same room, rather than isolated individually?

Underpinning and common to all these types of isolation management, with the exception of Strict Isolation (where additional specialized equipment and procedures are used), are good standards of hand hygiene prior to entering and leaving the facilities, safe disposal of sharps, decontamination of equipment, appropriate waste disposal, management of used laundry, good standards of daily environmental cleanliness and thorough terminal cleaning of the facility once the patient has been discharged.

Disinfection policy

Every institution must have a disinfection policy to ensure the environment is clean and safe for staff, patients and visitors. This policy will include several processes by which this is achieved – these are described below. It must be remembered that disinfection does not bring about the same level of reduction in microbial contamination as sterilization.

Sterilization

This is a process used to render an object free from all viable microorganisms, including viruses and spores. For example this process is used for theatre instruments and for instrumentation that enters a sterile body site such as needles and cannulae which are used to access the bloodstream.

It is essential to emphasize that prior to either process being employed, a thorough physical cleaning must be the first step to ensure that all dirt/soiling and organic matter is removed. Hand-hot water and detergent are usually all that is needed for cleaning processes; however, it is prudent to check with manufacturers of equipment to ensure that they are compatible with such.

Disinfectants and alcohol-based products are not effective in the presence of dirt and organic matter and therefore small but significant numbers of potentially pathogenic microorganisms may remain and pose an infection risk to subsequent patients.

Ward routine

Daily dusting and cleaning of the ward environment with hot water and detergent is a daily process and an essential part of ward routine. Occasionally, chlorine-releasing solutions are used to achieve a higher level of disinfection especially during outbreaks of viral gastro-enteritis, for example. Should the clinical environment not be subject to high standards of cleanliness it can play a significant part in contributing to the acquisition of health-care associated infections such as *Clostridium difficile*-associated diarrhoea, as highlighted in the UK Healthcare Commission's Report concerning Maidstone and Tunbridge Wells NHS Trust in 2007.

Commonly used pieces of equipment, such as stethoscopes and commodes for example, that are used between patients should also be subject to regular cleaning between each use. Wherever possible adequate numbers of frequently used equipment should be sought so that one particular piece of equipment is not in constant use, preventing cleaning and causing disinfection practices to fall by the wayside, thereby putting vulnerable patients at risk of potential cross infection.

It may be more prudent and cost effective to adopt a policy of single use disposable equipment than spending valuable nursing expertise cleaning equipment but this would need to be discussed and agreed locally. Regular visual checks and audits of cleaning and disinfection policies and practice will identify if cleanliness standards are being well maintained.

Other departments/units

More formal disinfection processes take place outside of wards in departments such as endoscopy and bronchoscopy units. The equipment used in these areas are usually constructed with materials that allow them to be very flexible and as such are made from heat-labile materials and therefore cannot be subjected to heat disinfection processes. They are thoroughly decontaminated in the first instance by staff wearing appropriate PPE and then processed in stand alone automated cold chemical washer disinfectors (WD) which are subject to regular disinfection programmes and monitoring of rinse water quality to ensure that the processes function optimally and that the machine does not pose a potential infection risk to subsequent instruments and patients.

Additionally, in departments such as sterile services units, surgical instrumentation undergoes heat disinfection processes in non-chemical WDs. These machines are monitored more closely via process assurance controls. If the system fails to provide water of an adequate temperature, for instance, the cycle will stop and the WD will require fixing and revalidation prior to being put back into operation. These machines utilize controlled programmes of 'time over temperature' processes and are therefore more efficient in killing bacteria and viruses than cold chemical processes or hand decontamination methods.

Waste management

Health-care waste comprises any waste that is produced by and/or is a consequence of health-care activities. Health-care facilities have a legal requirement (Duty of Care) to ensure that the disposal of all categories of waste (e.g. medical, chemical, radioactive, microbiological) occurs in an appropriate manner that does not pose a potential hazard to the public and the environment. Hospitals have received some adverse publicity as of late for poor waste disposal and leaving clinical waste in non-designated areas. Each health-care facility must have policies for disposal of all waste and to ensure that these waste streams are followed and adhered to.

CASE STUDY 12.1 Postoperative infection

It has been noted that two patients postoperatively have developed sternal wound infections following cardiac surgery. The patients are now pyrexial and the wound is showing a green exudate. A third patient operated on 3 days later has now developed a wound infection.

- **What samples would you take and from which health-care worker? What investigations would you request?**
- **Who would you involve in this process?**
- **What media would you use in sample processing?**
- **What nursing procedures would be implemented?**

After 24 h incubation all three patients' wound swabs grew a presumptive *Staphylococcus aureus*. Susceptibility tests show resistance to cefoxitin, erythromycin, trimethoprim and ciprofloxacin.

- **What are your conclusions about the organism involved?**
- **What further tests would you perform on the isolate?**

A total of 150 swabs were received at the request of the infection prevention and control team from all staff involved with the patient. It was shown that the nasal swab from the surgeon performing the operations on all three patients was heavily colonized with a presumptive *S. aureus*. Susceptibility tests show the organism has an identical susceptibility pattern.

- **What further tests would you perform on the isolate?**
- **What control of infection procedures would be implemented?**
- **Would any therapeutic regimes be considered?**

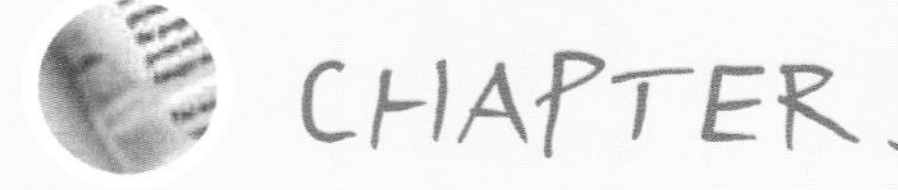

CHAPTER SUMMARY

After reading this chapter you should know and understand:

- The importance and role of the infection prevention and control team within the hospital
- The significance of MRSA and the impact it has on staff, patients and hospital resources
- How MRSA screening is conducted and the laboratory methods used to isolate the organism
- The importance of *C. difficile* infection and its impact on patients, staff and health resources

- Methods used for the detection of *C. difficile* toxins
- The importance of VRE and methods used for detection
- Consideration of other organisms which can cause problems in hospitals and strategies for their detection
- Hand hygiene and its importance in reduction of HCAI
- Disinfection strategies
- Management of cohorts and barrier nursing
- Monitoring air quality in theatres
- Importance of waste management in the hospital environment

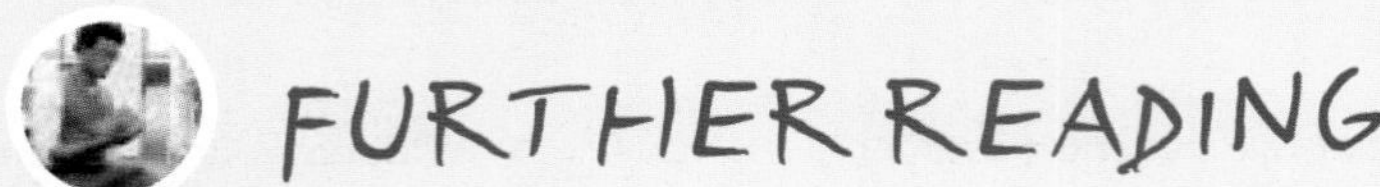

- **Centers for Disease Control (CDC). www.cdc.gov/index.htm**

  US website dealing with all aspects of infectious disease. Excellent resource for infectious disease information.

- **Finch RG, *et al*. eds. *Antibiotic and Chemotherapy: Anti-Infective Agents and their use in Therapy*, 8th edition. Churchill Livingstone, 2003.**

  Excellent text for use of antimicrobial agents in clinical practice.

- **Gould D, Brooker C. *Infection Prevention and Control: Applied Microbiology for Healthcare*, 2nd edition. Palgrave Macmillan, 2008.**

  Useful text describing the many aspects and problems associated with infection prevention and control.

- **Mims CA. *Medical Microbiology*, 3rd edition. Mosby, 2004.**

  Excellent text on infectious diseases and the process by which organisms initiate and spread infection.

12.1 Why do you think MRSA has become so prevalent in many hospital environments?

12.2 What do you think is making the biggest impact in the battle to reduce health-care acquired infection and why?

12.3 Why do you think *C. difficile* has become a problem for both the hospital and community? How can infection rates be reduced?

13

Laboratory investigations of viral infections

Jayne Harwood

Viral infections are among the most common and important causes of human disease. The word 'virus' comes from the Latin meaning 'a poison' and it is a microscopic organism, consisting mainly of nucleic acid in a protein coat. Viral infections affect every life form, from bacteria, fungi and plants to animals and man.

Viruses are metabolically inert, making them totally dependent upon living cells for replication and their existence. This means that until they enter a host cell capable of supporting their replication, they exist as inactive particles. This chapter describes the basic structure of a virus and the human host's immune response to it. It also describes the type of specimen needed to diagnose a viral infection, the laboratory investigations that are used in that diagnosis and some common viral illnesses.

Learning objectives

After studying this chapter you should be able to:

- Describe the basic organization, size range and structure of viruses
- Describe procedures used in the routine virology laboratory for the diagnosis of viral infections
- Describe the specimens required for the diagnosis of viral infections
- List viruses commonly isolated from pathological specimens
- Have a basic knowledge of antiviral therapy

13.1 General properties and structure of viruses

Viruses differ from other common infectious agents in a number of ways:

- Viruses are very small, measured in nanometers (nm), that is 10^{-9} of a metre. Viruses range from about 20 nm to 300 nm in diameter and this means they are not visible unless microscopy is used. Whilst conventional microscopes are used to visualize bacterial cells, viruses (apart from the largest viruses, e.g. poxviruses) can only be seen using specialized electron microscopes.
- Viral genomes consist only of DNA or RNA but not both.

The structure of viruses is important in their identification. Viruses are made up from the following basic components, which you can see in Figure 13.1:

- Capsid - protein coat or outer shell enclosing the nucleic acid. The role of the capsid is to aid entry into the host cell and to protect the viral nucleic acid.
- Capsomere - morphological subunit of which the capsid is composed, consisting of a small number of protein molecules.
- Nucleocapsid - the capsid together with the enclosed nucleic acid core.
- Envelope - a lipoprotein membrane which surrounds certain viruses and is acquired during the final stages of viral maturation by a process of budding through the host cell's outer cell membrane.

Whilst some viruses possess enzymes of their own they cannot reproduce and amplify the information in their genomes without the assistance of the host cellular components.

Key Point

Viruses are totally dependent upon living cells, either eukaryotic or prokaryotic, for replication and existence.

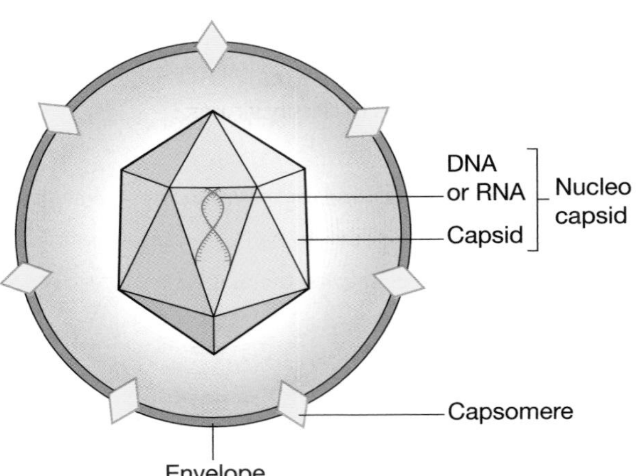

FIGURE 13.1
Structure of a virus. (Courtesy of J. Harwood.)

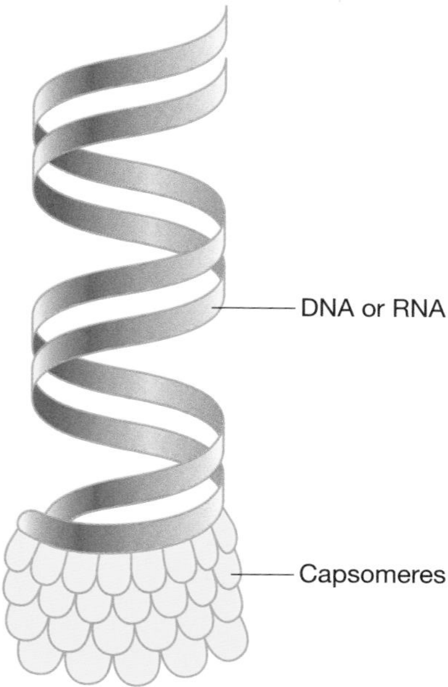

FIGURE 13.2
Structure of a helical virus. (Courtesy of J. Harwood.)

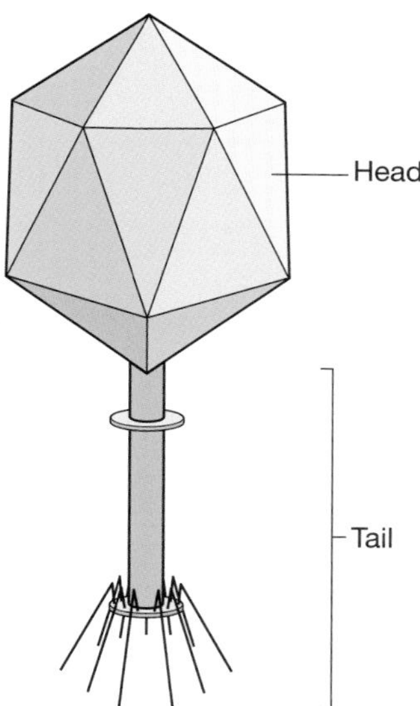

FIGURE 13.3
Structure of a complex virus (bacteriophage). (Courtesy of J. Harwood.)

Viral morphology

Viral structure shows that they tend to be simplistic and symmetrical in design because of the limited number of proteins available for their configuration. There are three distinct shapes:

- Helical viruses (Figure 13.2)
- Cubic or isometric (icosahedral) viruses
- Complex viruses (Figure 13.3)

Viral classification

The following criteria are used to classify viruses:

- The type of nucleic acid, i.e. DNA or RNA
- The number of strands of nucleic acid and their physical construction, i.e. single-stranded, linear, circular, circular with breaks, segmented
- Polarity of the viral genome, i.e. positive or negative stranded
- The symmetry of the nucleocapsid
- The presence or absence of a lipid envelope

For example, herpes simplex, which causes cold sores, is a DNA virus, whereas the influenza virus only contains RNA.

Key Point

Viruses, unlike bacteria, possess only one type of nucleic acid, either DNA or RNA.

SELF-CHECK 13.1

What are the main differences between a virus and a bacterium?

13.2 Viral pathogenesis

The consequences to the patient of a viral infection depend on a number of viral and host factors that affect the development of disease. Viral infection was long thought only to produce acute clinical disease (e.g. mumps); however, viruses can also cause chronic disease, such as liver cirrhosis as a result of hepatitis B infection. Other results of viral infection can include induction of various cancers, progressive neurological disorders and endocrine diseases such as diabetes.

Host cellular response to viral infection

Host cells can respond to a viral infection in three different ways:

1. No effect on the host cell - asymptomatic infection e.g. **cytomegalovirus** (CMV) infection
2. Cytopathic effect (CPE) and death - this is a morphological change in host cells as a result of the virus growing inside them
3. Loss of control of cell replication - this can result in the formation of tumours. This can occur with **Epstein–Barr virus** (EBV) infection which can cause **lymphoma**.

In a host, viruses can cause three basic patterns of infection: localized, **disseminated** and no apparent disease. In localized viral infection, replication remains localized near the site of entry, for example the skin or respiratory tract. Disseminated (that is systemic) infections usually take place through several steps:

1. Entry of virus into the body
2. Spread via the bloodstream (**viraemia**) to regional lymph nodes
3. Further spread to other susceptible organs such as liver or spleen

Key Point

Viraemia is the presence of virus in the bloodstream.

Many viral infections do not necessarily result in viral replication and cell death within the host. Instead following a primary infection, the viral genome enters into a latent state where it is maintained within the host cell in a repressed state which is compatible with the continued survival and normal functioning of that cell. Essentially the virus is 'hiding' in the host cell waiting for a trigger to 'awaken' (reactivate) it.

The virus can be reactivated, allowing active viral replication to occur and the production of numerous infectious virus particles. The events that trigger this reactivation of the **latent virus** include strong sunshine (UV light), trauma and immunosuppression. For example a patient can develop recurrent cold sores caused by the herpes simplex virus during the summer months as a result of exposure to sunshine.

SELF-CHECK 13.2

What is meant by the terms latency and reactivation in viral disease? Can you give an example?

13.3 Samples required for the diagnosis of viral infection

It is imperative that the clinicians know the correct samples required for the diagnosis of viral infections. In addition, it is essential that the samples are submitted in the required container, and transported in the appropriate manner.

The site and timing of specimens are possibly the most important factors for successful virus diagnosis. In general, specimens should be taken from the site of symptoms and as close to the onset of illness as possible.

Specimens for **cell culture** should be transported to the laboratory as soon as possible after being taken. Swabs should be put into a vial containing viral transport medium. There are various formulations of viral transport media but all of them generally contain: (a) a salt solution to ensure correct ionic concentrations; (b) a buffer to maintain pH; (c) a source of protein; and (d) antibiotics and antifungals to prevent overgrowth by bacteria and fungi which may be present in the sample. Bodily fluids and tissues should be placed in a sterile container. If delays of more than an hour are anticipated specimens are best stored at 4°C.

All specimens should be placed in secure transport containers and labelled in accordance with local policy. It is essential that the request form contains relevant clinical information. Such details must include the date of onset of illness, clinical signs and suspected diagnosis.

The specimen used for direct detection and virus isolation is very important. A positive result from the site of the disease would be of much greater diagnostic significance than those from other sites. For example, in the case of herpes simplex (HSV) encephalitis, a positive result from the cerebral spinal fluid (CSF) or the brain would be of much greater significance than a positive result from a cold sore, since as we have seen above reactivation of cold sores can occur during times of stress.

Table 13.1 summarizes the specimen types required for a range of viral infections.

SELF-CHECK 13.3

Why is it important for viral transport media to contain antibacterial and antifungal agents?

Swabs

The use of wooden stick swabs is not advised for the isolation of viruses as this material has been implicated as a source of cell culture toxicity. Plastic swab sticks are the required swabs

TABLE 13.1 Specimen types required for a range of viral diseases.

Disease	Specimen
Respiratory infection	Nasal or throat swabs, postnasal washing, BAL
Gastrointestinal infection	Faeces (rectal swab not satisfactory)
Vesicular rash	Vesicle swab, throat swab, faeces
Hepatitis	Clotted blood, faeces
Central nervous system	CSF, throat swab, faeces
AIDS, lymphadenopathy	Clotted blood

for isolation of viruses. The amount of material collected must be adequate for testing. Throat and skin swabs must be taken fairly vigorously to capture as many virus-infected cells as possible. The tip of the swab is broken off and placed into viral transport medium and forwarded as soon as possible to the laboratory.

Respiratory samples

Respiratory samples include nasopharyngeal aspirates, swabs, washes, broncheoalveolar lavages (BAL), throat swabs and sputum. Nasopharyngeal samples are extremely useful in the diagnosis of upper respiratory tract infections in young children (e.g. RSV). Respiratory samples are collected in sterile containers.

Cerebrospinal fluid (CSF)

CSF is often submitted to eliminate viral infection as a cause of aseptic meningitis or encephalitis. Other than collecting an adequate volume into a sterile container and prompt transport to the laboratory, no other special procedures are required.

Cross reference
Chapter 8, Section 8.1.

Vesicle fluid

Vesicle fluid for **electron microscopy** (EM) is collected on the tip of a scalpel blade, spread over an area of about 3–4 mm in diameter on an ordinary microscope slide and allowed to dry. Vesicle fluid for polymerase chain reaction (PCR) is taken in the same way but put into a vial containing viral transport medium.

Faeces

Faeces should be placed in a dry, sterile container. These are particularly useful for the diagnosis of enteroviruses and rotavirus.

Cross reference
Chapter 10.

Clotted blood

As a general rule, 5–10 mL of blood is taken. Clotted blood is used for serological tests.

EDTA blood

Blood is collected into EDTA blood tubes for any PCR work. Heparin tubes should be avoided for molecular work as the heparin can interfere with some assays.

13.4 Diagnostic methods in virology

Overview of diagnostic methods

In general, diagnostic tests can be grouped into three categories:

1. Direct detection
2. Indirect examination (virus isolation)
3. Serology.

In direct detection, the clinical specimen is examined directly for the presence of virus particles, viral antigen or viral nucleic acids. This is performed by PCR, electron microscopy and immunofluorescence. In indirect examination, the specimen is inoculated into **cell lines**, eggs or animals in an attempt to grow the virus: this is called virus isolation. Serology constitutes by far the bulk of the work of any virology laboratory. A serological diagnosis can be made by the detection of rising titres of antibody between acute and convalescent stages of infection, or the detection of **immunoglobulin M** (IgM). In general, the majority of common viral infections can be diagnosed by serology. Serology is also useful for determining the immune status of patients to viruses by detection of **immunoglobulin G** (IgG). For example determining the rubella (German measles) immune status of pregnant women as lack of immunity can lead to serious complications for the fetus if the mother comes into contact with the virus whilst pregnant.

Most serology is carried out using highly automated analysers, much like those used in biochemistry or haematology departments.

SELF-CHECK 13.4

Why do you think that serology is most commonly used for the diagnosis of viral infection?

Direct detection of viruses

Direct detection methods are often called rapid diagnostic methods because a rapid result is required for clinical management. For example respiratory syncytial virus (RSV) can spread rapidly through paediatric wards and the affected individual should be barrier nursed to stop the spread of infection. However, it is important to realize that not all direct examination methods are rapid, and, conversely, virus isolation and serological methods may sometimes give a more rapid result. Rapid diagnostic methods such as PCR have played a very important role in the diagnosis of the swine flu pandemic so patients and contacts can be given antiviral drugs in a bid to contain the severity and spread of the disease.

Electron microscopy (EM) morphology/immune EM

Virus particles can be detected and identified on the basis of morphology, that is how they appear under the electron microscope (Figure 13.4). A magnification of around 50 000 is

FIGURE 13.4
Photograph of the Phenom personal electron microscope. (Courtesy of FEI.)

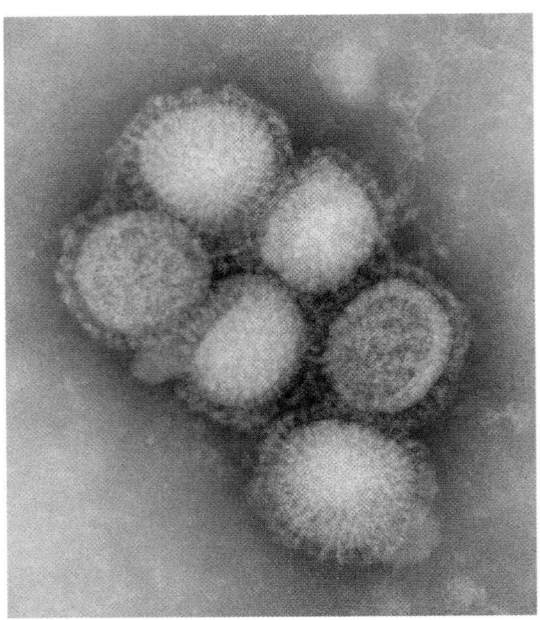

FIGURE 13.5
Electron micrograph of influenza A (H1N1) virus (swine flu virus). (Courtesy of CDC/C.S. Goldsmith and A. Balish.)

normally used. EM is now mainly used for the diagnosis of viral gastroenteritis by detecting viruses in faeces, for example rotavirus, adenovirus, astrovirus, calcivirus and Norwalk-like viruses (turn back to Figure 10.5). Respiratory viruses that can be detected using EM include influenza A, of which swine flu is an example (Figure 13.5). Occasionally, it may be used for the detection of viruses in **vesicles** and other skin lesions, such as herpesviruses and papillomaviruses. The sensitivity and specificity of EM may be enhanced by immune EM, whereby virus-specific antibody is used to agglutinate virus particles together, thus making them easier to recognize, or to capture virus particles onto the EM grid. The main problem with EM is the expense involved in purchasing and maintaining the facility. In addition, the sensitivity of EM is often poor, with at least 10^5 to 10^6 virus particles/millilitre in the sample required for visualization. Therefore the observer must be highly skilled. With the availability of reliable antigen detection and molecular methods for the detection of viruses associated with viral gastroenteritis, EM is becoming less widely used as a diagnostic aid.

Histopathological appearance

Replicating viruses often produce histological changes in infected cells. These changes may be characteristic or non-specific. Viral inclusion bodies are basically collections of replicating virus particles either in the nucleus or in cytoplasm of the host cell. Examples of inclusion bodies include the **Negri bodies** and **cytomegalic inclusion bodies** found in rabies

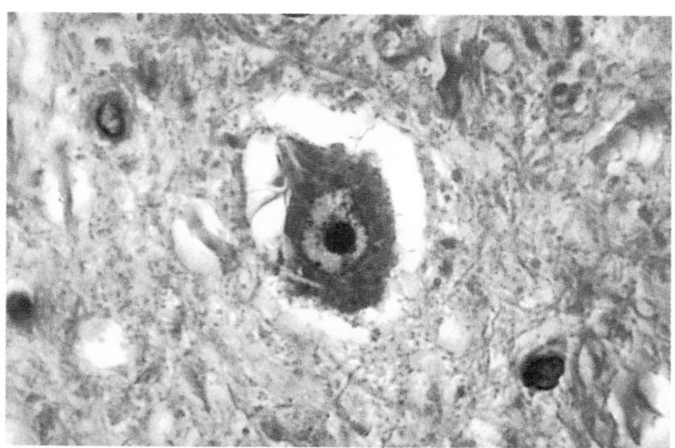

FIGURE 13.6
Micrograph of Negri bodies found in a patient with rabies encephalitis. (Courtesy of CDC/Dr Daniel P. Perl.)

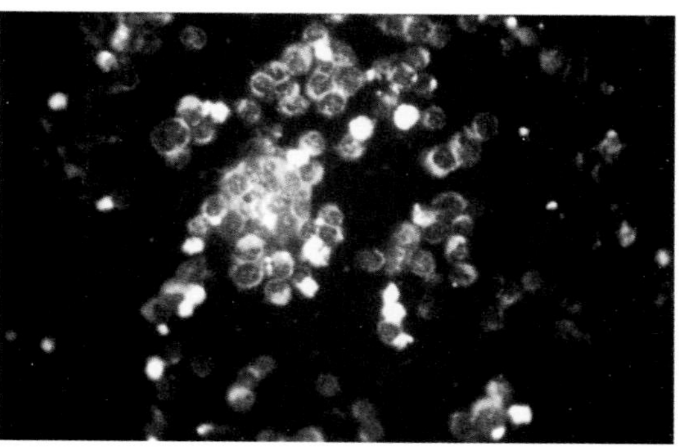

FIGURE 13.7
Photomicrograph of detection of respiratory syncytial virus (RSV) using an indirect immunofluorescence technique. (Courtesy of CDC/Dr H. Craig Lyerla.)

and cytomegalovirus (CMV) infections, respectively (Figure 13.6). Although not sensitive or specific, histology nevertheless serves as a useful adjunct in the diagnosis of certain viral infections.

Antigen detection

Examples of antigen detection include immunofluorescence testing of nasopharyngeal aspirates for respiratory viruses, for example RSV (Figure 13.7), influenza A, influenza B, detection of rotavirus antigen in faeces and detection of varicella zoster virus in skin scrapings. The main advantage of these assays is that they are rapid to perform with the result being available within a few hours. However, the technique is often tedious and labour intensive and the result can be difficult to interpret with often low sensitivity and specificity. The quality of the specimen obtained is of utmost importance in order for the test to yield good diagnostic results.

SELF-CHECK 13.5

What are the advantages and disadvantages of rapid detection methods for viral disease?

Molecular techniques for the detection of viral genomes

Methods for the detection of the viral genome are most commonly molecular methods. Molecular methods are now well established as the gold standard in non-clinical diagnosis of viral disease. However, in practice, although the use of molecular methods is widespread, the number of molecular tests is still low compared to serological tests.

Classical molecular techniques, such as dot blot and **Southern blot**, depend on the use of specific DNA/RNA probes for hybridization. The specificity of the reaction depends on the conditions used for hybridization. These techniques may allow for the quantification of

DNA/RNA present in the specimen. However, it is often found that the sensitivity of these techniques is comparable with other viral diagnostic methods such as serology.

Newer molecular techniques, such as the polymerase chain reaction (PCR), ligase chain reaction (LCR), nucleic acid based amplification (NASBA) and branched DNA (bDNA), depend on some form of amplification, either the target nucleic acid or the signal itself.

PCR is the only amplification technique which is in routine diagnostic use. PCR is an extremely sensitive technique; it is possible to achieve a sensitivity of one DNA molecule in a clinical specimen. However, PCR has many problems, most notably contamination, since only a very small amount of contaminating nucleic acid is needed to generate a false-positive result. In addition, because PCR is so sensitive compared to other techniques, a positive PCR result is often difficult to interpret as it does not necessarily indicate the presence of disease. This is a problem of particular concern in the detection of latent viruses such as CMV, since latent CMV genomes may be amplified from the blood of healthy individuals. Despite these drawbacks, PCR is being increasingly used for viral diagnosis, especially as the assays are becoming less expensive and the availability of highly automated systems is readily available, for example real-time PCR and Cobas Amplicor systems (Roche, Basel, Switzerland).

Other amplification techniques (such as LCR and NASBA) are just as susceptible to contamination as PCR but this is eliminated greatly by the use of proprietary closed systems. It is unlikely that other amplification techniques will challenge the dominance of PCR since it is much easier to set up an in-house PCR assay than other assays.

Polymerase chain reaction (PCR)

PCR allows the *in vitro* amplification of specific target DNA sequences by a factor of 10^6 and is therefore an extremely sensitive diagnostic technique. It is based on an enzymatic reaction involving the use of synthetic oligonucleotides flanking the target nucleic sequence of interest. These oligonucleotides act as primers for the thermostable Taq polymerase. Repeated cycles (usually 25 to 40) of denaturation of the template DNA (at 94°C), annealing of primers to their complementary sequences (50°C) and primer extension (70°C) result in exponential production of the specific target fragment. Look at Figure 13.8 to see how these cycles are carried out. Further sensitivity and specificity may be obtained by the nested PCR technique, whereby the DNA is amplified in two steps. In the first step, an initial pair of primers is used to generate a long sequence that contains the target DNA sequence. A small amount of this product is used in a second round of amplification, which employs primers to the final target DNA.

Detection of DNA sequence product of the PCR assay may be performed in several ways. The least sensitive and specific method is to size fractionate the reaction product on an agarose or acrylamide gel and stain the DNA with ethidium bromide. A more sensitive technique involves the attachment of DNA to a membrane through dot or slot-blot techniques followed by hybridization with a labelled homologous oligonucleotide probe. Alternatively, the PCR product may be probed directly by liquid oligomeric hybridization. However, these techniques provide no information on size of the amplified product and thus could not exclude the possibility that the product originated from a region of the human genome which exhibits homology with the target viral sequence. The most sensitive and specific detection methods result from combining the size information of gel electrophoresis with the improved sensitivity and specificity of hybridization techniques. This may be achieved by gel electrophoresis followed by Southern transfer and hybridization, or through liquid oligomeric hybridization followed by gel electrophoresis.

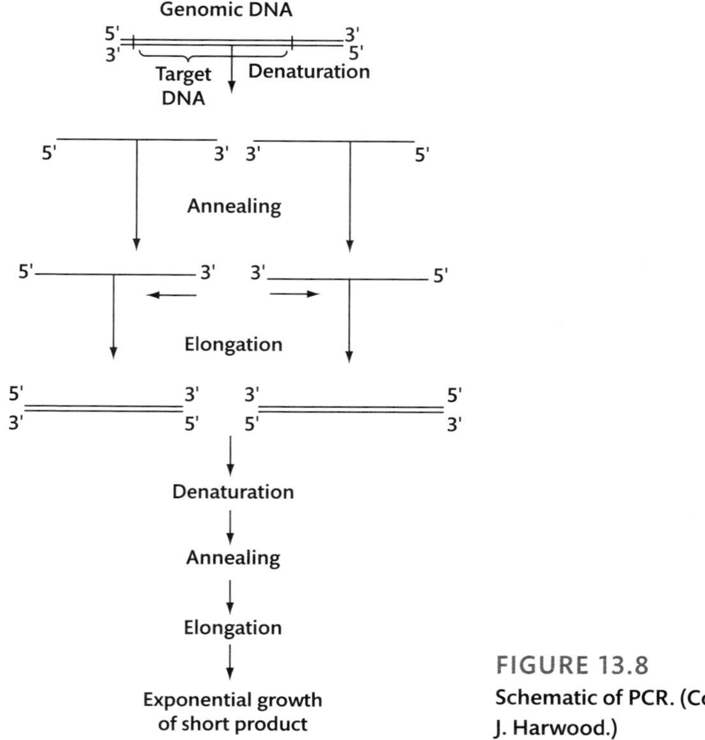

FIGURE 13.8
Schematic of PCR. (Courtesy of J. Harwood.)

The advantages of PCR include:

- Extremely high sensitivity – may detect down to one viral genome per sample volume
- Easy to set up
- Fast turnaround time

The disadvantages include:

- Liable to contamination
- High degree of operator skill required
- Not easy to quantitate results
- A positive result may be difficult to interpret, especially with latent viruses such as CMV, where any person previously infected with CMV will have virus present in their blood, irrespective of whether they have disease or not.

The first three problems have been addressed by the arrival of commercial closed systems such as Roche Cobas Amplicor, which requires minimum handling. The use of synthetic internal competitive targets in these commercial assays has facilitated the quantification of results. Otherwise, the same problems remain as with in-house PCR methods. The sensitivity would normally be sufficient if a single PCR reaction was used followed by hybridization with a specific oligonucleotide probe. This is the approach taken by commercial assays. The fourth problem is more difficult to resolve but it is generally found that patients with active CMV disease have a much higher viral load than those who do not. Therefore, it is simply a case of finding the appropriate cut-off level.

SELF-CHECK 13.6

Why have molecular methods largely replaced other direct detection methods?

Indirect examination

Typically, indirect examination is performed using the following three methods:

1. Cell culture
2. Fertile hens' eggs
3. Animals

Cell cultures, eggs and animals may be used for isolation of viruses from clinical samples. However, eggs and animals are difficult to handle and most viral diagnostic laboratories depend on cell culture only. To prepare cell cultures, tissue fragments are first dissociated, usually with the aid of trypsin or collagenase. The cell suspension is then placed in a flat-bottomed glass or plastic container together with a suitable liquid medium (e.g. Eagle's) and animal serum. After a variable lag phase, the cells will attach and spread on the bottom of the container and then start dividing, giving rise to a **primary culture**. Attachment to a solid support is essential for the growth of normal cells. Primary cultures are maintained by changing the fluid two or three times a week. When the cultures become too crowded, the cells are detached from the vessel wall either trypsin by or EDTA, and portions are used to initiate secondary cultures. In both primary and secondary cultures, the cells retain some of the characteristics of the tissue from which they are derived. Cells from primary cultures can often be transferred serially a number of times. The cells may then continue to multiply at a constant rate over many successive transfers. Eventually, after a number of transfers, the cells undergo culture **senescence** and cannot be transferred any longer. For human diploid cell cultures, the growth rate declines after about 50 duplications. During the multiplication of the cell strain, some cells become altered in that they acquire a different morphology, grow faster and become able to start a cell culture from a smaller number of cells. These cells are immortalized and have an unlimited life-span. However, they retain contact inhibition.

Senescence
The process of growing old and showing effects of increasing age.

There are three types of cell culture:

- Primary cells, e.g. monkey kidney. These are essentially normal cells obtained from freshly killed adult animals. These cells can only be **passaged** once or twice.
- Semicontinuous cells, e.g. human embryonic kidney and skin fibroblasts. These are cells taken from embryonic tissue, and may be passaged up to 50 times.
- Continuous cells, e.g. HeLa, Vero, Hep2, LLC-MK2, BGM. These are immortalized cells, i.e. tumour cell lines which may be passaged indefinitely.

Key Point

Cell cultures vary greatly in their susceptibility to different viruses. It is extremely important that the most sensitive cell cultures are used for a particular suspected virus.

Upon receipt, the specimen is inoculated into several different types of cell culture depending on the nature of the specimen and the clinical presentation. The maintenance media should be changed after 1 h or, if that is not practicable, the next morning. The inoculated tubes should be incubated at 35–37°C in a rotating drum. Rotation is optimal for the isolation of respiratory viruses and results in an earlier appearance of the CPE for many viruses. If stationary tubes are used, it is essential that the culture tubes be positioned so that the cell **monolayer** is bathed in nutrient medium.

The inoculated tubes should be read at least every other day for the presence of CPE. Certain specimens, such as urine and faeces, may be toxic to cell cultures and may produce a CPE-like effect. If toxic effects are extensive, it may be necessary to passage the inoculated cells. Cell cultures that are contaminated with bacteria should be either reinoculated or passed through a bacterial filter. Cell cultures should be kept for at least 1 to 2 weeks (longer in the case of CMV). Cell cultures should be re-fed with fresh maintenance medium at regular intervals or as required should the culture medium become too acidic or alkaline. When CPE is seen, it may be advisable to passage infected culture fluid into a fresh culture of the same cell type. For cell-associated viruses such as CMV and VZV, it is necessary to trypsinize and passage intact infected cells. Other viruses such as adenovirus can be **subcultured** after freezing and thawing infected cells.

Primary cell culture is widely acknowledged as the best cell culture system available since it supports the widest range of viruses. However, it is very expensive and it is often difficult to obtain a reliable supply. Continuous cells are the most easy to handle but the range of viruses supported is often limited.

The presence of growing virus is usually detected by:

- Cytopathic effect (CPE) - may be specific or non-specific, e.g. HSV and CMV produce a specific CPE, whereas enteroviruses do not.
- **Haemadsorption** - cultured cells acquire the ability to stick to mammalian red blood cells. Haemadsorption is mainly used for the detection of influenza and parainfluenza viruses.

Confirmation of the identity of the virus may be carried out using neutralization, haemadsorption-inhibition, immunofluorescence or molecular tests.

The main problem with cell culture is the long period (up to 4 weeks) required for a result to be available. Also, the sensitivity is often poor and depends on many factors, such as the condition of the specimen and the condition of the cell sheet. Cell cultures are also very susceptible to bacterial contamination and toxic substances in the specimen. Finally, many viruses will not grow in cell culture at all, including hepatitis B (HBV) and C (HCV), diarrhoeal viruses and parvovirus.

Rapid culture techniques are available whereby viral antigens are detected 2 to 4 days after the inoculation. Examples of rapid culture techniques include the CMV DEAFF test. In this test, the cell sheet (human embryonic fibroblasts) is grown on individual cover slips in a plastic bottle. After inoculation, the bottle is spun at a low speed for 1 h (to speed up the adsorption of the virus) and then incubated for 2 to 4 days. The cover slip if then taken out and examined for the presence of CMV early antigens by immunofluorescence.

The role of cell culture in the diagnosis of viral infections is being increasingly challenged by rapid diagnostic methods. Therefore, the role of cell culture is expected to decline in the future and is likely to be restricted to large central reference laboratories.

SELF-CHECK 13.7

Why is cell culture still used when there are so many more rapid diagnostic methods available?

13.5 Serology

Serology is the mainstay of viral diagnosis in many routine diagnostic laboratories. Following exposure to a virus, antibodies will start to appear in most patients, the first of which to appear is IgM (Figure 13.9). This is followed over time by production of IgG. Levels of IgG are often higher than those seen with IgM. The production of these immunoglobulins is most often how the viral infection is diagnosed.

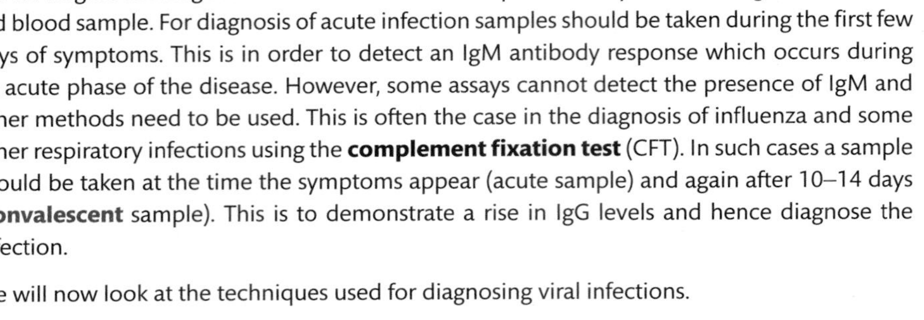

Key Point

Using serology techniques it must be remembered that it is not the virus that is detected but the patient's immune response to it in the majority of cases.

For serological investigations serum is used which is produced by the centrifugation of a clotted blood sample. For diagnosis of acute infection samples should be taken during the first few days of symptoms. This is in order to detect an IgM antibody response which occurs during an acute phase of the disease. However, some assays cannot detect the presence of IgM and other methods need to be used. This is often the case in the diagnosis of influenza and some other respiratory infections using the **complement fixation test** (CFT). In such cases a sample should be taken at the time the symptoms appear (acute sample) and again after 10–14 days (**convalescent** sample). This is to demonstrate a rise in IgG levels and hence diagnose the infection.

convalescent
Convalescent means the patient has often recovered from the infection or is showing an improvement in the condition. A convalescent sample is taken to demonstrate a rise in antibody levels.

We will now look at the techniques used for diagnosing viral infections.

Complement fixation test

The complement fixation test is used for the diagnosis of certain respiratory infections where IgM is difficult to detect and as such a rise in the IgG response is used to diagnose the infection using an acute and convalescent serum.

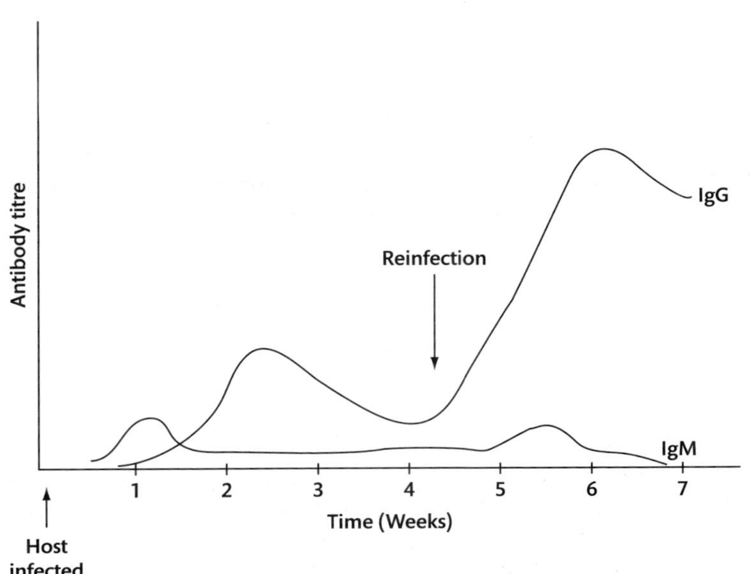

FIGURE 13.9
Serological events following primary infection and reinfection. Note that in reinfection, IgM may be absent or only present transiently at a low level. (Courtesy of J. Harwood.)

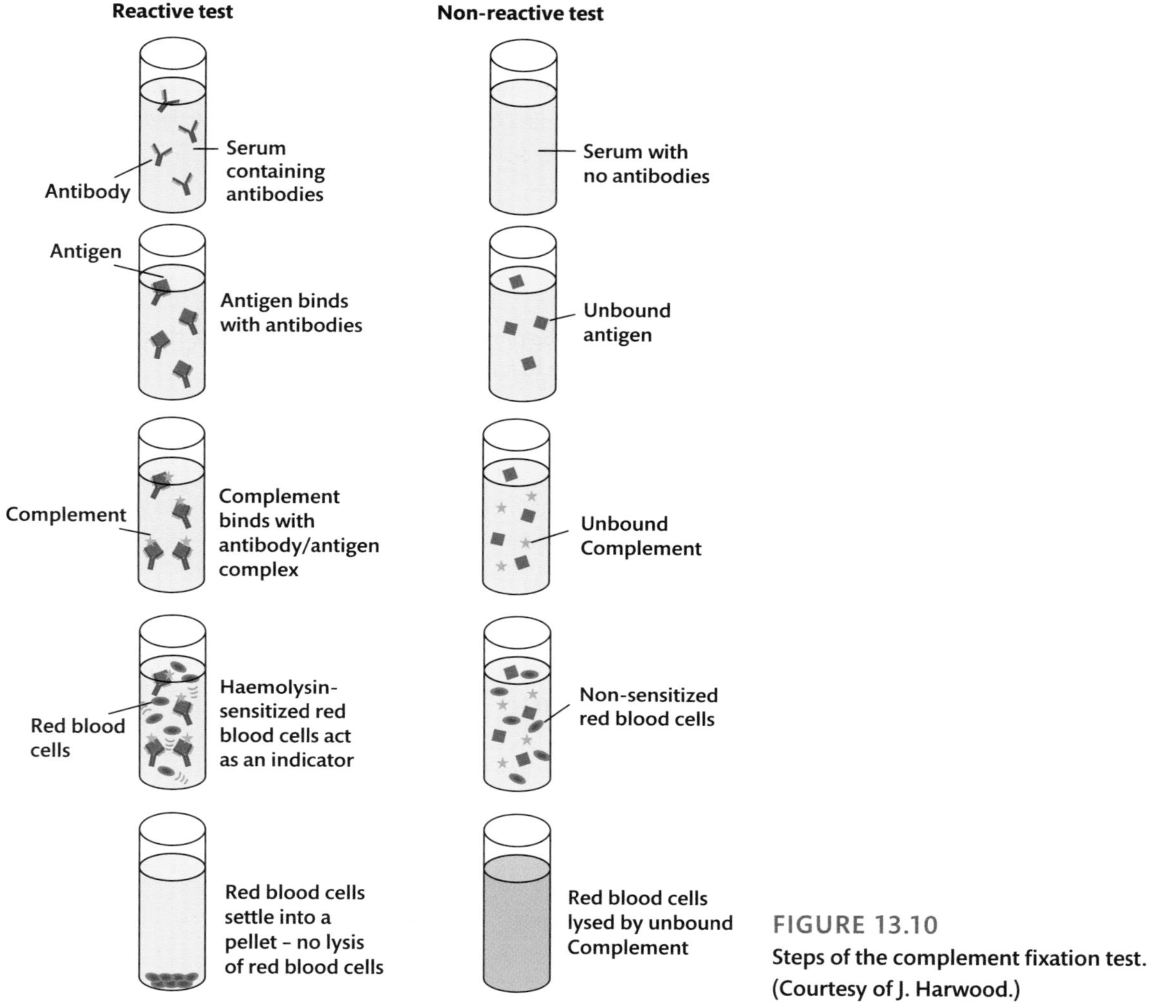

FIGURE 13.10
Steps of the complement fixation test. (Courtesy of J. Harwood.)

Although CFT is considered to be a relatively simple test, it is a very exacting procedure because five variables are involved, as you can see in Figure 13.10. In essence, the test consists of two antigen–antibody reactions, one of which is the indicator system. The first reaction, between a known virus antigen and a specific antibody, takes place in the presence of a predetermined amount of complement. The complement is removed or 'fixed' by the antigen–antibody complex. The second antigen–antibody reaction consists of reacting sheep red blood cells with haemolysin. When this indicator system is added to the reactants, the sensitized red blood cells will only lyse in the presence of free complement. In order for the CFT to perform correctly, the optimal concentration of haemolytic serum, complement and antigen must be determined in advance. The antigens used for CFT tend to be group antigens rather than type-specific antigens. For example influenza A antigens are antigens from many different strains of the virus. This allows for the diagnostic capacity of the test to be maximized.

The advantage of CFT is the ability to screen against a large number of viral (and bacterial) antigens at the same time. The tests are also relatively inexpensive to perform. The disadvantages are that the test has a low sensitivity and it is highly labour intensive to perform. The test can also be non-specific, for example cross-reactions can occur between the viral antigens used in the test.

Key Point

Many laboratories no longer offer a CFT service due to limitations of the test and the fact that many patients fail to submit a convalescent sample since they have recovered from the infection.

Detection of IgM

Several methods are available for the detection of IgM in serum. These include immunoassays using enzyme (EIA) or radiolabelled substrates (RIA). Immunofluorescence (IF) can also be used to detect IgM. These tests offer a rapid means of diagnosis and are routinely used for detection of IgM produced against viruses such as mumps and EBV. However, there are many problems associated with IgM assays, such as interference by rheumatoid factor, re-infection by the virus and unexplained persistence of IgM years after the primary infection. Problems with the use of radiolabelled substrates have rendered these assays unsuitable as a routine test. IF is relatively time consuming to perform and is generally only performed for virus detection in specialist reference centres. EIA is the most widely used method of IgM detection and is performed on highly automated platforms.

Enzyme-linked immunosorbent assay (ELISA)

ELISA was developed in 1970 and became rapidly accepted as the method of choice for the detection of IgG and IgM in serum samples for the diagnosis of viral infection. The principle of this test is that an antigen (or antibody) can be immobilized onto a solid-phase carrier, such as microtitre plates or beads, and then used to separate its **homologous antibody** (or antigen). This complex can be detected by an enzyme-mediated colour change (look at Figure 13.11 to see a diagram of the stages). In general the procedure is as follows:

homologous antibody
Homologous antibody is an antibody derived from an animal of the same species.

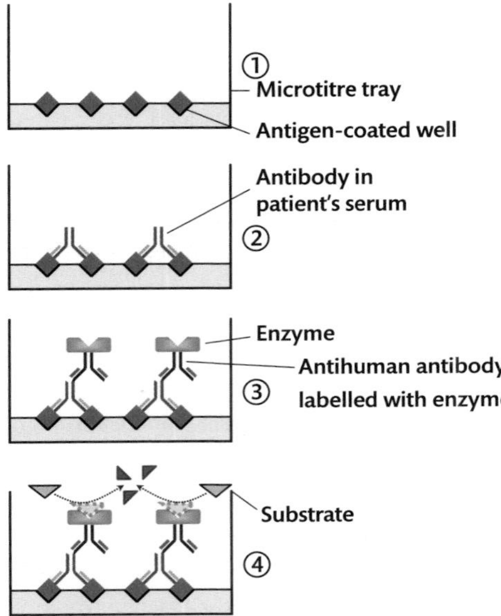

FIGURE 13.11
Example of an ELISA. The antigen to the antibody which is being tested for in the patient's serum is coated onto the well (1). The antibody in the serum will bind to the antigen (2). An antihuman antibody labelled with an enzyme is added (3). The substrate is added and will hydrolyse if the reaction has taken place (4) resulting in a colour change. (Courtesy of J. Harwood.)

- An antigen (or antibody) is linked to a solid phase surface, usually the wells of a 96-well microtitre tray, or on the surface of a polystyrene microparticle (usually a bead).
- This 'sensitized' solid phase surface captures its corresponding antibody (or antigen) if present in the specimen. This surface is washed to remove any unbound material.
- An enzyme-labelled antispecies antibody (conjugate) attaches to the bound antigen–antibody complex. This is again washed to remove any unbound conjugate.
- The attachment of the enzyme labelled conjugate can be detected by a colour change resulting from the hydrolysis of an added enzyme-specific substrate.
- The optical density of the colour reaction is directly proportional to the amount of unknown antibody (or antigen) in the test solution. The result is usually expressed as a quantitative optical density (OD) measurement.

SELF-CHECK 13.8

Why is it important for various washing steps to be performed in an ELISA test?

A wide variety of assay principles can be used in ELISA techniques. The most important of these are described in the following sections.

Competitive methods

The competitive ELISA is generally used to detect antibodies and is a quantitative assay over a narrow range. Unlike non-competitive methods, where the test serum is removed before adding the enzyme conjugate by washing (see above), the enzyme is in direct contact with antibodies in the serum. The method is simple, has only one incubation and wash step, does not **prozone** in the presence of high titre antibody concentrations and is particularly useful for detecting low-affinity antibodies. If antibodies in the test serum compete with the conjugate they block binding sites and prevent the hydrolysis of substrate, reducing the colour change. This lack of colour change indicates a positive reaction.

prozone
Prozone is the phenomenon exhibited by some sera, in which agglutination or precipitation occurs at higher dilution ranges, but is not visible at lower dilutions or when undiluted.

Key Point

A quantitative assay measures the exact amount of antibody present in the sample.

Sandwich methods

Indirect non-competitive ELISA is normally used to detect antibody. The antigen is coated onto the solid phase and incubated with the test serum. If homologous antibody is present, it will bind to the antigen. After washing, the conjugate (an antihuman IgG antibody covalently bound to the enzyme) is added and after incubation the plate is again washed. Substrate is added, the reaction stopped after a specific time period and the OD read at a specific wavelength defined by the substrate.

The double antibody sandwich is normally used to detect antigen, as shown in Figure 13.12. Specific antibody is bound to the solid phase and incubated with the test serum. If homologous antigen is present it will bind to the solid phase antibody. The conjugate is an enzyme linked to a specific IgG antibody homologous for the antigen being detected. Substrate is added and the reaction stopped after a specific time period. The OD is determined at a wavelength defined by the substrate.

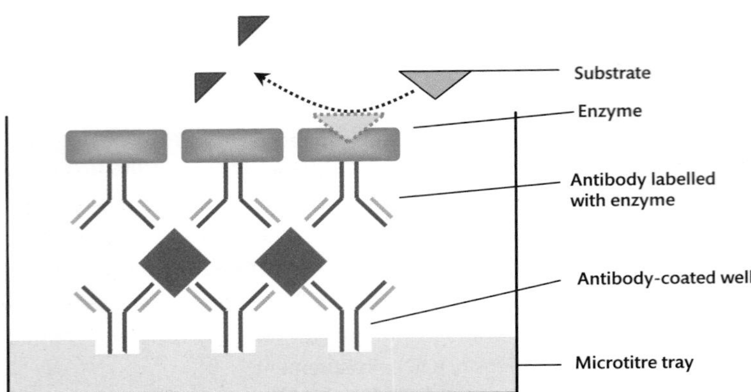

FIGURE 13.12
Sandwich method for ELISA. (Courtesy of J. Harwood.)

Antibody capture method

This method requires that the solid phase is coated with antibody to whichever class of immunoglobulin is being sought. For example, if IgM antibodies are to be detected, the solid phase is coated with anti-IgM to which any test IgM will bind. After washing, specific antigen is added followed by conjugate. The conjugate will only bind if the antigen is present which can itself only bind if the homologous IgM was present and captured to the solid phase. **Heterologous** IgM may be bound by capture to the solid phase, but will not bind antigen and therefore not react with the conjugate. Once again, coloration is indicative of a positive reaction which can be accurately assayed spectrophotometrically.

heterologous
Heterologous antibody is derived from an individual of a different species.

Assay characteristics

The use of monoclonal antibodies has lead to many improvements in ELISA systems.

- Higher sensitivity - either by selection of antibodies with an extremely high affinity, or by reduction of the height and variability of the background reaction, which makes very low concentrations of analyte more readily detectable.
- Higher specificity - by avoiding the presence of any antibody in the assay system with specific reactivity against non-analyte **epitopes**, and by selecting combinations of monoclonal antibodies which may further increase specificity.
- Higher practicality - by introducing simultaneous incubation of label, solid phase and sample without risk of prozone effect.
- The enzyme label - most of the assays employ horseradish peroxidase, alkaline phosphatase, or β-D-galactosidase. The most interesting recent developments have been in new methods to detect these enzymes rather than the use of new enzyme labels. Fluorimeters are used to detect alkaline phosphatase and β-D-galactosidase. Methods available for the detection of horseradish peroxidase are by means of **chemiluminescence**.
- Seroconversion - this is defined as changing from a previously antibody negative state to a positive state, e.g. seroconversion against human immunodeficiency virus (HIV) following a needle-stick injury, or against rubella following contact with a known case.
- A single high titre of IgG (or total antibody) - this is a very unreliable means of serological diagnosis since the cut-off is very difficult to define. This is because if the IgG level in the patient against the virus is unknown it is impossible to determine if the antibody titre has risen.

Criteria used for diagnosing a re-infection/re-activation

It is often very difficult to differentiate re-infection/re-activation from a primary infection. Under most circumstances, it is not important to differentiate between a primary infection and re-infection. However, it is very important under certain situations, such as rubella infection in the first trimester of pregnancy; primary infection is associated with a high risk of fetal damage whereas re-infection is not. In general, a sharp, large rise in antibody titres is found in re-infection whereas IgM is usually low or absent in cases of re-infection/re-activation.

Limitations of serological diagnosis

How useful a serological result is depends on the patient's immune response to an individual virus. For viruses such as rubella and hepatitis A, the onset of clinical symptoms coincides with the development of antibodies in the patient. The detection of IgM in the serum of the patient would be indicative of acute infection. Alternatively, rising titres of IgG in the serum of the patient who has previously been IgG positive would indicate re-activation or re-infection.

Many viruses often produce clinical disease well before the appearance of antibodies, such as respiratory and diarrhoeal viruses. In such instances, any serological diagnosis would be retrospective and therefore would not be of major clinical use. However, the diagnosis may well be useful for epidemiological purposes.

There are many viruses that produce clinical disease months or years after seroconversion (e.g. HIV and rabies). In the case of these viruses, the mere presence of antibody is sufficient to make a definitive diagnosis.

SELF-CHECK 13.9

Why are antibodies detected against HIV enough to make a definitive diagnosis?

Whilst a serological diagnosis is the mainstay in the diagnosis of viral infections, there are a number of problems associated with the serological diagnosis of infection. These problems are:

- The length of time required for diagnosis for paired acute and convalescent sera often results in the convalescent sample not being submitted as the patient may well have made a full recovery.
- Mild, local infections such as HSV genitalis (genital herpes) may not produce a detectable **humoral immune response** as antibodies may not be produced. A better sample would be a swab in viral transport media of the lesion.
- Extensive antigenic cross-reactivity between related viruses, e.g. HSV and VZV, may lead to false-positive results.
- Immunocompromised patients (e.g. transplant recipients) often give a reduced or absent humoral immune response. As such the antibody response may be too weak to detect.
- Patients with infectious mononucleosis (glandular fever) and those with connective tissue disease such as SLE may react non-specifically, giving a false-positive result.
- Patients given blood or blood products may give a false-positive result due to the transfer of antibody from the donor blood sample to the patient. This is particularly problematic for diagnosis of CMV. Samples from babies are also often problematic as IgG antibodies

will cross the placenta from the mother and will give positive results in IgG assays in the baby's sample.

SELF-CHECK 13.10

Why would swabs from a lesion be a better sample than a serum in the diagnosis of HSV genitalis?

13.6 Antiviral chemotherapy

Similar to antimicrobial chemotherapy, there are a number of agents available for the treatment of patients with viral infections. However, since most viral infections are mild and self limiting they are not as widely used as antibiotics.

Since antiviral agents are required to penetrate into host cells and viruses in general are capable of mutations, designing non-toxic agents is very difficult. As such, only a limited number of antiviral drugs are available and suitable for human use. There are few restrictions on the types of molecules that inhibit virus replication. These vary greatly in complexity and include natural products of plants, synthetic oligonucleotides, oligosaccharides, simple inorganic and organic compounds and nucleoside analogues. Examples of antiviral compounds in current use include:

- Nucleoside analogues - thousands of analogues of naturally occurring nucleosides have now been synthesized and tested in the laboratory, initially as herpesvirus inhibitors, and many now are retested as anti-HIV agents.
- Pyrophosphate analogues - foscarnet is an example of a pyrophosphate analogue. This specifically inhibits herpesvirus DNA polymerase at the pyrophosphate binding sites and it also has anti-HIV activity.
- Amantidine molecules - amantidine is licensed for the treatment of influenza A infection.

Only relatively recently have notable successes on a large scale been achieved with antiviral drugs such as acyclovir and azidothymidine (AZT). No antiviral compound tested has been able to inhibit completely the replication of any virus and a proportion of viral particles always seem to be able to circumvent the drug-induced blockade.

Resistance of viruses to inhibitors

A disappointing feature of antiviral chemotherapy has been the failure so far of any antiviral molecule to inhibit virus replication completely. Antiviral activity tends to produce a 100- to 1000-fold reduction in viral load which still allows some infective particles to survive. This may have important consequences in immunocompromised patients who may be unable to eradicate any residual virus by way of their immune response.

Site of action of common antiviral agents

The sites of action for common antiviral drugs can be seen in Table 13.2. These include:

- Interference with virus adsorption or attachment to the receptor-binding site on the cell wall. These drugs prevent the entry of the virus into the cell.
- Inhibition of virus uncoating. This drug will not allow the release of nucleic acid from the inside of the virus and so will not allow the reproduction of the virus.

TABLE 13.2 Common antiviral drugs showing the target and virus affected.

Target	Drug	Virus affected
Penetration and uncoating	Amantadine Rimantadine	Influenza A
Viral nucleic acid synthesis	Aciclovir Ganciclovir Ribavirin Lamivudine Zidovudine Foscarnet	Herpes simplex Varicella CMV Hepatitis C Hepatitis B HIV-1 Herpes simplex and HIV-1
Binding to intact virus particle	Disoxaril	Rhinovirus
Virus release	Oseltamivir Nefinavir	Influenza A and B HIV

- Inhibition of viral nucleic acid transcription and replication.
- Interference with cellular processing of viral polypeptides.
- Prevention of virus budding or maturation.

13.7 Common viral infections

Respiratory tract

Adenoviruses: these viruses were first isolated from human adenoids, hence their name. They give rise to about 5–10% of respiratory tract infections (RTIs). RTIs caused by adenovirus are mostly mild and require no therapy or only symptomatic treatment. Antigen detection, polymerase chain reaction assay and virus isolation can be used to identify adenovirus infections.

Rhinoviruses: usual cause of the 'common cold'. Symptoms involve nasal congestion, sneezing, sore throat and cough. Usually diagnosis from samples is not necessary. There are specific antibody tests available but these are not generally used.

Influenza: there are three genera: influenza A, B and C. Influenza C is much less important in human disease. Influenza B is almost exclusively a disease of humans. Influenza A viruses have been designated on the basis of the antigenic relationships of the external spike **haemagglutinin** (HA) and **neuraminidase** (NA) proteins. These are numbered sequentially H1–H15 and N1–N9. Of these, H1, H2 and H3 and N1 and N2 are known to infect humans and cause serious outbreaks. Symptoms include shivering, malaise, headache, aching in limbs and high temperature. In general, the severity of influenza is proportional to age.

In 2009, a new strain of influenza A, H1N1, was found to be spreading rapidly in Mexico and causing serious infection. The disease rapidly spread around the world and on 11 June 2009, the World Health Organization (WHO) raised the worldwide pandemic alert level to Phase 6 in response to an ongoing global spread of the influenza A (H1N1) virus. A Phase 6 designation indicates that a global pandemic is underway. More than 70 countries reported cases of human infection with novel H1N1 flu.

Diagnosis of influenza A and B is usually by the complement fixation test, however, in severe cases PCR on respiratory samples can be performed.

Childhood disease

Measles: this is an acute, highly infectious viral illness transmitted via droplet infection. Almost all who are infected develop symptoms. In temperate countries measles is a comparatively mild infection and serious complications are rare. Before the rash appears there is a prodromal stage which lasts for about 2-3 days, with running nose and eyes, cough and moderate fever. During this time, Koplik spots, which resemble grains of salt, can be seen in the buccal mucous membrane. The rash first appears on the face and spreads to the trunk and limbs within a couple of days. Within 2–3 days of onset, the rash will fade, the temperature will decrease and the child will feel better.

Measles is usually diagnosed by looking for specific IgM antibodies in the patient's serum. Vaccination is available as a preventative measure. The vaccine is one of the components of MMR vaccine which is given in the national immunization programme at 12–15 months and at 3–5 years of age.

Mumps: this is an acute viral illness transmitted by direct contact with saliva or droplets from the saliva of an infected person. Humans are the only known host of the mumps virus. The onset is marked by malaise and fever followed within 24 h by a painful enlargement of one or both parotid glands. In most cases the swelling subsides within a few days and recovery is uneventful. Diagnosis is usually based on clinical symptoms but confirmation of the diagnosis can be made by looking for the IgM antibodies in the serum. As with measles, vaccination is available.

Common viral gastroenteritis

Cross reference
Chapter 10, Section 10.8.

Rotavirus: this primarily infects young children; babies under the age of 2 years are the main victims. The symptoms are mainly diarrhoea, vomiting and fever. Dehydration is a major problem and must be treated promptly. In the Third World, dehydration resulting from rotavirus infections causes many millions of deaths in young children annually.

Adenovirus: enteric adenoviruses 40 and 41 cause gastroenteritis, usually in children. Most of the adenoviruses identified in stool specimens are not typed.

Norovirus: the most common cause of infectious gastroenteritis (diarrhoea and vomiting) in England and Wales. The illness is generally mild and people usually recover fully within 2–3 days; there are no long-term effects resulting from infection. Infections can occur at any age because immunity is not long lasting.

Several other viruses can cause gastrointestinal infection including astrovirus and calcivirus.

Herpesviruses

The herpesviruses form a large and important group of infective agents. They cause a wide range of syndromes, varying from trivial mucocutaneous lesions to life-threatening infections. An important property of herpesviruses is their ability to cause latent infections that may subsequently become reactivated.

Herpes simplex (HSV): HSV-1 and HSV-2 have many features in common but, as a general rule, HSV-1 affects the upper body and HSV-2 is more usually, but not exclusively, the cause of genital infections. HSV can cause cold sores, painful skin lesions, ophthalmic infections, meningitis and **encephalitis**. HSV-2 can cause infections in newborn babies caused by passage through an infected birth canal. HSV can cause severe and persistent infection in immunocompromised hosts. Diagnosis is usually made by PCR on a swab of the lesion.

Varicella zoster virus (VZV, chickenpox): a common childhood infection. The rash is usually the first sign of disease. The rash is usually more pronounced on the trunk than on the limbs, which first appears as flat macules which rapidly become raised into fluid-filled vesicles. Complications are rare but can include bacterial infection of the spots.

In humans, VZV can reactivate, causing an attack of herpes zoster with a characteristic distribution of vesicular lesions along the affected **dermatome**.

Laboratory diagnosis is rarely sought in cases of varicella because a diagnosis can be reliably made on clinical grounds.

Cytomegalovirus (CMV): CMV can affect all ages. CMV is transmissible to the fetus via the placenta, young children can be affected with overt disease, adolescents and young adults can be infected when the infection is spread by kissing. In this age group, individuals can have an infectious mononucleosis-like infection (glandular fever). The last category of patient to be affected by CMV is not age-related but is due to a depleted immune system, either through transplantation or an immune-deficient disease such as AIDS. In these patients the disease can be exogenous, that is deriving from the donor's tissue or blood transfusion, or endogenous, that is reactivation of an existing infection in the recipient. Reactivations are less serious than primary infections (i.e. exogenous) and may result only in mild symptoms. Primary infections in these patients, however, are extremely serious and may result in death if appropriate antiviral therapy is not instigated quickly.

Hepatitis viruses

Hepatitis A (HAV): primarily spread by the faecal-oral route and by ingestion of contaminated water or food (e.g. contaminated shellfish). HAV is a benign and self-limiting illness with an incubation period of 2–6 weeks and does not cause chronic hepatitis or cirrhosis. It is the most common cause of infective hepatitis in the world. The appearance of IgM type anti-HAV antibodies in the serum is used to diagnose the infection. A vaccination is available for those at risk of contracting HAV, for example, those travelling to endemic countries and health-care workers.

Hepatitis B (HBV): a blood-borne viral infection that can be prevented through vaccination. Patterns of infection are variable and are influenced by age, sex and the state of the immune system. Many people have no symptoms while others experience a flu-like illness. Other symptoms may include nausea and vomiting. Acute infection can be severe causing abdominal discomfort and jaundice. Mortality during the acute phase of infection is less than 1%.

It is common in South-East Asia, the Middle and Far East, southern Europe and Africa. The World Health Organization estimates that one-third of the world's population has been infected at some time and that there are approximately 350 million people who are infected long term. In Europe, there are estimated to be one million people infected every year.

In the UK, approximately one in 1000 people is thought to have the virus. In some inner-city areas with a high percentage of people from parts of the world where the virus is common, as many as one in 50 pregnant women may be infected.

Several markers are looked for in the laboratory to distinguish between the different phases of the disease (i.e. acute, resolved or chronic). Hepatitis B is implicated in chronic hepatitis, cirrhosis of the liver and **hepatocellular carcinoma**. HBV is transmitted in blood and other body fluids, and is prevalent in men who have sexual intercourse with other men and intravenous drug abusers (IVDAs). A vaccination is available for at-risk individuals, such as IVDAs and health-care workers. HBV immunoglobulin can be administered for temporary passive protection from HBV. It is commonly used in babies born to HBV-positive mothers, sexual partners of individuals with acute HBV and unvaccinated health-care workers who sustain a needlestick injury from a patient with HBV.

Individuals with chronic HBV infection should be evaluated for the potential need for treatment. Drugs currently licensed for the treatment of chronic HBV infection include adefovir dipoxil, alpha interferon and lamivudine.

Hepatitis C (HCV): a major cause of chronic hepatitis, cirrhosis and hepatocellular carcinoma. Hepatitis C infection affects different people in different ways; many experience no symptoms at all while others experience extreme tiredness and can feel very unwell, with symptoms such as fatigue, weight loss, abdominal pain and jaundice. The incubation period varies from a few weeks to 6 months.

It is estimated that around 15–20% of infected people clear their infections naturally within the first 6 months of infection. For the remainder, hepatitis C is a chronic infection that can span several decades and can be life-long.

Estimates suggest over 250 000 people in the UK have been infected with hepatitis C, but eight out of ten don't know that they have it because they have no symptoms. Approximately, 75% of these people go on to develop a chronic hepatitis, but because it can take years, even decades, for symptoms to appear, many people (possibly 100 000 or more) remain unaware that they have the infection. By the time they become ill and seek help, considerable damage has been done to the liver.

HCV is blood-borne and many infections are transmitted in IVDAs sharing needles. HCV is diagnosed serologically by detecting antibodies specific to the HCV. There is currently no vaccine for HCV. The ability of HCV to mutate is in part the main difficulty in developing an effective vaccine. Prevention of HCV is focused mainly on education in IVDAs (not sharing needles), screening of blood and blood products and effective universal precautions when performing venepuncture. The treatment for individuals with chronic hepatitis C infection is a combination of two drugs: interferon and ribavirin. This combination therapy is successful in clearing virus from the blood of around 40% of those treated; however, not everybody is suitable for treatment or can tolerate it. Factors such as age, sex, duration of infection, the strain of the virus and the degree of existing liver damage determine the effectiveness of treatment.

Human immunodeficiency virus (HIV)

HIV continues to be one of the most important communicable diseases in the UK. It is associated with serious morbidity, high costs of treatment and care, significant mortality and high number of potential years of life lost. Each year, many thousands of individuals are diagnosed with HIV for the first time. HIV infection is still regarded as stigmatizing and has a prolonged 'silent' period during which it often remains undiagnosed.

CASE STUDY 13.1

Clinical history

A 28-year-old man attends his GP with a history of weight loss (30 lb in the last few months), poor appetite, night sweats and intermittent fevers. He admits to having used intravenous drugs in the past and having sexual intercourse with both men and women. On examination, he has enlarged lymph nodes in his neck and groin.

This patient falls into a high risk category for HIV, hepatitis B and hepatitis C. He should be tested for all three viruses after being counselled and having consented for the tests to be performed. Patients having HIV tests should always be counselled as to the implications of the test result being positive.

His blood test results reveal that he is HIV positive and his $CD4^+$ lymphocyte count is 375 cells/mm^3. He is given a prescription for anti-HIV drugs and is given an appointment for the Infectious Disease Clinic. The patient fails to attend his clinic appointment and does not take the prescribed drugs.

He is next seen 18 months later when he attends the casualty department of the local hospital complaining of shortness of breath. He is admitted to a ward and chest X-rays show bilateral pulmonary infiltrates. A preliminary diagnosis of *Pneumocystis jiroveci* pneumonia (PCP) is made and the patient is started on antibiotics. He is also found to be very thin, suffering from memory loss and diarrhoea. His $CD4^+$ count is 146 cells/mm^3. Although the patient is treated for his symptoms and infections, he dies the following week.

Discussion

The patient is in a high risk category for HIV and hepatitis B and C. The highest risk patients of contracting HIV are: homosexual or bisexual males, intravenous drug abusers, patients with haemophilia, recipients of blood transfusions and heterosexual contacts of the aforementioned. On initial diagnosis of HIV, his $CD4^+$ count was not low enough ($<$200 cells/mm^3) to be given a diagnosis of AIDS, nor did he have documented AIDS-related infections (e.g. *Candida*, toxoplasmosis), tumours (e.g. Kaposi's sarcoma, B-cell lymphoma) or severe wasting. However, on his second attendance at the casualty department his HIV infection had progressed to AIDS: his $CD4^+$ count was $<$200 cells/mm^3 and he had a diagnosis of PCP. The diarrhoea was also probably attributable to an infectious cause such as *Cryptosporidium* spp. or *Salmonella* spp. The memory loss which was witnessed was probably caused by HIV encephalitis resulting in brain atrophy.

In cases of reinfection, the level of specific IgM either remains the same or rises slightly. However, the level of IgG increases rapidly and far more earlier than in a primary response. Many different types of serological tests are available. With some assays such as EIA and RIA, one can look specifically for IgM or IgG, whereas with other assays such as CFT and HAI, one can only detect total antibody, which compromises mainly IgG. Some of these tests are much more sensitive than others: EIAs and RIAs are the most sensitive tests available, whereas CFT and HAI tests are not so sensitive. Newer techniques such as the EIA offer better sensitivity, specificity and reproducibility than classical techniques such as CFT and HAI. The sensitivity and specificity of the assays depend greatly on the antigen used. Assays that use recombinant protein or synthetic peptide antigens tend to be more specific than those using whole or disrupted virus particles.

CHAPTER SUMMARY

After reading this chapter you should know and understand:

- The general properties and structure of viruses
- Viral pathogenesis and the consequences of a viral infection on the host

- Samples required for the diagnosis of viral infection
- The commonly used diagnostic methods in virology and how they are performed
- The sites and modes of action of antiviral agents
- The names of common viral infections and how they are diagnosed

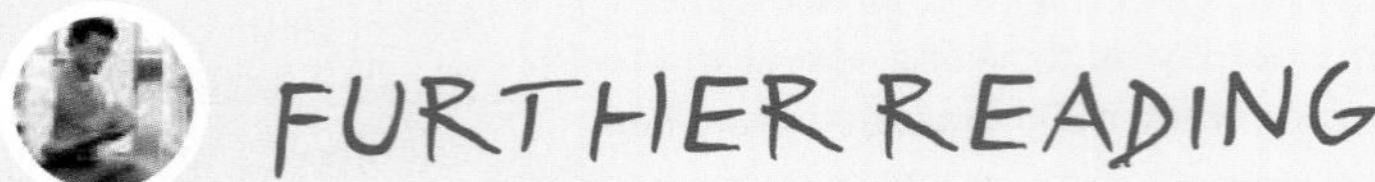

- **Fields BN, *et al.* *Virology*, 3rd edition. Lippincott-Raven, 1996.**

  An excellent reference book for any virologist. The book covers just about every family of viruses. The book is structured well and has sections discussing both the epidemiology, molecular biology and pathology of the viruses contained within the book.

- **Granoff A, Webster RG, eds. *Encyclopaedia of Virology*, 2nd edition. Academic Press, 1999.**

  An excellent reference book which covers all aspects of virology.

- **Timbury MC. *Notes on Medical Virology*, 11th edition. Churchill Livingstone, 1997.**

  A good basic book for students learning virology.

- **World Health Organization. http://www.who.int**

  Website dedicated to health. Monitors disease outbreaks and assesses performances of health systems around the globe.

- **Health Protection Agency. http://www.hpa.org.uk**

  Details on viral infections, outbreaks and diagnostic algorithms can be found on this website

- **Clinical Virology network. http://www.clinical-virology.org**

  Website dedicated to virology. News, clinical tests and current viruses in circulation are all contained within the pages.

DISCUSSION QUESTIONS

13.1 Discuss the provision of a diagnostic virology service to a district general hospital.

13.2 'There is no longer a need for diagnostic laboratories to attempt isolation of viruses in tissue culture.' Comment on this statement.

13.3 What factors would you consider before selecting a commercially available diagnostic assay?

13.4 Discuss the quality assurance of serological assays.

Hints and tips for discussion questions

Chapter 1

1.1 Think of the sites of the body that swabs are used to take samples from - this should indicate in what way they need to be different.

1.2 Are blood tubes for haematology and biochemistry samples sterile? If blood is taken from a patient on antibiotics what will be the concentration? What would be the consequences of a high concentration of antibiotics in the blood stream on isolating a pathogen?

1.3 Consider how long bacteria survive outside the body. Are there any legal or health and safety issues associated with sample transport?

1.4 Look at why organisms are categorized into groups and the effect on the laboratory worker if there is a spillage, for example.

Chapter 2

2.1 Think about the number and types of organism that are isolated on a daily basis in microbiology laboratories. Are all of them pathogens - if not what are they? What is the cost of bacterial identification both in terms of staffing and reagents? Does it matter clinically if an organism is called an *E. coli* or just a 'coliform'?

2.2 Consider if automated systems perform all the different types of identification required. What do automated systems identify? Think about when an immunological test is required, and would it still be required if an automated system was used?

Chapter 3

3.1 Consider the advantages of chromogenic media in differentiating organisms in mixed culture. Think about the savings in biomedical scientists' time. Are less organisms forwarded for identification? Are there certain sample types where chromogenic media are becoming more popular based on clinical need?

3.2 In your laboratory has PCR replaced routine culture in any areas? If not why do you think this is? You need to consider workload and cost. What pathogens are isolated from clinical samples? Is it always a single pathogen? Urine samples would be a good area to consider in your answer.

3.3 Consider quality of the media - are commercially prepared media of a higher quality? You need to discuss how the media are quality assured. Are any problems associated with commercial deliveries? If so what would be the consequences of this? Consider the flexibility of having media production facilities on-site.

Chapter 4

4.1 Think about your answer in terms of ease of use, reproducibility and standardization. What methods are used routinely - will this always be the case? Remember cost and staff required need to be considered in your answer, as well as the scientific reasons for performing susceptibility tests using both these methods.

4.2 This is quite controversial as some laboratories don't perform direct susceptibility tests - why would this be the case? Discuss the situations in which they can be performed and is there an obvious clinical benefit? Are costs and turnaround times important or are these potential benefits outweighed by the lack of standardization?

4.3 Consider how many people are on antibiotics at any one time worldwide. What antibiotics are monitored and why are they more 'important' than others? Are different patient groups of any significance?

Chapter 5

5.1 Think about the situation if the clinician is correct and consequences if not. What can be done with a positive blood culture isolate to aid patient management?

5.2 Does the Gram stain tell you exactly what the organism is? Does it tell you its significance? What does the Gram stain *really* tell you?

5.3 Think about the organism isolated and what species is it. Are some species more important than others? What about the patient - how ill are they?

5.4 Are some organisms more important than others? Think about contaminants, the cost and time of identifying all isolates. Are suspect contaminants always so - do patients ever develop other conditions later?

5.5 You need to understand how blood cultures are taken and the techniques used. What are the consequences of using a poor technique? Will it have any clinical consequences when the results are relayed back to the wards?

Chapter 6

6.1 For both parts of the question think about the differences between the two locations and what type of patients they are likely to be. Are they likely to be simple or complicated infections? What organism would you expect to commonly isolate?

6.2 You need to consider what is meant by bacteriuria and what is significant - is it the type of organism, patient type, number of bacteria present or other factors?

6.3 Consider the types of method available. Think about tests that are performed outside of the laboratory. Consider the advantages and disadvantages of these.

6.4 How many antibiotics can be used? Consider which are the best and for what reason. Think about the organisms being treated and what they are likely to be susceptible to. Why are some better than others, and why are some prescribed more than others?

6.5 Are all urine samples sent because the patient has a suspected UTI or can it be for other reasons? Are all causes of a UTI likely to grow on the media used? Consider factors present in the urine which may cause the organism not to grow.

6.6 Consider the different types of media used and how susceptibility tests can be performed. Does the identification have any relevance and is it better with certain media? Is cost important?

Chapter 7

7.1 Why is it only some laboratories perform such tests? Is it staffing, resources, complexity or some other reason? Would a more rapid result make any difference?

7.2 Would a molecular method be better than culture - if so why? Think about how you would perform a susceptibility test before deciding on your answer.

7.3 Where does *S. aureus* commonly reside on the human body? Would this explain why it occurs so frequently as a cause of wound infections? Are there other reasons, apart from colonization, why its incidence is so high? What properties does the organism possess to enable it to cause wound infections?

7.4 Remember the human body is covered in normal flora - think about the types of flora involved in your answer and the pathogens you would look for.

7.5 How many methods are there currently? How different are they? What are the obstacles to having only one method?

Chapter 8

8.1 Think about the age of the patient and perhaps their surroundings when they develop meningitis. Which bacteria colonize humans and at what age? Think of immunization - is it always effective? Has everyone been vaccinated?

8.2 What are the other departments in laboratory medicine (pathology)? Think about the tests they perform. Can these tests rule out meningitis?

8.3 Think about how easy it is to see small numbers of organism in a smear. If you got it wrong what consequences would this have for patient management?

Chapter 9

9.1 Does smoking damage the respiratory tract? If so, how will this promote bacterial growth? What other things can damage the respiratory tract (think seasonal)?

9.2 Think about the patient types and how quickly organisms can change from normal flora to being involved in disease. Consider environmental factors also.

9.3 Think about the cell properties of certain mycobacteria, think of how they are transmitted. What is it that gives some organisms an advantage over others?

9.4 Consider the patient types they infect. What advantage do they have over other, similar organisms? What are their characteristic properties?

Chapter 10

10.1 Consider the causes of bacterial gastroenteritis and how the disease progresses. Are cases always so severe that the patient would seek medical advice? Would it be worldwide?

10.2 What culture media are used to isolate the various pathogens? You will need to consider the problems associated with culture of faeces samples in general as well as what is likely to grow. Do other organisms 'interfere' with the isolation of target pathogens and is this important?

10.3 First of all, do you believe gastrointestinal infections will ever be eradicated - what are the barriers to this? Consider the source(s) of the pathogens involved and how they could be eliminated. Your answer must be broad enough to consider all causes of infective diarrhoea.

Chapter 11

11.1 Think about how many specialized mycology staff are in routine microbiology departments and what this means to service provision. What services would be required? Also, think of other things that are required to provide the service.

11.2 How many spores does one *Aspergillus* head contain? What happens when one spore lands on a culture plate? Think of the consequences of this.

11.3 In what patients are fungal infections seen? Are these conditions likely to decrease? Think about what improvements have been seen in fungal diagnosis and therapies.

Chapter 12

12.1 Consider the factors that allow the organism to spread - are they environmental or human? What unique properties does the organism have that enable it to become established?

12.2 Consider the environment and the staff within it. How have both of these changed to enable health-care acquired infections to reduce and what have been the drivers for such change?

12.3 Evaluate why spread of the organism occurs and what patient groups are affected. What is happening to the age of the population and what effect is this having both in hospital and community patients? Does the organism have any specific pathogenic properties that make it such a problem organism? Are there any mechanisms that can be applied to reduce infection rates of any organism?

Chapter 13

13.1 You need to consider the range of tests you would offer clinicians - what tests would they want? Would you perform all serological investigations, for example? What use can be made of reference laboratories and other referral centres? Consider financial as well as space considerations in your answer.

13.2 Your answer must include the advantages and disadvantages of cell culture in comparison with other methods. What other methods are available? Are cost, turnaround times and staffing issues important? Which methods are considered as gold standard?

13.3 In your answer you should consider

- actual requirements
- practical considerations
- cost

What sensitivity and specificity would you require? Is the assay a screening assay or a confirmatory one? You must also consider other factors important to requesting clinicians - what would these be? Also, think about how the assay will be performed, by whom and when.

13.4 Is quality control important? How is it performed and when is it performed? How do you know you have achieved the required level of quality? Consider how you record this and make reference to external methods of quality assurance.

Self-check answers

Chapter 1

1.1 The main problems are taking the incorrect sample, delays in transport, unlabelled samples, delays in inoculation of samples and taking samples after the patient has received antibiotics. Selective media may need to be used to eliminate contaminating organisms.

1.2 The main types of sample containers are sterile containers (with or without spoon), swabs, blood tubes and blood culture bottles.

1.3 The skin may contain numerous organisms which can grow on the culture media used to isolate the target pathogen. By cleaning it, their numbers are reduced significantly and this will allow pathogens to grow unhindered.

1.4 Probably the best method of sample transport is the fastest, although all methods have limitations. Air tube systems are most likely to allow samples to get to the laboratory quickly.

1.5 Since selective media contain inhibitory substances it is best to inoculate non-selective media first to allow maximum isolation of target pathogens.

1.6 Chemical treatment will not kill the mycobacteria but will kill or inhibit the growth of any contaminating organisms.

1.7 The highest is Level 4 and the lowest Level 1. Level 2 is most commonly used for processing routine microbiology samples.

Chapter 2

2.1 Firstly, there are health and safety issues if it is not performed correctly as aerosols are produced which could be very hazardous to the operator. The test should not be performed from blood agar as this can lead to a false-positive result and therefore a mis-identification.

2.2 Carbohydrates can be linked to chromogenic or fluorogenic substrates so that upon enzymic hydrolysis a coloured or fluorescent compound is released, which is easily visualized. Examples include *o*-nitrophenol and 4-methyl umbelliferone.

2.3 In the main, chromogenic substrates are used, as the reaction product is easily visualized and there is no need for a UV light source. However, fluorescent substrates are more sensitive and, as a result, a positive test will be determined much sooner. Substrates are therefore chosen depending upon the rapidity of the result required and the ease by which the tests are set up and results obtained.

2.4 (a) Oxidase. (b) Indole or tryptophan deaminase; also swarming on blood agar could be used. (c) Hydrogen sulphide production or fermentation of several carbohydrates. Serological tests can also be used. (d) Catalase would be the most useful differentiator.

2.5 The tube coagulase test is considered the gold standard for the detection of coagulase activity. However, it is impractical to set up one of these tests every time a *Staphylococcus* is grown. As a result, slide coagulase tests have been introduced which make detection of coagulase activity simple and rapid.

2.6 Several groups of organisms fail to ferment carbohydrates (e.g. *Pseudomonas* spp.) and as a result other tests, such as carbon source utilization tests, need to be used to differentiate between species.

2.7 Generally, biochemical tests would be better to differentiate between species. In general, there are a wide variety of tests that can be used compared to substance susceptibility tests. Even with established tests, such as Optochin susceptibility, there are still problems with other closely related strains showing susceptibility to this agent.

2.8 Commercial kits have the advantages of utilizing a wide range of substrates to provide an accurate identification and differentiation from organisms which can be closely related. They are easily inoculated and results are generally easy to read. Powerful software packages are available containing databases of thousands of strains to allow for accurate identification.

Chapter 3

3.1 Like most living things bacteria require a 'balanced' diet in order to survive and multiply. The nutrients supplied are required for cell wall production, enzymes, etc.

3.2 Since pathogenic organisms require a host in order to survive and multiply the pH for growth reflects that of the host.

3.3 Horses are large animals and as such a large amount of blood can be withdrawn. Human blood is not available in such quantities, and there are ethical issues also. Additionally, the blood would need to be screened for HIV and hepatitis before it could safely be used. Horse or sheep blood is therefore used.

3.4 For the selective isolation of Gram-positive organisms, bacitracin and aztreonam are commonly used agents. For Gram-negative strains, vancomycin is commonly used to suppress the growth of Gram-positive organisms and allow the Gram-negative strains to flourish.

3.5 The main advantages are that it eliminates the need for further confirmatory testing, compared to conventional media such as DCA or XLD. This reduces labour and lowers turnaround times for results. In the majority of cases, a coloured colony on a *Salmonella* chromogenic medium is highly likely to be a *Salmonella* spp.

3.6 The susceptibility profile of *Candida* spp. can vary widely with some species often 'intrinsically' or naturally resistant to antifungal agents. Finding these species can alter patient management. Differentiation of mixed *Candida* spp. is very difficult if not impossible on non-chromogenic media.

3.7 In general serum or blood would break down if added to media that was too hot. More importantly, antibiotic supplements for selective media would be totally inactivated by adding to media not cooled to the right temperature. If this was to happen the media would no longer be selective and important pathogens could be missed.

3.8 Media are one of the most important products used in a microbiology laboratory and as such ensuring high quality media is essential. The consequences of poor quality or contaminated media could be disastrous in that target pathogens from clinical samples may not grow or may be covered in contaminants. The consequences for the patient could be significant.

Chapter 4

4.1 If the agar was too thin, larger zones may be encountered. With the medium too thick, the zones of inhibition would be reduced. Remember the antibiotic will diffuse downwards as well as across the plate.

4.2 If the temperature of incubation is incorrect the expected growth of the test organism will not occur. This could result in reduced or larger zone sizes depending upon the conditions. Incubation in carbon dioxide can affect the zone sizes of macrolide antibiotics such as erythromycin. Incubation of aminoglycosides anaerobically can also result in reduced zones.

4.3 Briefly, the MIC for the organism is determined at the same time as a zone size is produced. Since, from pharmokinetic studies, the serum concentration of the antibiotic is produced there is a relationship between them. From these data a zone size is produced which allows an organism to be called sensitive, resistant or intermediate.

4.4 Potentially, the same consequences as for any incorrect result issued. For antibiotics it would mean the patient may be on inappropriate therapy. For a patient with life-threatening septicaemia, for example, this could result in serious consequences, even death.

4.5 The main problem is that the inoculum is not standardized and could be incorrect by several orders of magnitude. As a result, it can be difficult, if not impossible, on some occasions to interpret the susceptibility test result. The benefit is that for certain clinical conditions, such as septicaemia, a 'direct' result can confirm the patient is on appropriate therapy, or indicate antibiotic therapy should be changed more quickly than setting up the test after the culture result is known.

4.6 Broth tests are often performed in tubes or in microtitre plates and are often more labour intensive than conventional agar disc tests. Broth tests are considered more accurate due to the many problems that can be encountered with agar methods. Unless the process is automated, laboratories often prefer the convenience and cost benefits of agar susceptibility methods.

4.7 There are a number of species that appear to be susceptible to the antibiotic *in vitro* but are known to produce inducible enzymes which are expressed *in vivo*, for example *AmpC* enzymes. It is also important to recognize unusual patterns of resistance. For example, Group A streptococci are very unlikely to show resistance to penicillins and *S. aureus* strains are very rarely resistant to glycopeptides. This can be highly useful in stopping the spread of such organisms if they are identified fully.

4.8 Most antibiotics are relatively non-toxic to the host in the correct dose. Some, however, such as the aminoglycosides, can have serious side effects such as ototoxicity if the level is too high.

4.9 Generally, the automated immunoassay method is considered the best method for routine diagnostic use as it is relatively simple to perform with accurate results available in less than 30 minutes.

Chapter 5

5.1 Septicaemia is a fatal condition if not treated early and effectively. Clinicians will always treat if the indications are the patient has septicaemia.

5.2 Virtually any organism can cause septicaemia and some are highly nutritionally dependent. As a result, the blood culture systems use highly nutritious media to maximize isolation of the pathogen.

5.3 The Gram stain can provide information as to the possible identity of the infecting organism and allow appropriate antibiotic treatment to start before the culture result is obtained.

5.4 Having an indication of the susceptibility pattern of the organism growing in a blood culture bottle can aid significantly in the management of the patient.

5.5 Whilst numerous organisms cause septicaemia, the commonest are *S. aureus*, Enterobacteriaceae, *Enterococci*, *Pseudomonas* and *Candida spp.*

5.6 If multiple organisms are involved then finding a single antibiotic to which they are all susceptible is difficult.

5.7 They may receive antibiotics which they do not need, and these may be costly, cause side effects, reduce the chance of a correct diagnosis and increase length of hospital stay.

5.8 Yes - remember the most nutritious media are used in blood culture systems and one organism will rapidly grow into millions and a positive bottle will result. Some contaminants will grow faster than pathogens and as a result can outgrow and even inhibit their growth.

5.9 Most commonly the oropharynx in cases of native endocarditis. For prosthetic valve endocarditis it is contamination of the valve during surgery.

5.10 Symptoms can be quite vague and non-specific, such as the patient feels generally unwell and lethargic with low-grade pyrexia. Other symptoms can be heart murmurs, nodules in the eye or fingernails as well as a raised white cell count.

Chapter 6

6.1 The commonest cause of a UTI is *E. coli*. The body has a number of mechanisms to prevent a UTI, such as urinary flow to flush the organism out. Other mechanisms include antibodies (IgA) and high osmolarity.

6.2 Whilst essential for patients with bladder problems and those post op there are disadvantages also. Colonization by bacteria and trauma, which can result in bacteraemia and septicaemia, are two major problems associate with urinary catheters.

6.3 A complicated urinary tract infection means an infection in a patient who has a structural abnormality, who is immunocompromised or who may be infected with a highly resistant organism.

6.4 In most cases upper urinary tract infection starts from infection of the lower tract. It can progress to bacteraemia in over one third of cases. Upper urinary tract infections are considered clinically very important and essential to treat effectively.

6.5 Excretion of mycobacteria into the urine is intermittent and may not be picked up in a single sample.

6.6 The first few millilitres of a urine sample are likely to be contaminated with skin organisms. If these are present in the sample then it may be sent out as contaminated or contaminants may overgrow the causative organism of the UTI.

6.7 Microscopy, especially for the presence of WBC and RBCs, is very important alongside the culture result. The lack of WBCs in a sample may indicate asymptomatic bacteruria if significant numbers of organisms are present. The presence of RBCs (haematuria) may indicate renal infection. The presence of epithelial cells suggests the sample may be contaminated.

6.8 Automated methods are useful particularly when there is a high workload. Since most urine samples submitted are likely to be negative, these samples can be screened out without the need for culture. This can save labour, time, money and patients being put on antibiotics they don't need.

6.9 Small numbers of organisms are likely to be insignificant or possibly represent contamination. Apart from special categories of patients, $>10^5$ organisms/mL is considered as significant.

6.10 Culture methods can vary between laboratories. Whilst CLED agar is commonly used, these plates are being superseded by chromogenic agar media, as such plates can give an accurate identification of bacterial colonies without the need for further work.

6.11 Generally, contamination is assessed by the number and types of organisms present in the sample. The majority of UTIs are caused by a single organism and the presence of thee or more organisms is suggestive of contamination. Remember that low WBC counts and the presence of epithelial cells also indicate a contaminated sample.

6.12 For the patient an oral antibiotic is much preferred to an IV agent. Often IV antibiotics are administered in hospital which is more of an inconvenience than the patient being at home. Preferred choices would depend on the age of the patient and other factors such as pregnancy, and the common resistance patterns prevailing. Ideally, it would be oral agents such as amoxicillin, trimethoprim, nitrofurantoin or a first-line cephalosporin such as cefalexin.

Chapter 7

7.1 Some organisms have more virulence mechanisms which makes them more likely to cause disease. *S. aureus* colonizes the nose, so skin infections are likely to be more common due to its occurrence on human skin.

7.2 In the tissue necrosis which occurs with gangrene there is very limited oxygen. This is an ideal environment for rapid-growing anaerobes such as *C. perfringens* to thrive.

7.3 Most throat infections are caused by a virus and antibiotics are of no use. Also, most infections get better on their own. Using antibiotics unnecessarily will only encourage resistance and limit the use of antibiotics in the future.

7.4 Group A streptococci produce numerous toxins which cause tissue damage, haemolysis and evasion of phagocytosis. Serious infections caused by this organism are monitored epidemiologically so that the spread can be monitored.

7.5 Firstly, it can be difficult to get good 'sample material' from a swab and as such it may be difficult to see any organism. Secondly, it is impossible to say with any accuracy that Gram-positive cocci seen, for example, are pathogenic strains or part of the normal flora.

7.6 Some organisms may only grow in CO_2 or under anaerobic conditions. Since organisms that grow under such conditions may be present, plates are incubated under a variety of conditions. In some cases β-haemolysis is best observed under anaerobic conditions and this can be very important in determining the identity of β-haemolytic streptococci in particular.

7.7 The biofilm will consist of organisms with a very low metabolic rate; thus antibiotics which are designed to kill rapidly growing cells will have a much lower effect on these organisms. Secondly, the organism in the biofilm will be encapsulated in a polysaccharide matrix making penetration of antibiotics difficult.

7.8 The number of organisms causing an infection in such sites may be very low. Culturing the samples direct may fail to yield the causative organism. Secondly, antibiotics may have damaged the organisms present. It is therefore essential to use broth enrichment to ensure any organisms present are isolated.

7.9 *N. gonorrhoeae* may be difficult to isolate because of the large numbers and diversity of the normal flora which may be present. In addition, the target organism may only be present in small numbers.

7.10 This is complex to answer and may be due to the fact that many infections are asymptomatic with no obvious signs – people wrongly think that protection is therefore not needed.

7.11 The organism cannot be grown by conventional culture methods, hence the need for alternative methods of detection. Screening tests for syphilis look for IgA or IgM antibodies depending upon the length of the disease.

7.12 The use of endocervical swabs is most common in females or urethral swabs in males. Culture is still used as the most common method of detection. Isolation media include antibiotics such as vancomycin, colistin and antifungal agents such as amphotericin. Molecular methods (combined with detection of *Chlamydia*) are now proving to be useful for detection of the organism but culture still needs to be performed if susceptibility tests are required.

Chapter 8

8.1 The commonest causes of bacterial meningitis are *Neisseria meningitidis*, *S. pneumoniae* and *S. agalactiae* (Group B).

8.2 Polymorphonuclear cells are often associated with bacterial meningitis, whereas lymphocytes are often associated with viral infections.

8.3 Bacteria will not always be seen, especially if it is an early infection where there may be few organisms infecting the CSF. Obviously the biomedical scientist won't be able to see viruses in a Gram stain!

8.4 It is important to identify any pathogen infecting a sterile site as it can have an impact upon patient management and in the case of meningitis it may be important to determine precisely what the organism is in an outbreak situation. The latter allows contacts to be traced and ensures prophylaxis is given promptly. Also, remember the problems associated with incorrect identification of *Listeria* and Group B streptococci.

8.5 Definitely not! This is one of the most serious clinical conditions which has a very high mortality rate if not treated promptly. Clinicians would always treat promptly if suspected.

8.6 In general it depends on the site. For example, respiratory organisms are likely to be encountered in pleural fluid. In peritoneal fluid it is more likely to encounter 'faecal' organisms, although any sterile site can be infected with virtually any bacteria.

8.7 Bacteria may be inside white blood cells - lysis causes them to be released allowing them to grow on culture media.

Chapter 9

9.1 Patients who are immunocompromised fail to produce as many white cells as a normal healthy individual. As a result the secretions from these patients may be mucoid.

9.2 Normal sputum samples should contain a mixture of mouth flora and the pathogen(s) in question. If diluted the mouth flora will diminish, making it easier to see the target pathogens. Other respiratory samples, for example pleural fluid, will contain no 'normal flora' organisms so there is no need to dilute the sample.

9.3 PCP is a disease only associated with patients who are immunocompromised.

9.4 Pathogens and 'normal flora' organisms will grow on basic isolation media. Selective media are used so it is easier to find the pathogen.

9.5 Bcc organisms can indicate a poor prognosis in transplant patients if they are colonizing or infecting them. As a result, it is vitally important to identify these organisms, which can be very difficult.

9.6 Decontamination agents are used to kill off the contaminating organisms in the sample so that the mycobacteria can grow unhindered. Mycobacterial cells are resistant to the decontamination procedure.

9.7 The most common species are *M. tuberculosis*, *M. bovis* and *M. africanum*. These are Hazard Group 3 organisms.

Chapter 10

10.1 There are many such mechanisms including the pH of the stomach. White blood cells and immunoglobulins such as IgA are abundant which also provides defence against infection. Digestive enzymes and the presence of normal colonizing bacteria are also important mechanisms.

10.2 In the western world these infections are generally less severe, which is often due to the better health status of the individuals. Poor nutrition and sanitation contribute to the severity and length of disease in the developing world.

10.3 Think about the number of different types of gastrointestinal infection that an individual can acquire. Most have viral causes and it is very difficult to detect these routinely. Infections caused by parasites are often associated with foreign travel and it is not worth looking if there is no supporting clinical history.

10.4 For both the answer would be viral infections. If bacterial infections are considered then in developing countries it is mainly *Salmonella* and *Campylobacter* infections, although *Shigella* can be a significant cause. The same infections also occur in developing countries plus *Vibrio* and parasite infections.

10.5 The use of sodium selenite, the main ingredient of selenite broth, inhibits the growth of other interfering bacteria and allows preferential growth of *Salmonella* sp. There are no selective agents that would allow the enrichment of *Shigella* sp. at the expense of other intestinal flora.

10.6 It is highly suggestive but other organisms that are not causes of gastrointestinal disease, such as *Proteus* sp. and *Citrobacter freundii*, would produce similar colonies on XLD.

10.7 Whilst cell culture is often considered the gold standard it does not detect strains that only produce toxin A. In addition, faeces may contain chemicals and other toxin-producing bacteria, which produce a similar cytotoxic effect. The method is considered to be quite labour intensive.

10.8 Other Gram-negative organisms can have similar antigens to both *Salmonella* and *Shigella*, and cross reactions (false positives) do occur. It is generally considered that biochemical tests provide the most definitive identification in routine laboratories.

10.9 Most definitely not. This is because a patient with diarrhoea could potentially infect a whole ward or other unit before the laboratory confirmed the cause of the infection.

10.10 Most viral illnesses are mild and self limiting and it would be very difficult and expensive for routine laboratories to look for all potential viral causes. Generally, most laboratories would look for viruses such as adenovirus and rotavirus which cause significant disease in children. Laboratories would often look for viruses such as norovirus in outbreak situations.

10.11 Since most gastrointestinal infections are self limiting, generally rest and fluid replacement are sufficient.

Chapter 11

11.1 The main difference is that the bacteria are prokaryotic organisms and fungi are eukaryotic organisms.

11.2 More patients are immunosuppressed, there is better diagnosis of fungal infections and there are better treatments, which has prompted a demand for earlier diagnosis and better diagnostic aids.

11.3 Factors predisposing to fungal infections are immune state of the patient, age, use of broad spectrum antibiotics and diabetes. These are the same factors that will predispose to many bacterial infections.

11.4 They are very unlikely to cause systemic infections as they require a lower temperature for optimal growth, hence they cause superficial infection. They are also of low virulence.

11.5 The result is available several weeks before culture. This allows therapy to be started earlier if the microscopy is positive.

11.6 The respiratory tract is the most common site of invasive *Aspergillus* infections because spores are infective and present in the air and are commonly breathed into the lung.

11.7 Allergic bronchopulmonary aspergillosis, aspergilloma, non-invasive aspergillosis and acute invasive aspergillosis are associated with *Aspergillus*.

11.8 They are often labour intensive, require specialist media and there is a lack of standardized protocols. They are also more costly to perform than bacterial susceptibility tests.

Chapter 12

12.1 Most commonly, the nose is the site of carriage of MRSA. In order to ensure that the best attempt has been made to locate MRSA, a nasal swab must be sent.

12.2 Because on mannitol salt agar the acid generated from fermentation of mannitol changes the pH throughout the medium, so the whole plate can become yellow, making the individual MRSA colonies difficult to detect in mixed culture. On chromogenic media there is no diffusion of the reaction product, making the colonies easier to detect.

12.3 The molecular methods offer significant advantages in the rapidity of results (<1 hour) and the simplicity of the methodology is now evident. However, the cost of each assay is more than 20-fold that for a chromogenic medium so its use for widespread screening is difficult to imagine financially. If large-scale screening is required the space must also be available to house multiple platforms. Several institutions advocate the use of molecular for certain situations combined with chromogenic/enrichment culture for the majority of 'routine' screens.

12.4 Screening is important for picking up MRSA carriage and such information is useful to help stop spread of infection. Most important are ward procedures such has hand washing, general cleanliness and avoiding poor practice when inserting IV lines. These will have the most impact on MRSA bacteraemia rates.

12.5 These organisms can be highly drug resistant and vancomycin is an extremely important antibiotic for treatment of Gram-positive infections in general. If these organisms become common in a hospital environment, clinicians will be faced with difficult therapeutic decisions.

12.6 In winter months, patients often succumb to respiratory infections (especially elderly patients) and antibiotics can alter the gastrointestinal flora and allow overgrowth of *C. difficile*. In general, more antibiotics are used in the winter than in summer months.

12.7 It must be remembered that no single method is optimal for detection of *C. difficile* infection. New algorithms are being developed which initially screen for GDH as this has a high NPV. Samples positive for GDH would then be tested for the presence of toxins A+B. Some authors advocate the use of culture as part of any diagnostic algorithm. More recently, molecular methods are becoming popular due to the speed of results (<1 hour).

12.8 No. This is because *C. difficile* is likely to exist as spores in a ward environment and these gels have poor activity against spores.

12.9 If a group of infected individuals all have the same infection then it is prudent to manage them in the same environment. It would be very difficult to allocate such individuals a separate ward environment as most health-care institutions have only limited resources of this nature.

Chapter 13

13.1 Viruses have a characteristic very small size when compared to bacteria. They also have a very different genome (DNA or RNA but not both) and they are metabolically inert. In addition, they will not grow on artificial agar media.

13.2 Latency is a state of inactivity and reactivation is to become active again after a period of quiescence or as in a viral infection, latency. An example of a virus that can be latent and then reactivate is herpes simplex.

13.3 Viral transport medium should contain antibacterial and antifungal agents to prevent the overgrowth of contaminating bacteria and fungi. The presence of bacteria and fungi, particularly in cell culture, can lead to contamination of the culture and render it impossible to interpret.

13.4 Serology tests are easy, quick and relatively cheap to perform.

13.5 These tests are rapid and results can be obtained in relatively short time spans. However, they can be tedious and labour intensive. These tests can be difficult to interpret and have a low sensitivity and specificity. The quality of the sample must be good to produce optimal results.

13.6 Molecular methods are much more sensitive and specific than other direct detection methods. They are also quick and easy to perform.

13.7 Cell culture is still used for many viral infections because it is a relatively inexpensive way to diagnose such infections. PCR is expensive to set up and so many smaller laboratories may not be financially in a position to set up a molecular laboratory.

13.8 Washing steps in an ELISA test rid the reaction well of any excess antibody or antigen that may be present. If excess antibody or antigen is present in the well, false-positive reactions will occur giving rise to inaccurate results.

13.9 HIV infection does not necessarily produce symptoms/clinical disease for many years after the initial infection; therefore although the patient may be completely free of clinical disease, they may still be infected with HIV, and as such will produce antibodies.

13.10 Mild local infections such as HSV genitalis (genital herpes) may not produce a detectable humoral immune response as such and antibodies may not be produced.

Glossary

α- or β-haemolytic streptococci bacterial colonies can be described as α-haemolytic or β-haemolytic depending on their action on the red blood cells in the culture medium.

acid-fast microorganisms that resist decolorization with acid; typically applies to mycobacteria.

acyclovir an antiviral agent with activity against herpesvirus types 1 and 2 and varicella zoster virus. Acyclovir is converted by a herpesvirus enzyme into a molecule that inhibits the synthesis of deoxyribonucleic acid (DNA) molecules in the virally infected cells. This has the effect of inhibiting viral replication.

adenylate cyclase an enzyme that catalyses the conversion of adenosine triphosphate (ATP) to inorganic pyrophosphate and cyclic adenosine monophosphate (cAMP).

adhesins appear to mediate bacterial adherence to host epithelial cells thereby assisting in pathogenesis.

aglycone a molecule that is not a sugar but is linked to a natural sugar (e.g. galactose) via a glycosidic bond (e.g. nitrophenol in *o*-nitrophenyl-β-D-galactopyranoside).

AIDS disease of the immune system caused by HIV.

aminoglycosides a group of antibiotics, including gentamicin and tobramycin. They are considered broad-spectrum agents and are particularly active against *S. aureus* and aerobic Gram-negative bacilli. These agents can be both nephrotoxic and ototoxic.

aminopeptidase a peptidase enzyme (see above) that targets amino acids with a 'free' amino group.

amnionitis infection of the membrane sac that surrounds and protects the embryo.

AmpC β-lactamase an inducible enzyme occurring in Gram-negative species such as *Enterobacter* and *Serratia* spp. They hydrolyse broad- and extended-spectrum cephalosporins.

amphotericin B a polyene antifungal drug; it shows high patient toxicity.

anaerobes organisms that only grow when oxygen is absent.

antibiotic a chemical compound that has the ability to kill or inhibit the growth of microorganisms. They are sufficiently non-toxic to be administered to the host.

arthritis the term used to describe inflammation in a joint. Septic arthritis is the name used to describe inflammation of a joint caused by infection.

arthropathy a disease or abnormality of a joint.

arthroplasty surgical procedure where an arthritic or problem joint is replaced.

asepsis the practice of trying to eliminate contaminants from entering samples; also used in surgical and other situations to ensure organisms don't enter the patient during the procedures used.

aseptic technique the use of sterile equipment and procedures to avoid possible contamination.

aspergilloma a ball of fungi found in the lungs; typically occurs with *Aspergillus* spp.

aspiration pneumonia occurs when the epiglottis fails to cover the larynx during swallowing. Food and saliva, which would normally be directed down the oesophagus, enter the trachea and can lead to pneumonia.

asymptomatic the presence of an organism but without symptoms developing.

atypical pneumonia organisms that cause pneumonia without characteristic symptoms, e.g. lack of sputum production.

auramine–phenol stain auramine O is a fluorescent dye and this is a common stain used to detect mycobacteria in clinical samples.

autolysis when an organism produces enzymes that lead to the breakdown of its own cells.

β-lactam antibiotics a family of antibiotics which contains penicillins, cephalosporins and some other antibiotics. They are composed of a β-lactam ring and some, e.g. penicillin, are used mainly to treat Gram-positive infections. Cephalosporins are more broad-spectrum agents.

β-lactamase enzymes that break down penicillin and related antibiotics (β-lactams) such as ampicillin and destroy their antibacterial activity.

bacitracin a naturally produced antibiotic that targets bacterial cell walls.

bacteraemia the presence of bacteria in the blood stream.

bacteristatic refers to an antibiotic that inhibits the growth of an organism but doesn't necessarily kill it. Bactericidal antibiotics are those that are likely to kill the organism.

bacteriuria presence of bacteria in the urine.

balanitis inflammation of the glans penis.

barrier nursing this involves putting the patient in a side room or with other individuals with exactly the same infection in the same room. The nursing staff wear disposable aprons and gloves when dealing with these patients to minimize any spread of the infection to other patients.

Bartholin's gland two glands situated below the vagina.

biofilm a collective community of microorganisms encapsulated in a polysaccharide matrix.

blind subculture subculture of the bottle regardless of the presence or absence of turbidity or indications of growth.

breakpoint this refers to when only certain antibiotic concentrations are tested, normally two. These concentrations would be appropriate for differentiating between susceptible, intermediate and resistant.

broad-spectrum antibiotics drugs that show activity against a wide range of organisms, typically both Gram-negative and Gram-positive species.

bronchiectasis disease that causes a localized and irreversible dilatation of the bronchi.

***Burkholderia cepacia* complex** a group of bacteria that are most often associated with patients who have cystic fibrosis.

Buruli ulcer an infectious disease caused by *Mycobacterium ulcerans*.

butyrous resembling butter.

capsule a polysaccharide coat that covers some organisms such as *C. neoformans*. Capsules help resist ingestion by white blood cells.

carbapenemase an enzyme that hydrolyses not only penicillins but also cephalosporins, monobactams and carbapenems. Strains that produce these enzymes are highly drug resistant.

carbuncles a large abscess made up of several skin boils, filled with fluid and pus.

carrier state this is where an individual, who is not ill, can harbour an infectious organism which may cause disease if it is passed on.

casein a protein precipitated from cow's milk.

catalase an enzyme that catalyses the decomposition of hydrogen peroxide, forming water and oxygen.

catheter a hollow tube that can be inserted into a body cavity. The procedure is known as catheterization.

cell culture the growth of cells *in vitro*.

cell line one derived from a primary culture which has been cultured for the first time.

cellulitis a spreading infection of the skin and connective tissue.

cerebal palsy a name for a group of conditions which are non-progressive and result in physical disability.

cerebrospinal fluid a clear fluid that fills the spinal column.

chancroid an uncommon sexually transmitted infection which presents as painful sores on the genitalia.

chemiluminescence the emission of light by a substance as a result of a chemical reaction that does not involve an increase in its temperature. Chemiluminescence occurs when a highly oxidized molecule, such as a peroxide, reacts with another molecule. The bond between the two oxygen atoms in a peroxide is relatively weak, and when it breaks the atoms must reorganize themselves, releasing energy in the form of light.

chromogenic media differential culture media that contain chromogenic enzyme substrates. When such substrates are metabolized (typically by hydrolysis), they release a coloured dye or 'chromogen' indicating the presence of a specific enzyme.

chromogenic substrate may be natural or synthetic and is metabolized (usually by a hydrolase) to release a coloured dye.

cilia hair-like beating structures which protrude from respiratory epithelial cells.

Clostridium difficile an organism involved in outbreaks of gastrointestinal disease which can be fatal, especially in the elderly. Production of two protein toxins A+B is responsible for production of disease.

cmax is the peak concentration obtained following standard dosing.

coliform a loose term used for bacteria that are '*Escherichia coli*-like'. They are members of the Enterobacteriaceae family which ferment lactose to produce acid and gas at 35–37°C.

colonization an organism that is found at a particular body site but is playing no role in disease. Colonization can be transient in that individuals can be colonized with MRSA at the start of a shift and be clear at the end and *vice versa*.

commensal organisms organisms that make up the normal flora of the skin.

complement a series of proteins found in blood; when activated the proteins have an antibacterial action and help the body to fight infection.

complement fixation test an immunological test that can be used to detect the presence of either specific antibody or specific antigen in a patient's serum. However, in clinical diagnostics labs it has been largely superseded by other serological methods such as ELISA and by DNA-based methods of pathogen detection, particularly PCR.

computed tomography scan (CT scan) a medical imaging technique which generates a 3D image.

conjunctivitis an infection of the outermost layer of the eye (the conjunctiva).

containment levels describes the facilities and practices that are appropriate for the control of microorganisms.

contaminant an organism that is not deemed to be pathogenic. This can also mean an organism that was not present in the original sample.

cord factor one of the mycolic acids in mycobacteria. This promotes the formation of strings of cells that give a cord-like appearance microscopically.

C-reactive protein (CRP) a protein found in blood; levels of CRP increase when there is inflammation or infection.

crepitus the medical name for sounds under the skin. Typically this is popping or cracking sounds.

crystalluria presence of crystals in the urine.

cycloheximide a chemical that has antifungal properties. It suppresses growth of fungi such as *Aspergillus* and *Penicillium* spp. which are common in the environment and may contaminate skin and nail samples.

cystitis acute cystitis is acute inflammation of the bladder.

cytokine a regulatory protein, released by cells of the immune system.

cytology the study of cells.

cytomegalic inclusion bodies nuclear or cytoplasmic aggregates of stainable substances, usually proteins. They typically represent sites of viral multiplication in a bacterium or a eukaryotic cell and usually consist of viral capsid proteins.

cytomegalovirus (CMV) a member of the herpesvirus family and one of the most common human viruses. The virus occurs worldwide, and most people become infected with CMV sometime during their lives.

cytopathic effect the presence of the virus often gives rise to morphological changes in the host cell. Any detectable changes in the host cell due to infection are known as a cytopathic effect. Cytopathic effects (CPE) may consist of cell rounding, disorientation, swelling or shrinking, death, or detachment from the surface.

cytotoxic agents a chemical that is toxic to host cells.

cytotoxin a toxin that kills cells.

decontamination when samples are treated with chemicals to kill or reduce the numbers of other organisms leaving the target pathogen undamaged.

dermatome an area of skin that is mainly supplied by a single spinal nerve.

diabetic nephropathy damage to the small blood vessels within the kidney seen in patients with longstanding diabetes mellitus.

differential media culture media that include an indicator system for detection of biochemical activity, thus enabling differentiation of species or groups of microbes. A typical example would be inclusion of a sugar plus a pH indicator to demonstrate oxidation/fermentation.

dimorphic a fungus that can exist in two forms, a yeast or a filamentous hyphal form.

diphtheria an acute upper respiratory tract disease caused by *Corynebacterium diphtheriae*.

diptheroids a loose term to describe a range of Gram-positive rods that appear 'coryneform' on Gram stain.

direct susceptibility tests a sensitivity test carried out directly on a sample rather than on an isolated bacterial culture.

disseminated spread throughout the body.

dysuria pain on passing urine; patients often describe this as a burning or stinging feeling on passing urine.

echinocandin antifungal drugs which inhibit cell wall synthesis.

EDTA ethylenediaminetetraacetic acid is an anticoagulant to prevent blood samples from clotting.

electron microscope a type of microscope that uses a particle beam of electrons to illuminate a specimen and create a highly magnified image. An electron microscope has a much greater resolving power than a light microscope and can obtain a much higher magnification of up to 2 million times, while the best light microscopes are limited to magnifications of 2000 times.

emetic toxin a toxin that causes severe, often projectile vomiting; such a symptom can be characteristic of *B. cereus* food poisoning.

empiric therapy the commencement of therapy before it is known what the infective organism is or what its susceptibility is.

empyema a collection of pus and necrotic tissue in a body cavity, e.g. the pleural cavity.

encephalitis inflammation of the brain usually caused by a viral infection.

endocarditis an infection of the heart valves.

endocytosis a process where substances are taken into the cell.

endotoxin is a component of Gram-negative bacterial cell walls; when released into blood it causes fever.

enrichment culture a method that utilizes a nutrient liquid medium so that very small numbers of organism(s) are encouraged to multiply to large numbers. This method is used so that small numbers of potential pathogenic organisms can be detected where they may be missed on direct culture.

enrichment media culture media (typically broth based) that are designed to favour or 'enrich' the growth of specific target pathogens.

enterotoxin a toxin released by an organism in the intestine. They are protein in nature and frequently cytotoxic and kill host cells.

eosinophils a type of WBC associated with allergy. They may be detected in sputum microscopically.

epidemiology the study of the distribution of disease and its impact upon the population.

epiglottitis inflammation of the epiglottis which when inflamed can cause serious breathing difficulties.

epithelium cellular tissue which covers the external body surface, or the lining of internal surfaces.

epitopes the part of a macromolecule that is recognized by the immune system.

Epstein-Barr virus (EBV) a member of the herpesvirus family and one of the most common human viruses. The virus occurs worldwide, and most people become infected with EBV sometime during their lives.

erysipelas a diffuse infection of the upper layers of skin; often has a defined edge.

erysipeloid a cutaneous form of an infection caused by *Erysipelothrix rhusiopathiae*.

esculin a naturally occurring glucoside consisting of glucose linked via a glycosidic bond to 6,7-dihydroxycoumarin; a substrate for detecting β-glucosidase.

esterase a hydrolase enzyme that splits esters into an acid and an alcohol.

exocytosis the passage of material to the cell surface and subsequent release from the cell.

exopolysaccharide high-molecular-weight compounds composed of sugar residues. These are produced by microorganisms and released into their surrounding environment.

exotoxin a component of Gram-negative bacterial cell walls; when released into blood it causes fever.

extended-spectrum β-lactamases enzymes that can be produced by bacteria making them resistant to penicillins and cephalosporins, e.g. cefuroxime, cefotaxime and ceftazidime. These antibiotics are some of the most important antibiotics available for the treatment of infectious disease.

faecal lactoferrin presence of leucocytes in a stool sample. Can be used as a marker for intestinal inflammation; not specific for *C. difficile*.

fastidious organisms that require additional nutrient in order to grow.

fatty acids acids found in fats and oils. They consist of long hydrocarbon chains with a carboxyl acid group at one end.

fish tank granuloma skin lesions caused by *Mycobacterium marinum*, also known as swimming pool granuloma.

flagella hair-like projections that extend from some bacteria which aid in their movement.

fluorogenic substrate may be natural or synthetic and is metabolized (usually by a hydrolase) to release a fluorescent compound.

folliculitis inflammation of a hair follicle anywhere on the skin.

free radicals an atom or chemical species that has at least one unpaired electron and is therefore highly unstable/reactive.

fumigation describes the process of sterilizing a safety cabinet with the use of formaldehyde vapour.

galactomannan a polysaccharide component of the cell wall of *Aspergillus* spp.

galactosidase a glycosidase that hydrolyses a glycosidic bond between galactose and a second sugar molecule, e.g. glucose.

gangrene a complication of cell death (necrosis) where tissues of the body start to decay.

gas gangrene the decay or necrosis of body tissues infected with a *Clostriduim* spp. where gas is formed, which results in extreme swelling of the infected area.

genetic probe a nucleic acid sequence that is complementary to the target sequence.

glomerulonephritis inflammation of the glomeruli, or small blood vessels of the kidney.

glucuronidase a glycosidase that hydrolyses a glycosidic bond between glucuronic acid and a second sugar molecule.

glutamate dehydrogenase (GDH) marker for *C. difficile* strains, present in large amounts. Samples screened for the presence of GDH have a high negative predictive value (NPV); as such, the test is becoming widely used as part of a *C. difficile* detection algorithm for faeces samples.

glycopeptides these antibiotics bind to cell wall peptidoglycan which affects cell wall synthesis. They are mainly used for the treatment of serious staphylococcal and enterococcal infections.

glycopeptides these are a class of antibiotics that are used for the treatment of Gram-positive infections, e.g. teicoplanin and vancomycin.

glycosidase a hydrolase enzyme that hydrolyses a glycosidic bond, e.g. linking sugar units.

Guillian-Barré syndrome an autoimmune disease usually triggered by an acute infection which affects the peripheral nervous system causing paralysis.

HACEK a group of organisms, namely *Haemophilus* spp., *Aggregatibacter actinomycetemcomitans* (previously named *Actinobacillus*), *Cardiobacterium hominis*, *Eikenella corrodens* and *Kingella* spp. whose initial letters form the abbreviation. They are CO_2 dependent and initial infections can lead to endocarditis.

haemadsorption when cultured cells acquire the ability to stick to mammalian red blood cells.

haemagglutinin a substance that causes red blood cells to agglutinate.

haematogenous spread an agent that spreads through the blood stream.

haemocytometer a device for counting blood cells.

haemoptysis literally means the 'spitting of blood'.

hazard groups microorganisms belong to certain hazard groups depending upon the seriousness of the disease caused. The highest level is 4.

headspace the gas in a blood culture bottle above the liquid.

health-care-acquired infections these occur when pathogenic organisms are transferred from health-care workers, patients or the hospital environment, to a previously uninfected patient, who then develops the infection.

helminths worm parasites of the intestines, e.g. roundworm or tapeworm.

hepatocellular carcinoma (HCC) a primary carcinoma of the liver.

heterologous antibody is an antibody derived from an individual of a different species.

heterotrophic requiring complex organic compounds (growth factors) for growth.

homologous antibody an antibody derived from an animal of the same species.

hospital-acquired infection also called a nosocomial infection, it is an infection that appears more than 48 h after a patient has been admitted to hospital or other health-care facility.

humoral immune response an immune response mediated by secreted antibodies produced in B lymphocytes.

hydrocephalus literally means water on the brain. It is a build up of CSF around the brain and it causes enlargement of the head in neonates.

hydrolase an enzyme that catalyses the hydrolysis of a chemical bond.

hydrophobic water hating.

hyphae a long, branching filamentous cell of a fungus, hyphae are the main mode of vegetative growth, and are collectively called a mycelium.

immunocompromised or immunodeficient a state where the patient's immune function is absent or impaired so the ability to fight infection is impaired.

immunoglobulin G (IgG) an immunoglobulin, built of two γ heavy chains and two light chains. Each IgG has two antigen binding sites. It is the most abundant immunoglobulin and is approximately equally distributed in blood and in tissue liquids. IgG antibodies are predominantely involved in the secondary antibody response which occurs approximately 1 month following antigen recognition; thus the presence of specific IgG generally corresponds to maturation of the antibody response.

immunoglobulin M (IgM) an immunoglobulin built of μ heavy chains. It is the first immunoglobulin to appear on the surface of B cells and the first to be secreted. IgM is by far the physically largest antibody in the human circulatory system.

impetigo generally appears as scabs on a superficial skin infection.

in vitro in an artificial environment such as the laboratory.

in vivo inside the living organism.

incubation period the time between the receipt of infection and the onset of illness.

index patient the individual who first shows signs of the disease.

indicator antibiotics agents that are generally the least microbiologically active member of a family of antibiotics and are therefore more likely to detect low-level resistance. Examples of agents that have been used for this purpose include oxacillin for the detection of penicillin resistance in *Streptococcus pneumoniae* and methicillin resistant *S. aureus*.

induced sputum sample that is induced in the patient by inhaling nebulized saline to promote the production of sputum.

inoculum the portion of the specimen that is put onto the culture plate or into the fluid medium.

inspissation a process for heating and thus solidifying high-protein media via dehydration. The process, classically used for the coagulation of egg-based media, typically involves heating at 80–85°C and is not a process of sterilization, but most vegetative bacteria are killed.

interstitial fluid is the fluid that bathes and surrounds cells; produced in capillaries it is also known as tissue fluid.

intracellular organisms organisms that can be seen or multiply inside the host's cells.

intraluminal the inside space of a tubular structure (lumen).

intrathecal refers to an event that happens inside the spinal column.

intrauterine contraceptive device a device placed in the uterus for birth control.

intrinsic/inherent resistance resistance to antibiotics which occurs naturally within most or all species of a particular organism.

Kauffmann–White scheme the definitive listing of all essential serological characteristics of *Salmonella*.

keratin a protein found in the outer layers of skin; it serves to protect skin.

keratitis inflammation of the cornea.

lactic acid bacilli bacteria that produce lactic acid as the main metabolic end product of carbohydrate fermentation.

lactophenol blue a good background dye to visualize fungal structures.

latent virus any virus that remains in its host organism without undergoing replication. Latent viruses may be induced to replicate and cause cell lysis some time after the initial infection, for example when the host's immunity is reduced.

lecithinase a type of phospholipase enzyme that acts upon lecithin.

Lemierre's disease a disease mainly caused by *Fusobacterium necrophorum* but other fusobacteria can be involved. It mainly affects young adults. Lemierre's syndrome follows a streptococcal throat infection which has led to a peritonsillar abscess. *Fusobacterium necrophorum* can flourish in this abscess and can penetrate into the neighbouring jugular vein. Bacteria can then disseminate into the blood stream. It is associated with a high mortality rate.

lenticules and vitroids solid matrices designed for preserving microbes. They are typically presented as small discs containing microbial cells and spontaneously re-hydrate and dissolve on addition to culture media thus delivering a specified microbial inoculum. May be used for quality control of culture media.

leucopenia a reduction in the total number of white blood cells.

lipids fat soluble, naturally occurring compounds such as fats, oils or waxes. Cells of mycobacteria have high lipid content.

lipooligosaccharide a large molecule which has both a lipid and a polysaccharide component.

lipophilic lipid loving; some fungi require the addition of lipids to the medium in order to grow.

lumbar puncture the procedure used to obtain a CSF sample from a patient.

lumen the inside of any hollow, tube-like structure.

lymphoma a form of cancer that originates in the lymphocytes of the immune system.

lysosyme a family of enzymes that damage bacterial cell walls.

macrolides a group of antibiotics containing erythromycin and clindamycin. They are most often used for the treatment of staphylococcal and streptococcal infection, particularly in patients allergic to penicillins.

macrophages white blood cells (WBCs) that engulf then digest foreign bodies, including organisms.

malabsorptive state a characteristic state marked by subnormal intestinal absorption of dietary constituents, and thus excessive loss of nutrients in the stool, with chronic diarrhoea and weight loss.

mandatory reporting this is the obligatory reporting of a particular condition to local or national authorities. This often applies to communicable disease. Reporting of MRSA bacteraemia and *C. difficile* infection is mandatory in the UK.

mecA this is a gene that codes for an altered penicillin binding protein (PBP2 or PBP2a). These proteins have a lower affinity for β-lactam antibiotics and confer resistance to the organism.

meningitis inflammation of the meninges.

metachromatic describes granules (also known as volutin granules) that are cytoplasmic inclusions composed mainly of polyphosphates. They appear red-purple when stained with methylene blue.

metronidazole an antimicrobial agent with excellent activity against anaerobic bacteria and protozoa.

micro and macroconidia names given to the spores produced by dermatophyte fungi. They differ in size; microconidia are small often spherical or tear drop shaped, 2–4 μm in diameter; macroconidia are much larger.

microaerophillic a term applied to organisms such as *Campylobacter* spp. whereby the organisms require an environment containing a reduced oxygen level for optimal growth. Many microphiles also require an elevated concentration of carbon dioxide (>5%).

minimum bactericidal concentration the lowest concentration of an antibiotic which is cidal to ≥99.9% of the test inoculum.

minimum inhibitory concentration the lowest concentration of an antibiotic needed to inhibit bacterial growth.

monolayer a single layer of cells growing on a surface.

monomicrobial bacteraemia bacteraemia where only one significant organism is isolated.

mucinase an enzyme that catalyses the breakdown of mucopolysaccharides such as mucin.

mycolic acids these are long fatty acid molecules with one long chain and one short chain. They are commonly found in the cell wall of mycobacteria.

nasogastric feed a tube, generally made of soft rubber or plastic, that is inserted through a nostril and into the stomach.

nebulized antimicrobials antibiotics are inhaled using a device which creates a cloud of small particles. High drug concentration in the lung is achieved and minimal systemic absorption reduces toxicity.

necrotizing fasciitis a rare infection often known as 'flesh eating disease' affecting the deeper layers of skin and subcutaneous tissue.

necrotizing pneumonia when multiple abscesses can be found in the lungs.

needlestick injury when a needle used to take a blood sample from a patient accidentally pierces the skin of a health-care worker.

Negri bodies eosinophilic inclusion bodies found in the cytoplasm of certain nerve cells containing the rabies virus.

neuraminidase frequently found as an antigenic determinant on the surface of influenza virus. It is also a drug target for the prevention of influenza infection.

neutropenia a reduction in the number of phagocytic white blood cells.

non-dermatophyte fungi other filamentous fungi that can cause infections. There are several species that can cause infections of nails.

non-selective agar media that support the growth of a wide range of organisms; they do not contain inhibitory substances.

normal flora (human microbiotica) the collective term for the organisms colonizing a given body site.

nosocomial infections infection associated with health care which develops more than 48 h after the patient is admitted. For example, a urinary tract infection occurring as a complication of a procedure such as urinary catheter insertion.

novobiocin a naturally produced antibiotic and an inhibitor of DNA gyrase enzyme.

number needed to treat a statistical term that assesses the effectiveness of a health-care intervention, for example seven pregnant women with asymptomatic bacteriuria would have to be treated to prevent one case of pyelonephritis.

'O' antigen an antigen that is part of the lipopolysaccharide layer of the wall of Gram-negative bacteria.

oligonucleotide a short nucleic acid polymer, typically with 20 or fewer bases.

onychomycosis term given to a fungal infection of the toe or finger nails.

ophthalmia neonatorum also known as neonatal conjunctivitis. An eye infection of newborn infants contracted during delivery.

opportunistic pathogen an infection caused by a pathogen(s) that doesn't normally cause disease in healthy individuals.

optochin a shorthand term for ethylhydrocupreine hydrochloride, a chemical that inhibits the growth of most strains of *Streptococcus pneumoniae* but not other α-haemolytic streptococci.

orbital cellulitis infection of orbital tissue

otitis externa infection of the outer ear; often called 'swimmer's ear'.

otitis media inflammation of the inner ear.

ovaritis inflammation of the ovaries.

oxidase shorthand for cytochrome oxidase which oxidizes compounds and therefore utilizes oxygen with an electron transfer chain.

pandemic an epidemic that occurs over a wide geographical area such as a region or even worldwide.

passage the subculture of cells from one culture vessel to another.

pathogen an organism that has the ability to grow in the human and animal body and initiate disease.

pelvic inflammatory disease describes any infection in the lower female reproductive tract that spreads to the upper female reproductive tract.

peptidase a hydrolase enzyme that hydrolyses a peptide bond, e.g. linking amino acids.

peptones a protein derivative which is produced upon partial hydrolysis of a protein.

pericarditis inflammation of the pericardium (the fibrous sac surrounding the heart).

perineum the area between the anus and the vagina.

peristalsis the rhythmic, wavelike motion of the walls of the alimentary canal. These motions move the contents of the alimentary canal onwards.

pernasal swab a thin flexible swab that passes through the nose to the back of the throat, often used to diagnose whooping cough.

personal protective equipment equipment that creates a physical barrier between a worker and potential hazards; in its simplest form this could be a laboratory coat.

Peyer's patches oval elevated patches of closely packed lymph follicles on the mucosa of the small intestines.

phagocytosis the cellular process of engulfing solid particles such as bacteria.

pharmacodynamics the study of the biochemical and physiological effects of drugs and the mechanisms of their actions.

pharyngitis inflammation of the throat or pharynx.

phenol formerly carbolic acid. Acts like a detergent; it reduces the hydrophobic effect of lipids, enabling the dye to penetrate the cell wall.

phenotype a typical characteristic of an organism. This could be seen as its colonial appearance or biochemical properties. Atypical *E. coli*, for example, may show different biochemical properties but it is still an *E. coli*.

phosphatase an enzyme that removes a phosphate group from its substrate.

photophobia a dislike of bright lights.

pigbel a fatal type of food poisoning caused by a specific type of *C. perfringens* toxin (β-toxin).

pleural effusion fluid that accumulates in the pleural cavity.

pleural fluid the fluid between the two layers of pleura.

point of care testing (POCT) these are tests that can be done at the point of care, for example the GP surgery, the outpatient clinic, on the ward or even in the patient's own home. It is often used as a rapid assay to help immediate management and to decide which patients need further laboratory work up.

polymerase chain reaction (PCR) a technique involving amplification of specific DNA sequences which can be used to rapidly identify pathogens.

polymicrobial bacteraemia bacteraemia where two or more significant organisms are isolated.

pouch of Douglas the extension of the peritoneal cavity situated between the uterus and rectum.

primary culture any culture derived from cells, tissues or organs taken directly from an organism.

primary culture plates those plates onto which the clinical samples are directly inoculated. Organisms from primary plates can be inoculated onto further plates; these secondary plates are called subcultures.

productive cough this results in the expectoration of sputum.

prophylaxis the administration of a treatment to prevent disease developing.

prosthesis a mechanical or electronic device placed in or on the patient's body to help it function, e.g. artificial knees and hips, and pacemakers.

proteolytic describes the action of protein digestion by microbial enzymes such as proteases.

prothrombin glycoprotein precursor of thrombin that is produced in the liver and is necessary for the coagulation of blood.

prozone the phenomenon exhibited by some sera, in which agglutination or precipitation occurs at higher dilution ranges but is not visible at lower dilutions or when undiluted.

pseudomembranous colitis an infection of the colon, the symptoms of which are severe diarrhoea, abdominal pain and fever. The disease can often be fatal, particularly in elderly patients.

puerperal sepis also known as childbed fever. Infection of the female reproductive system following childbirth or abortion.

purpuric rash the appearance of red or purple discolorations on the skin, caused by bleeding underneath the skin.

purulent containing a lot of pus.

pus a discharge made up of white cells and fluid which is present in infected wounds.

pyelonephritis an infection of the kidney, which is a complication of urine infection.

pyrexial an increase in the patient's body temperature.

quinolones a group of antibiotics containing nalidixic acid. The newer agents such as ciprofloxacin are known as fluoroquinolones because of a fluorine atom added to the quinolone nucleus. They are generally well tolerated by the patient and display a broad spectrum of activity.

radiolabelled the tagging of a compound with a radioisotope.

rapid-growing mycobacteria mycobacteria that produce colonies within 7 days.

redox potential this is measured in volts under standardized conditions and expresses the propensity of a chemical to acquire electrons and thereby be reduced or release electrons and undergo oxidation. A lower redox potential will favour reduction and *vice versa*.

resident organisms (also known as normal flora or commensals) organisms deeply seated within the epidermis in skin crevices, hair follicles, sebaceous glands and beneath fingernails. Their purpose is protecting the skin against colonization by transient organisms. They are not readily removed by hand washing.

ribotyping the process of 'fingerprinting' an organism by detection of specific genomic DNA.

ringworm the common name for fungal infections of the skin and hair.

salpingitis inflammation of the fallopian tubes; can be acute or chronic.

saponin compounds found generally in plants which have a detergent-like action.

satellitism growth of *H. influenzae* on blood agar around colonies of other organisms producing V factor, e.g. *S. aureus*.

search and destroy a programme of screening all individuals on admission and discharge from hospital for the presence of MRSA.

selective media culture media that contain growth inhibitors (often antibiotics) to preclude the growth of unwanted commensal bacteria and facilitate the isolation of target pathogens.

sensitivity that proportion of true positives which the test correctly identifies, calculated as the number of true-positive results divided by the number of true-positive and false-negative results.

septicaemia an infection of the blood stream caused by toxic microorganisms.

seroconversion the development of detectable specific antibodies to viruses in a patient's serum as a result of infection or vaccination.

serogroup a grouping of microorganisms based on cell surface antigens.

skin antisepsis preparing the skin with disinfectant to remove bacteria before a blood sample is taken.

slow-growing mycobacteria colonies of mycobacteria that take longer than 7 days to appear.

Southern blot a method used to check for the presence of a DNA sequence in a sample. Southern blotting combines agarose gel electrophoresis for size separation of DNA with methods to transfer the size-separated DNA to a filter membrane for probe hybridization.

specificity the probability that an individual who does not have the particular disease being tested for will be correctly identified as negative, expressed as the proportion of true-negative results to the total of true-negative and false-positive results.

splenectomy removal of the spleen.

staphylocoagulase a protein secreted by some staphylococci. It activates prothrombin to cause the conversion of flbrinogen to fibrin, thus stimulating coagulation or clotting.

subarachnoid haemorrhage bleeding into the area between the pia mater and the arachnoid membrane around the brain.

subculture the transplantation of cells from one culture vessel to another.

superficial fascia the soft tissue part of the connective tissue.

superoxide a compound containing the highly reactive superoxide ion O^{2-}, an important killing mechanism generated in the lysosomes of phagocytes.

synergy testing tests to determine if antibiotic combinations would have an additive, indifferent or antagonistic effect.

synthetic substrate designed for detection of enzymes and not usually occurring in nature, synthetic substrates link sugars, fatty acids or amino acids to indicator molecules such as coloured or fluorescent dyes that are released to indicate enzyme activity.

syphilis a sexually transmitted disease caused by the bacterium *Treponema pallidum*.

therapeutic index the ratio between the toxic dose and the therapeutic dose of a drug, used as a measure of the relative safety of the drug for a particular treatment. A drug with little difference between toxic and therapeutic doses is referred to as having a narrow therapeutic index.

thermonuclease a nuclease is an enzyme capable of cleaving the phosphodiester bonds between the nucleotide subunits of nucleic acids. Staphylococci, and some other species, produce a thermonuclease capable of withstanding heat (e.g. 100°C).

time to positivity the time taken from when a blood culture bottle is inoculated to when it is identified as being positive.

tinea the name given to a fungal infection of the body; it is usually followed by a term to describe the body site, e.g. tinea capitis.

toxic shock syndrome a serious, often life-threatening syndrome caused by bacterial toxins, commonly from *S. aureus* or *S. pyogenes*.

transient colonization colonization by a microorganism that occurs for a short time only.

transmissible plasmids small, circular molecules of DNA that are separate from the bacterial chromosome. These can be transferred between bacteria and may carry resistance genes.

traumatic tap where a capillary or blood vessel is 'nicked' when a CSF sample is taken. Samples become blood stained. Serial samples can help identify whether the blood is due to a traumatic tap.

Triton X a non-ionic detergent used to lyse cells.

tuberculin skin test involves the subcutaneous inoculation of a purified protein derivative of MTB.

urea produced by the liver and excreted in the kidney, it is the means by which the body removes potentially toxic ammonia, a breakdown product of proteins.

urease a hydrolase that decomposes urea to form water and ammonia.

urethral syndrome patient experiences symptoms of a UTI but significant bacteruria is not detected.

uropathogens organisms that can infect the urinary tract.

vaccination the administration of material (an antigen) to produce immunity to a disease, usually by production of antibodies.

venepuncture taking blood via a needle placed into a vein.

vesicle a small, clear blister-like eruption on the skin that eventually ruptures, ulcerates, crusts over and heals without scarring unless it becomes infected with a bacterial pathogen.

vesicoureteric reflux failure of the one way valve mechanism that usually prevents bladder urine flowing back in to the ureter.

Vincent's angina a polymicrobial infection of the gums which leads to their necrosis.

viraemia presence of viruses in the bloodstream.

virucidal a substance capable of killing or neutralizing a virus.

virulence the ability of an organism to produce disease; highly virulent organisms are able to produce disease easily.

Wood's light a long-wave ultraviolet light used as a diagnostic tool by dermatologists to distinguish dermatophyte infections from other causes of scalp irritation and hair loss.

xanthachromia the yellow coloration of CSF samples caused by lysed blood.

Ziehl-Neelsen (ZN) used to stain mycobacteria using fuchsin combined with phenol. Malachite green is used as a counter stain. In modified form can be used to stain some intestinal parasites.

References

Chapter 1

Advisory Committee on Dangerous Pathogens. *Biological Agents: Managing the Risks in Laboratories and Healthcare Premises*. London: Health and Safety Executive, 2005.

Advisory Committee on Dangerous Pathogens. *Categorisation of Biological Agents According to Hazard and Categories of Containment*, 4th Edition. HSE Books, 1995.

Carriage of Dangerous Goods (Classification, Packaging and Labelling) and Use of Transportable Pressure Receptacles Regulations 2004, SI No 568. The Stationery Office, 2004.

Church D. The seven principles of accurate microbiology specimen collection. *Microbiology Newsletter*, Calgary Laboratory Services. Vol. 6, 2005.

Clinical Practice Committee, Great Ormond Street Hospital for Children. *Microbiological Specimen Collection*, version 2.0, 2006.

Control of Substances Hazardous to Health Regulations 2002. The Stationery Office, 2002.

Evans EGV, Richardson MD, eds. *Medical Mycology. A Practical Approach*. Oxford University Press, 1989.

Gilchrist B. Wound infection – sampling bacterial flora: a review of the literature. *Journal of Wound Care* 1996; **5**: 386–388.

Health and Safety Executive. *Biological Agents: Managing the Risks in Laboratories and Healthcare Premises*, 2005.

Higgins C. An introduction to the examination of specimens. *Nursing Times* 1994; **90**: 29–32.

Isenberg HD, ed. *Clinical Microbiology Procedures Handbook*, Vol 1. American Society for Microbiology, 1992.

Garcia LS. *Diagnostic Medical Parasitology*, 4th edition. ASM Press, 2001.

Miller MJ. *A Guide to Specimen Management in Clinical Microbiology*. American Society for Microbiology, 1999.

Richardson MD, Warnock DW. *Fungal Infections. Diagnosis and Management*. Blackwell Scientific Publications, 1993.

Standards Unit, Evaluations and Standards Laboratory, Centre for Infections. *Inoculation of Culture Media*. QSOP 52. Issue date 2007.

Standards Unit, Evaluations and Standards Laboratory, Centre for Infections. *Investigation of Blood Cultures (for Organisms other than Mycobacterium Species)*, BSOP 37, 2005.

Weinbaum FI. Doing it right first time: quality improvement and the contaminant blood culture. *Journal of Clinical Microbiology* 1997; **35**: 563–565.

Chapter 2

Barrow GI, Feltham RKA, eds. *Cowan and Steel's Manual for the Identification of Medical Bacteria*, 3rd edition. Cambridge University Press, 2004.

Bascomb S. Enzyme tests in bacterial identification. In: Colwell RR, Grigorova R, eds. *Methods in Microbiology*, Vol. 19, pp. 105–160. London and New York: Academic Press, 1987.

Collier L *et al*, eds. *Topley and Wilson's Microbiology and Microbial Infections*, 9th edition. Arnold, 1998.

Collins CH, Lyne PM, Grange JM, Falkinham JO, eds. *Collins and Lyne's Microbiological Methods*, 8th edition. Arnold, 2004.

Isenberg HD, ed. *Clinical Microbiology Procedures Handbook*. American Society for Microbiology, 2004.

Manafi M, Kneifel W, Bascomb S. Fluorogenic and chromogenic substrates used in bacterial diagnostics. *Microbiology and Molecular Biology Reviews* 1991; **55**: 335–348.

Perry JD, Ford M, Hjersing N, Gould FK. Rapid conventional scheme for biochemical identification of antibiotic resistant *Enterobacteriaceae* isolates from urine. *Journal of Clinical Pathology* 1988; **41**: 1010–1012.

Chapter 3

Barrow GI, Feltham RKA, eds. *Cowan and Steel's Manual for the Identification of Medical Bacteria*, 3rd edition. Cambridge University Press, 2004.

Collier L *et al.*, eds. *Topley and Wilson's Microbiology and Microbial Infections*, 9th edition. Arnold, 1998.

Difco Laboratories. *Difco Manual: Dehydrated Culture Media and Reagents for Microbiology*. 1953.

Madigan M, Martinko J, eds. *Brock Biology of Microorganisms*, 11th edition. Prentice Hall, 2005.

Snell JJS, Brown DFJ, Roberts C, eds. *Quality Assurance: Principles and Practice in the Microbiology Laboratory*. Public Health Laboratory Service, 1999.

Chapter 4

Andrews J, Brenwald N, Brown DFJ, Perry J, King A, Gemmell C. Evaluation of a 10 μg cefoxitin disc for the detection of methicillin resistance in *Staphylococcus aureus* by BSAC methodology. *Journal of Antimicrobial Chemotherapy* 2005; **56**: 599–600.

Andrews JM, Brown DFJ, Wise R. A survey of antimicrobial susceptibility testing in the United Kingdom. *Journal of Antimicrobial Chemotherapy* 1996; **37**: 187–204.

Andrews JM. Determination of minimum inhibitory concentrations. *Journal of Antimicrobial Chemotherapy* 2001; **48** (Suppl. 1): 5–16.

Andrews JM. The development of the BSAC standardized method of disc diffusion testing. *Journal of Antimicrobial Chemotherapy* 2001; **48** (Suppl.1): 29–42.

Andrews JM. BSAC standardized disc susceptibility testing method (version 6). *Journal of Antimicrobial Chemotherapy* 2007; **60**: 20–41.

Olsson-Liljequist B, Larsson P, Walder M, Miörner H. Antimicrobial susceptibility testing in Sweden. *Scandinavian Journal of Infectious Diseases* 1997; **105** (Suppl.): 3.

Berenbaum MC. A method for testing for synergy with any number of agents. *Journal of Infectious Diseases* 1978; **137**: 122–130.

Berenbaum MC. Correlation between methods for measurement of synergy. *Journal of Infectious Diseases* 1980; **142**: 476–480.

Ericsson H, Sherris JC. Antibiotic sensitivity testing. Report of an International Collaborative study. *Acta Pathologica et Microbiologica Scandinavica Section B* 1971; **217** (Suppl.): 1–90.

Hakanen AJ, Linggren M, Huovinen P, Jala J, Kotilainen P. New quinolone resistance phenomenon in *Salmonella enterica* nalidixic acid-susceptible isolates with reduced fluoroquinolone susceptibility. *Journal of Clinical Microbiology* 2005; **43**: 5775–5778.

Hoffner SE, Svenson SB, Kallenius G. Synergistic effects of antimycobacterial drug combinations of *Mycobacterium avium* complex determined radiometrically in liquid medium. *European Journal of Clinical Microbiology* 1987; **6**: 530–555.

Lang BJ, Aaron SD, Ferris W, Herbert PC, MacDonald NE. Multiple combination bactericidal antibiotic testing for patients with cystic fibrosis infected with multiresistant strains of *Pseudomonas aeruginosa*. *American Journal of Respiratory and Critical Care Medicine* 2000; **162**: 2241–2245.

Louie L, Matsumura SO, Choi E *et al*. *Performance Standards for Antimicrobial Susceptibility Testing*; Fifteenth Informational Supplement, M100-S15, Vol. 25. Wayne, PA, USA: Clinical and Laboratory Standards Institute, 2005.

MacGowan AP, Wise R. Establishing MIC breakpoints and the interpretation of in vitro susceptibility tests. *Journal of Antimicrobial Chemotherapy* 2001; **48** (Suppl. 1): 17–28.

MacGowan AR, Bowker K, Bedford KA, Holt HA, Reeves DS. Synergy testing of macrolide combinations using the chequerboard technique. *Journal of Antimicrobial Chemotherapy* 1993; **32**: 913–915.

Metzler CM, DeHaan RM. Susceptibility tests of anaerobic bacteria: statistical and clinical considerations. *Journal of Infectious Diseases* 1974; **130**: 588–594.

Muller-Serieys C, Andrews J, Vacheron F, Cantalloube C. Tissue kinetics of Telithromycin, the first ketolide antibacterial. *Journal of Antimicrobial Chemotherapy* 2004; **53**: 149–157.

Ragunathan PL, Ison C, Livermore DM. Nalidixic acid-susceptible, ciprofloxacin-resistant *Neisseria gonorrhoeae* strain in the UK. *Journal of Antimicrobial Chemotherapy* 2005; **56**: 437.

Nightingale CH, Murakawa T, Ambrose PG, eds. *Antimicrobial Pharmacodymanics in Theory and Clinical Practice*. New York, USA: Marcel Dekker, 2002.

Ragunathan PL, Ison C, Livermore DM. *Methods for Antimicrobial Susceptibility Testing of Anaerobic Bacteria; Approved Standard*, 4th edition, M11-A4, Vol. 17, No. 22. Wayne, PA, USA: Clinical and Laboratory Standards Institute, 1997.

Comité de l'Antibiogramme de la Société Français de Microbiologie. Technical Recommendations for in vitro susceptibility testing. *Clinical Microbiology and Infectious Diseases* 1996; **2** (Suppl. 1): S11–S25.

Snell JJS, Brown DFJ, Perry SF, George R. Antimicrobial susceptibility testing of enterococci: results of a survey conducted by the United Kingdom National External Quality Assessment Scheme for Microbiology. *Journal of Antimicrobial Chemotherapy* 1993; **32**: 401–412.

Working Party on Antibiotic Sensitivity Testing of the British Society for Antimicrobial Chemotherapy. A guide to sensitivity testing. *Journal of Antimicrobial Chemotherapy* 1991; **27** (Suppl. D): 1–50.

Yeo SF, Livermore DM. Effect of inoculum size on the in-vitro susceptibility to beta-lactam antibiotics of *M. catarrhalis* isolates of different beta-lactamase types. *Journal of Medical Microbiology* 1994; **40**: 252–255.

Chapter 5

Elliott TSJ, Foweraker J, Gould FK, Perry JD, Sandoe JAT. Guidelines for the antibiotic treatment of endocarditis in adults: report of the Working Party of the British Society for Antimicrobial Chemotherapy. *Journal of Antimicrobial Chemotherapy* 2004; **54**: 971–981.

Finch RG, *et al*. eds. *Antibiotic and Chemotherapy: Anti-Infective Agents and Their Use in Therapy*, 8th edition. Churchill Livingstone, 2003.

Freeman R, Gould FK. *Infection in Cardiothoracic Intensive Care*. Hodder Arnold, 1987.

Shanson DC. *Sepicaemia and Endocarditis: Clinical and Microbiological Aspects*. Oxford University Press, 1990.

Standards Unit, Evaluations and Standards Laboratory, Centre for Infections. *Investigation of Blood Cultures (for Organisms other than Mycobacterium Species)*, BSOP 37, 2005.

Chapter 6

Fallon D, Ackland G, Andrews N, Frodsham D, Howe S, Howells K, Nye KJ, Warren RE. A comparison of the performance of commercially available chromogenic agars for the isolation and presumptive identification of organism from the urine. *Journal of Clinical Pathology* 2003; **56**: 608–612.

Graham JC, Galloway A. The laboratory diagnosis of urinary tract infection. *Journal of Clinical Pathology* 2001; **54**: 911–919.

Leigh DA, Williams JD. Method for detection of significant bacteriuria in large groups of patients. *Journal of Clinical Pathology* 1964; **17**: 498–503.

National Institute for Health and Clinical Excellence. *UTI in Children: Urinary Tract Infection in Children: Diagnosis, Treatment and Long-term Management*, 2007. Available at http://www.nice.org.uk/nicemedia/pdf/CG54NICEguideline.pdf.

Smyth M, Moore JE, McClurg RB, Goldsmith CE. Quantitative colorimetric measurement of residual antimicrobials in the urine of patients with suspected urinary tract infection. *British Journal of Biomedical Science* 2005; **62**: 114–119.

Chapter 7

Adler MW, Cowan F, French P, Mitchell H, Richens J, eds. *ABC of Sexually Transmitted Infections*, 5th edition. Wiley-Blackwell, 2004.

Beers MH, Robert Berkow R. *The Merck Manual of Diagnosis and Therapy*, Centennial Edition, 17th edition. John Wiley & Sons, 2004.

Hay PE, Morgan DJ, Ison CA, Bhide SA, Romney M, McKenzie P *et al*. A longitudinal study of bacterial vaginosis during pregnancy. *British Journal of Obstetrics and Gynaecology* 1994; **101**: 1048–1053.

Health Protection Agency. *Investigation of Genital Tract and Associated Specimens*. BSOP 28. National Standard Method. http://www.hpa-standardmethods.org.uk/documents/bsop/pdf/bsop28.pdf.

Health Protection Agency. *Investigation of Skin, Superficial, and Non-surgical Wound Swabs*. BSOP 11. National Standard Method. http://www.hpa-standardmethods.org.uk/documents/bsop/pdf/bsop11.pdf.

Holmes KK, Sparling PF, Stamm WE *et al*. *Sexually Transmitted Diseases*, 4th edition. McGraw-Hill Medical, 2008.

Nugent RP, Krohn MA, Hillier SL. Reliability of diagnosing bacterial vaginosis is improved by a standardised method of Gram stain interpretation. *Journal of Clinical Microbiology* 1991; **29**: 297–301.

Chapter 8

Bain BJ. *Blood Cells – A Practical Guide*, 4th edition. Blackwell Publishing, 2006.

Finch RG *et al*. *Antibiotic and Chemotherapy: Anti-Infective Agents and their Use in Therapy*, 8th edition. Springer, 2004.

Merck online manual. http://www.merck.com/mmhe/index.html.

Standards Unit, Evaluations and Standards Laboratory, Centre for Infections. *Investigation of Fluids from Normally Sterile Sites*. National Standard Method BSOP 26.

Standards Unit, Evaluations and Standards Laboratory, Centre for Infections. *Investigation of Continuous Peritoneal Dialysis Fluids*. National Standard Method BSOP 25.

Standards Unit, Evaluations and Standards Laboratory. Centre for Infections. *Investigation of Cerebrospinal Fluid*. National Standard Method BSOP 27.

Tunkel A. *Bacterial Meningitis*. Lippincott Williams and Wilkins, 2001.

Chapter 9

BTS Pneumonia Guideline Committee. *BTS Guidelines for the Management of Community Acquired Pneumonia in Adults – 2004 update*. http://www.brit-thoracic.org.uk/Portals/0/Clinical%20Information/Pneumonia/Guidelines/MACAPrevisedApr04.pdf.

Cystic Fibrosis Trust. *Antibiotic Treatment for Cystic Fibrosis*, 2nd edition. http://www.cftrust.org.uk/aboutcf/publications/consensusdoc/C_3200Antibiotic_Treatment.pdf.

Drake R *et al*. *Gray's Anatomy for Students*, 2nd edition. Churchill Livingstone, 2009.

Goering R, Dockrell H, Zuckerman M, Wakelin D. *Mims' Medical Microbiology*, 4th edition. Mosby, 2007.

Murray PR *et al*. *Manual of Clinical Microbiology*, 9th edition. Blackwell Publishing, 2007.

Ratledge C, Dale JW. *Mycobacteria*. Wiley Blackwell, 1999.

Chapter 10

Ash LR, Orihel TC. *Atlas of Human Parasitology*, 5th edition. ASM Press, 2007.

Collier L *et al*. eds. *Topley and Wilson's Microbiology and Microbial Infections, 9th edition*. Arnold Publication, 1998.

Lynn Shore Garcia. *Diagnostic Medical Parasitology*, 4th edition. ASM Press, 2001.

Mims C, Nash A, Stephen J. *Mims' Pathogenesis of Infectious Disease*, 5th edition. Academic Press, 2000.

Standards Unit, Evaluations and Standards Laboratory, Centre for Infections. *Investigation of Faecal Specimens for Bacterial Pathogens*. National Standard Method BSOP 30. http://www.hpa-standardmethods.org.uk/documents/bsop/pdf/bsop30.pdf.

Standards Unit, Evaluations and Standards Laboratory, Centre for Infections. *Investigation of Specimens other than Blood for Parasites*. National Standard Method BSOP 31. http://www.hpa-standardmethods.org.uk/documents/bsop/pdf/bsop31.pdf.

Chapter 11

Anaissie J *et al. Clinical Mycology*, 2nd edition. Churchill Livingstone, 2009.

Bancroft D, Gamble M. *Theory and Practice of Histological Techniques*, 6th edition. Churchill Livingstone, 2007.

Dr Fungus. http://www.doctorfungus.org.

Odds FC *et al. Principles and Practice of Clinical Mycology*. Wiley Blackwell, 1996.

Stevens DA. Diagnosis of fungal infections: current status. *Journal of Antimicrobial Chemotherapy* 2002; **49** (Suppl S1): 11–19.

Chapter 12

Care Quality Commission. Several publications from the Commission and previous bodies on health-care-acquired infection. www.cqc.org.uk.

Casewell M, Philpott-Howard J. *Hospital Infection Control. Policies and Practical Procedures.* W.B. Saunders, 1994.

Centre for Evidence Based Purchasing, NHS. *Evaluation Report: Clostridium difficile Toxin Detection Assays.* CEP08054, 2009.

Conterno LO *et al.* Real-time polymerase chain reaction detection of methicillin-resistant *Staphylococcus aureus*: impact on nosocomial transmission and costs. *Infection Control and Hospital Epidemiology* 2007; **28**: 1134–1141.

Department of Health, Chief Medical Officer. *Winning Ways: Working Together to Reduce Health Care Infection*, 2003. http://www.dh.gov.uk/en/Publicationsandstatistics/Publications/PublicationsPolicyAndGuidance/DH_4064682.

Department of Health. HTM 07-01. *Safe Management of Healthcare Waste*, 2006. http://www.dh.gov.uk/en/Publicationsandstatistics/Publications/PublicationsPolicyAndGuidance/DH_063274.

Kuijper EJ *et al.* Emergence of *Clostridium difficile*-associated disease in North America and Europe. *Clinical Microbiology and Infection* 2006; **12** (Suppl. 6): 2–18.

National Audit Office. Health-care-acquired infection audits. www.nao.org.uk.

Planche T, Aghaizu A, Holliman R, Riley P, Poloniecki J, Breathnach A, Krishna S. Diagnosis of *Clostridium difficile* infection by toxin detection kits: a systematic review. *Lancet Infectious Diseases* 2008; 8: 777–784.

Standards Unit, Evaluations and Standards Laboratory, Centre for Infections. *Investigation of Specimens for Screening for MRSA.* National Standard Method B29. http://www.hpa-standardmethods.org.uk/documents/bsop/pdf/bsop29.pdf.

Standards Unit, Evaluations and Standards Laboratory, Centre for Infections. *MRSA Selective Agar.* National Standard Method MSOP 30. http://www.hpa-standardmethods.org.uk/documents/msop/pdf/msop30.pdf.

Standards Unit, Evaluations and Standards Laboratory, Centre for Infections. *Processing of Faeces for Clostridium difficile.* National Standard Method BSOP10. http://www.hpa-standardmethods.org.uk/documents/bsop/pdf/bsop10.pdf.

Chapter 13

Desselberger U. *Medical Virology – A Practical Approach.* IRL Press, 1995.

Lennette EH, Smith TF, eds. *Laboratory Diagnosis of Viral Infections*, 3rd edition. Marcel Dekker, 1999.

Mahy BWJ, Collier L, eds. *Topley and Wilson's Microbiology and Microbial Infections*, Vol. 1 Virology, 9th edition. Arnold, 1998.

Schachter J, Stamm WE. *Chlamydia.* In: Murray PR ed. *Manual of Clinical Microbiology*, 7th edition. ASM Press, 1999.

White DO, Fenner FJ. *Medical Virology*, 4th edition. Academic Press, 1994.

Zuckerman AJ *et al. Principles and Practice of Clinical Virology*, 4th edition. Wiley, 1999.

Index

N

O

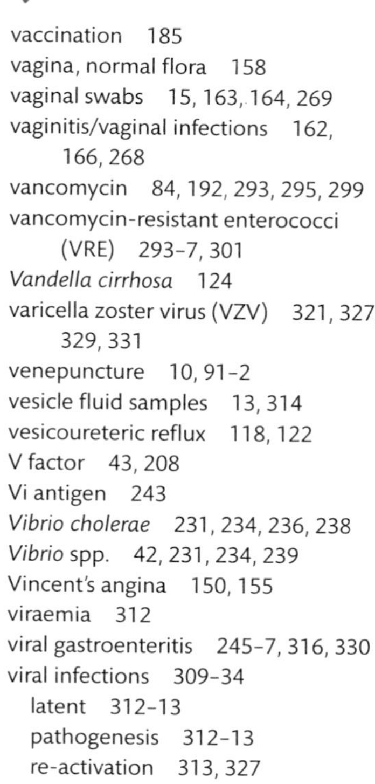

V

W

X

Y

Z